Amorphous Drugs

Marzena Rams-Baron • Renata Jachowicz •
Elena Boldyreva • Deliang Zhou •
Witold Jamroz • Marian Paluch

Amorphous Drugs

Benefits and Challenges

Springer

Marzena Rams-Baron
Institute of Physics
University of Silesia
Chorzów, Poland

Renata Jachowicz
Chair of Pharmaceutical Technology and
Biopharmaceutics, Faculty of Pharmacy
Jagiellonian University - Medical College
Kraków, Poland

Elena Boldyreva
Institute of Solid State Chemistry and
Mechanochemistry, Siberian Branch
of the Russian Academy of Sciences
Novosibirsk State University
Novosibirsk, Russian Federation

Deliang Zhou
AbbVie
North Chicago, USA

Witold Jamroz
Chair of Pharmaceutical Technology and
Biopharmaceutics, Faculty of Pharmacy
Jagiellonian University - Medical College
Kraków, Poland

Marian Paluch
Institute of Physics
University of Silesia
Chorzów, Poland

ISBN 978-3-319-72001-2 ISBN 978-3-319-72002-9 (eBook)
https://doi.org/10.1007/978-3-319-72002-9

Library of Congress Control Number: 2017962001

Printed on acid-free paper

This Springer imprint is published by Springer Nature
The registered company is Springer International Publishing AG
The registered company address is: Gewerbestrasse 11, 6330 Cham, Switzerland

Acknowledgment

The authors, Marzena Rams-Baron, Renata Jachowicz, Witold Jamroz and Marian Paluch are grateful for the financial support received within the Project No. 2015/16/W/NZ7/00404 (Symfonia 3) from the Polish National Science Centre.

Contents

Chapter 1
Why Amorphous Drugs?

Low aqueous solubility of active pharmaceutical ingredients (APIs) is one of the most important challenges facing drug development researchers today [1, 2]. With the development of computational chemistry and high throughput screening methods it is possible to obtain a large number of compounds with attractive therapeutic activity. However, at the same time the selection of novel active molecules with suitable biopharmaceutical properties (like solubility, intestinal permeability) becomes a great challenge and a bottleneck in drug development. Statistically, more than 40% of approved drugs and even 70–90% of those under investigations are poorly water-soluble and additional efforts are required to improve their water solubility [3–5].

The rate of drug absorption depends on the complex interplay of various physicochemical and physiological conditions [6]. Among them Amidon et al. have distinguished membrane permeability and drug solubility/dissolution rate as those of fundamental importance for oral drug absorption [7]. In the framework of the biopharmaceutical classification of drug products, depending on their aqueous solubility and gastrointestinal permeability, these are divided into four groups distinct in terms of expected in vivo performance. The APIs with poor water solubility are classified as class II (with low solubility but high permeability) and class IV (with low solubility and low permeability). The currently observed trend in drug discovery indicates the rapid and continuous growth of class II compounds and the corresponding decrease of class I drugs which due to high solubility and high permeability are much easier to formulate [4].

The increasing amount of poorly water soluble chemical entities appearing during development research motivates pharmaceutical companies to search for novel solubilizing approaches able to overcome the urgent problem of their inefficient biopharmaceutical performance [3, 8–10]. In the case of drugs which can penetrate the intestinal mucosa easily, like class II drugs, the insufficient solubility will be a factor limiting their bioavailability. To trigger biological response the drug has to dissolve in biological fluids sufficiently enough to exert the desired therapeutic response. If the drug cannot be dissolved fast enough, it might pass the absorption

M. Rams-Baron et al., *Amorphous Drugs*,
https://doi.org/10.1007/978-3-319-72002-9_1

site without appropriate action [11]. Increasing the dose may induce the desired outcome, however, it raises other problems relating to the proper patient compliance.

Although the nature of solubility and dissolution process are different, the former is a purely thermodynamic phenomenon while the latter is a kinetic event, they are closely related to each other. This relationship can be rationalized by modified Noyes-Whitney equation [12, 13]:

$$\frac{dC}{dt} = K(C_s - C_t) \tag{1.1}$$

where dC/dt is the dissolution rate, C_s is the drug solubility at saturated equilibrium condition and C_t denotes the concentration of drug dissolved at time t. The constant $K = AD/h$ depends on the diffusion coefficient D, value of surface area available for dissolution A and the thickness of diffusion layer h. Various physicochemical and structural factors may tune the parameters in Eq. (1.1). It is difficult to alter the diffusion coefficients of drug in biological fluids or thickness of diffusion layer since both quantities are governed by viscosity or hydrodynamics inside the gastrointestinal track [14]. Thus, one can deduce the following possibilities to enhance the dissolution profile of a drug, i.e. increasing the particle surface area and/or improving the drug saturated solubility in the gastrointestinal fluids. These solutions can be realized in a number of different ways giving raise to different formulating approaches which are summarized in Table 1.1. Each approach has its own advantages and weak points that need to be considered. Matching the

Table 1.1 Formulation approaches on the basis of BCS classification

BCS class	I	II	III	IV
Solubility	High	Low	High	Low
Permeability	High	High	Low	Low
Examples	Verapamil hydrochloride, warfarin sodium,	Diazepam, ibuprofen, glibenclamide, nevirapine, nifedipine, ritonavir	Cimetidine, amoxicillin, captopril, chloramphenicol	Dapsone, paracetamol, sulfamethoxazole
Formulation strategy	Capsule or tablet	*Physical modifications:* – particle size reduction – solid state modifications (polymorphs, co-crystals, amorphous forms) – complexation – solubilization by surfactants – drug dispersion in appropriate carrier *Chemical modifications:* – prodrag application – salt formation	Capsule or tablet, absorption enhancers	The same as for BCS II, absorption enhancers

Adopted from [5]. Examples taken from [15]

optimal formulation strategy to the drug development is a time and cost consuming task. To make a rational decision several factors need to be taken into account, for example, the physicochemical properties (e.g. pKa, log P, solubility, stability etc.) or the targeted profile of developing product (e.g. required dose, preferential administration route) [5].

Among the available approaches aimed at improving the dissolution behavior of poorly-water soluble drugs amorphization has been considered. Conversion of crystalline drugs into the amorphous form has been recognized as an effective way to achieve the longstanding goal of pharmaceutical science and drug developments, i.e. beneficial drug dissolution in vivo [16, 17]. It is possible due to the unique nature of the amorphous state (disordered nature and high energy) which differs substantially from the crystalline state. On the market a few examples of products containing amorphous API can be found, for instance Accolate® (zafirlukast), Ceftin® (cefuroxime axetil), and Accupril® (quinapril hydrochloride) [18].

The differences between crystalline and amorphous solids are schematically depicted in Fig. 1.1. When we cool a liquid slowly, allowing nucleation and crystal growth to occur, the drop of enthalpy and volume observed at the melting temperature (T_m) is due to the presence of a first order liquid-crystal transition. Contrary, when the liquid will be cooled fast enough to avoid crystallization, its liquid-like properties will be preserved below T_m in the supercooled liquid state. As we continue decreasing the temperature the liquid-glass transition will take place. The observed change in the slope of V(T) or H(T) determines the glass transition

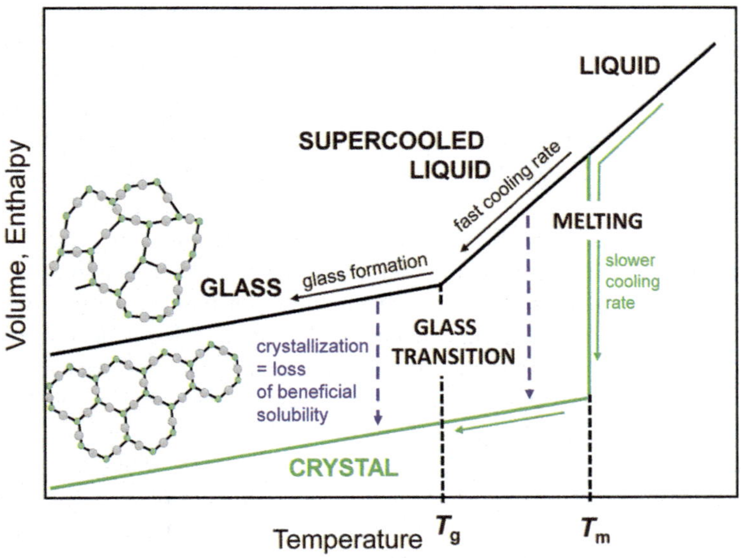

Fig. 1.1 Temperature dependence of volume and enthalpy at constant pressure. Fast cooling may lead to glass formation, while for slower cooling rates the crystallization may likely occur. Besides, crystallization may be observed from glassy or supercooled liquid states canceling any improvements in drug dissolution properties

temperature (T_g). Then, the material becomes an amorphous solid with macroscopic properties distinct from equilibrium values. It is worth mentioning that the cooling process is accompanied by a huge change in rheological properties. Below T_g in the glassy state the system is so viscous that its inhibited molecular mobility in relation to cooling rate is responsible for its fall out of equilibrium [19]. From a pharmaceutical perspective both glassy and supercooled liquid states are relevant. Usually, we keep the drugs at room temperature which corresponds to the glassy state of most pharmaceuticals. However, it is necessary to study amorphous drugs both below and above T_g since higher-temperature conditions corresponding to the supercooled liquid state may be applied during drug manufacturing. Due to higher molecular mobility the risk of drug conversion to crystalline form increases.

In general, the proper processing of crystalline material (e.g. by mechanical activation during milling, fast melt cooling, rapid precipitation from solution) [16] allows for material transformation into the amorphous form. Instead of three-dimensional ordering typical for crystalline lattice we obtain a structure with random atomic arrangement. Amorphous solids, in comparison to crystals, do not exhibit the long-range ordering (LRO). Instead, short range ordering (SRO) relevant only over few molecular dimensions can be found [19]. The faster dissolution and beneficial absorption of drugs in amorphous state is related to their higher free energy in comparison to crystals [3, 20]. The thermodynamics of solubilization is driven by the difference in the Gibbs free energy of initial state formed by undissolved components and final dissolved state. Since the amorphous state has higher free energy when compared to crystalline state less energy is necessary to dissolve when the amorphous form is applied. Thus, better dissolution rates should be expected. Attempts to estimate the actual solubility benefits arising from the application of drug in the amorphous form were made by Hancock and Parks [21]. Based on simple thermodynamic considerations they estimated that in the case of the amorphous forms 10- to 1600-fold improvement of drug water solubility in comparison to crystals should be expected. However, the measured values are usually significantly lower which was explained by difficulties in their experimental determination.

The excess thermodynamic properties of the amorphous state, like its greater entropy, enthalpy and free energy as well as its higher molecular mobility make the amorphous drugs more prone to crystallization. So far finding an effective stabilization approach is the major challenge related to the development of drugs in the amorphous form. The crystalline drug forms, more stable and easier to handle, dominate the pharmaceutical market for practical and economical reasons. However, the importance of the problem of insufficient APIs solubility encourages pharmaceutical companies to support and invest in new solutions, even those requiring additional efforts to obtain the beneficial drug absorption in vivo. Therefore, drug compositions based on amorphous active ingredients attract particular interest despite their unstable and problematic nature. The emphasis in developing amorphous formulations is put on searching drug compositions providing stability at each stage of drug processing—from its manufacturing to administration. The progress we are witnessing today, reflected in the growing number of amorphous products available on the market is related to the successful implementation of amorphous solid dispersions technology. This concept, substantially improving water solubility and effectively protecting against drug recrystallization, allows for successfully entering of amorphous products onto the pharmaceutical market and secured their stable position in the offer of pharmaceutical companies.

Searching for stable amorphous drug formulations is a complex issue which cover a variety of challenges that we have to face at each stage of the drug product lifecycle. All should be predicted and resolved at the initial stage of research and drug development. Without complete understanding of the theoretical principles which are responsible for the recrystallization behavior, achieving the desired goal of producing efficient, well-soluble and safe amorphous drug product will be unattainable. It is well known that the process of drug discovery is extremely costly and highly risky. To save time and money unnecessarily lost during verification of ineffective solutions, a rational approach to the problem of amorphous drug instability is required. Such an approach requires interdisciplinary knowledge, skills and insights into the problems. The recrystallization of amorphous content might be promoted by elevated temperature, mechanical stress or humidity at each stage of drug processing, storage or even administration. The systematic investigations of crystallization behavior of amorphous drugs at different thermodynamic conditions and a comprehensive insight into manufacturing procedures allows one to establish processing conditions minimizing the risk of drug recrystallization. Only in-depth understanding of factors controlling crystallization kinetics allows for design of effective stabilizing solutions. The lack of such knowledge makes it impossible to understand the reasons of unexpected failure at a formulation stage.

A large number of reports concerning amorphous pharmaceuticals reflects the amount of work and efforts that have been made to advance this field in the last decades. This motivates us to summarize the current state of the art and indicate the paramount perspectives for future development. This book is addressed to people who are motivated to work with amorphous pharmaceuticals, but do not understand in detail what truly impacts their behavior. Our goal is to increase their awareness by improving understanding of the benefits and challenges associated with the application of high energy amorphous forms as active pharmaceutical ingredients. Based on our own experience, we refer to common problems that one may experience when starting work with amorphous drugs, but also we present here the robust solutions how these potential difficulties may be predicted and overcome. Practical and technological aspects are presented along with theoretical background allowing a rational approach to the task of amorphous drug preparation. Content of the book should guide those who are interested in amorphous formulations through the process of new product development decreasing the risk of failure. This book is not only dedicated to those who are actually involved in the implementation of amorphous formulations. Students and scientists who are simply interested in learning the subject, through many examples contained in the book, can understand the phenomenon of amorphous pharmaceuticals also. This book covers all key theoretical and practical issues related to working with pharmaceutical materials in the amorphous form. The particular chapters were prepared by experts from different fields—physicists, pharmacists and representatives of pharmaceutical companies, which allows discussion of the problems from different perspectives.

To fully understand the properties of amorphous pharmaceuticals, one must have a thorough knowledge about theoretical concepts standing behind them. Thus, at the beginning the fundamental aspects concerning order-disorder transition and structure-property relationship for amorphous and crystalline phases will be outlined. In Chap. 2 by introducing the physics of disordered systems will provide a theoretical background for further considerations contained in the book. In Chap. 3 we focus on bioavailability advantage of amorphous formulations. A short review of the current state of the art methods of drug amorphization is provided in Chap. 4. Various manufacturing technologies are discussed there, from those applied in the laboratory environment to the most common approaches in the pharmaceutical industry like hot melt extrusion or spray drying. In Chap. 5 we will focus on the biggest challenge associated with amorphous drug application, i.e. their tendency for recrystallization. Understanding which factors are responsible for the recrystallization behavior is crucial to fully exploit and commercialize the potential of amorphous formulations in the future. The chapter will cover some fundamental aspects concerning the mechanism of nucleation and crystal growth, their resultant kinetics and methods of their experimental determination. From an industrial perspective finding drug properties that correlate with its recrystallization behavior is extremely important to facilitate the process of amorphous drug development. The experimental opportunities and existing models of drug long-term stability prediction are discussed extensively. Finally, we discuss the most long-standing issue in the field concerning methods of amorphous drug stabilization. Various well-established strategies and the most recent experimental results are presented and comprehensively discussed to give insight into the actual state of the art and to point out the most exciting research topics in the field. The last chapter gives some basic insight into various practical aspects of amorphous drug formulation and manufacturing (Chap. 6). To properly select the formulation composition and processing technology, the effect of different variables on quality and performance of the final product must be thoroughly understood. We hope that issues carefully chosen by us and described herein provide an in-depth understanding of the various aspects of working with amorphous products which will translate into further progress in this field. We believe that our expertise and interdisciplinary experience which we share with the readers will enable them to confidently and consciously enter the world of amorphous drug formulations in the future.

References

1. Savjani KT, Gajjar AK, Savjani JK (2012) Drug solubility: importance and enhancement techniques. ISRN Pharm 2012:195727. https://doi.org/10.5402/2012/195727
2. Lipinski CA, Lombardo F, Dominy BW, Feeney PJ (1997) Experimental and computational approaches to estimate solubility and permeability in drug discovery and development setting. Adv Drug Deliv Rev 23:3–25. https://doi.org/10.1016/S0169-409X(96)00423-1

3. Kalepu S, Nekkanti V (2015) Insoluble drug delivery strategies: review of recent advances and business prospects. Acta Pharm Sin B 5(5):442–453. https://doi.org/10.1016/j.apsb.2015.07.003
4. Ku MS (2008) Use of the biopharmaceutical classification system in early drug development. AAPS J 10(1):208–212. https://doi.org/10.1208/s12248-008-9020-0
5. Fridgeirsdottir GA, Harris R, Fischer PM, Roberts CJ (2016) Support tools in formulation development for poorly soluble drugs. J Pharm Sci 105(8):2260–2269. https://doi.org/10.1016/j.xphs.2016.05.024
6. Dahan A, Miller JM, Amidon GL (2009) Prediction of solubility and permeability class membership: provisional BCS classification of the world's top oral drugs. AAPS J 11 (4):740–746. https://doi.org/10.1208/s12248-009-9144-x
7. Amidon GL, Lenneras H, Shah VP, Crison JR (1995) A theoretical basis for a biopharmaceutical drug classification: the correlation of in vitro drug product dissolution and in vivo bioavailability. Pharm Res 12(3):413–420. https://doi.org/10.1023/A:1016212804288
8. Babu NJ, Nangia A (2011) Solubility advantage of amorphous drugs and pharmaceutical cocrystals. Cryst Growth Des 11:2662–2679
9. Kawabata Y, Wada K, Nakatani M, Yamada S, Onoue S (2011) Formulation design for poorly water-soluble drugs based on biopharmaceutics classification system: basic approaches and practical applications. Int J Pharm 420(1):1–10. https://doi.org/10.1016/j.ijpharm.2011.08.032
10. Leuner C, Dressman J (2000) Improving drug solubility for oral delivery using solid dispersions. Eur J Pharm Biopharm 50(1):47–60. https://doi.org/10.1016/S0939-6411(00)00076-X
11. Horter D, Dressman JB (1997) Influence of physiochemical properties on dissolution of drugs. Adv Drug Deliv Rev 25(1):3–14. https://doi.org/10.1016/S0169-409X(96)00487-5
12. Newman A (2015) Pharmaceutical amorphous solid dispersions. Wiley, Hoboken, NJ
13. Costa P, Lobo JMS (2001) Modeling and comparison of dissolution profile. Eur J Pharm Sci 13:123–133. https://doi.org/10.1016/S0928-0987(01)00095-1
14. Janssens S, Van den Mooter G (2009) Review: Physical chemistry of solid dispersions. J Pharm Pharmacol 61(12):1571–1586. https://doi.org/10.1211/jpp/61.12.0001
15. Kasim NA, Whitehouse M, Ramachandran C et al (2004) Molecular properties of WHO essential drugs and provisional biopharmaceutical classification. Mol Pharm 1(1):85–96. https://doi.org/10.1021/mp034006h
16. Hancock BC, Zografi G (1997) Characteristics and significance of the amorphous state in pharmaceutical systems. J Pharm Sci 86(1):1–12. https://doi.org/10.1021/js9601896
17. Murdande SB, Pikal MJ, Shanker RM, Bogner RH (2011) Aqueous solubility of crystalline and amorphous drugs: challenges in measurement. Pharm Dev Technol 16:187–200. https://doi.org/10.3109/10837451003774377
18. Shah N, Sandhu H, Choi DS, Chokshi H, Malick AW (2014) Amorphous solid dispersions: theory and practice. Springer, New York
19. Zografi G, Newman A (2017) Interrelationships between structure and the properties of amorphous solids of pharmaceutical interest. J Pharm Sci 106:5–27. https://doi.org/10.1016/j.xphs.2016.05.001
20. Grohganz H, Priemel PA, Löbmann K et al (2014) Refining stability and dissolution rate of amorphous drug formulations. Expert Opin Drug Deliv 11(6):977–989. https://doi.org/10.1517/17425247.2014.911728
21. Hancock BC, Parks M (2000) What is the true solubility advantage for amorphous pharmaceuticals? Pharm Res 17(4):397–404. https://www.ncbi.nlm.nih.gov/labs/articles/10870982/

Chapter 2
Order vs. Disorder in the Solid State

2.1 Perfect Order (Crystalline Materials)

2.1.1 Periodic Structures

According to the definition given by the International Union of Crystallography, "*by "crystal" is meant any solid having an essentially discrete diffraction diagram*" [1]. A typical diffraction pattern corresponding to a "classical" periodic, perfect crystal looks like the one shown in Fig. 2.1 [2]. This pattern corresponds to the inner structure of the material, which can be represented as an array of periodically repeating fragments. The whole structure can be described by defining the repeating fragment (basis) and a set of three non-coplanar unit vectors. The three unit vectors can be used to build a parallelepiped: a unit cell. Translations are not the only symmetry elements that can be used to describe a periodic structure. Combinations of mirror reflections, rotations, inversions, glides and screw rotations form groups, that are termed space symmetry groups [3]. A periodic structure can then be described only by defining the crystallographic[1] coordinates of an asymmetric unit and the symmetry operations of the space symmetry group.

2.1.2 Aperiodic Structures

For a periodic crystal structure the positions of the diffraction patterns can be expressed by:

[1]Crystallographic coordinates are defined in the coordination system related to the three primitive translation vectors.

M. Rams-Baron et al., *Amorphous Drugs*,
https://doi.org/10.1007/978-3-319-72002-9_2

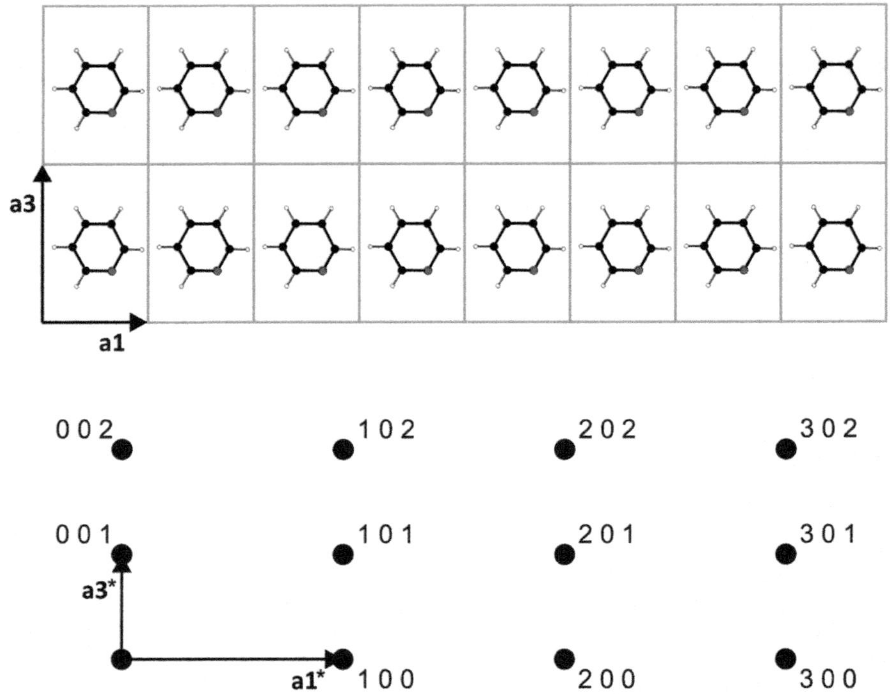

Fig. 2.1 Above: a schematic presentation of an imaginary periodic structure which can be represented as a three-dimensional array of periodically repeated fragments (only a 2D layer is shown for clarity), $\mathbf{a_1}$ and $\mathbf{a_3}$—unit translation vectors defining an elementary cell; Below: simplified periodic diffraction pattern corresponding to the imaginary periodic structure shown above, $\mathbf{a_1}$* and $\mathbf{a_3}$*—unit vectors in the reciprocal space corresponding to the $\mathbf{a_1}$ and $\mathbf{a_3}$ vectors in the direct space. Ratio $\mathbf{a}^*_1$: $\mathbf{a}^*_3$ is inverse to $\mathbf{a}^*_1$: $\mathbf{a}^*_3$. Numbers show the indices of reflections equal to h_1, h_2, h_3 in Eq. (2.1) [2]. Reproduced with permission of the "International Union of Crystallography" from [2]. http://journals.iucr.org/

$$\mathbf{H} = h_1\mathbf{a_1}^* + h_2\mathbf{a_2}^* + h_3\mathbf{a_3}^*, \tag{2.1}$$

where the three vectors $\mathbf{a_i}^*$ in reciprocal space are related to the basic translations $\mathbf{a_i}$ as:

$$\mathbf{a_1}^* = [\mathbf{a_2} \times \mathbf{a_3}]/\mathbf{a_1}[\mathbf{a_2} \times \mathbf{a_3}]; \quad \mathbf{a_2}^* = [\mathbf{a_3} \times \mathbf{a_1}]/\mathbf{a_1}[\mathbf{a_2} \times \mathbf{a_3}];$$
$$\mathbf{a_3}^* = [\mathbf{a_1} \times \mathbf{a_2}]/\mathbf{a_1}[\mathbf{a_2} \times \mathbf{a_3}]. \tag{2.2}$$

There are other structures for which three translation vectors are not sufficient to describe all the diffraction maxima, and additional terms must be added to Eq. (2.1):

$$\mathbf{H} = h_1\mathbf{a_1}^* + h_2\mathbf{a_2}^* + h_3\mathbf{a_3}^* + h_4\mathbf{a_4}^* + \ldots + h_n\mathbf{a_n}^*, \tag{2.3}$$

where $\mathbf{a_i}^*$ and h_i are the reciprocal lattice vectors and integer coefficients, respectively, and the number n is the minimum number for which the positions of the

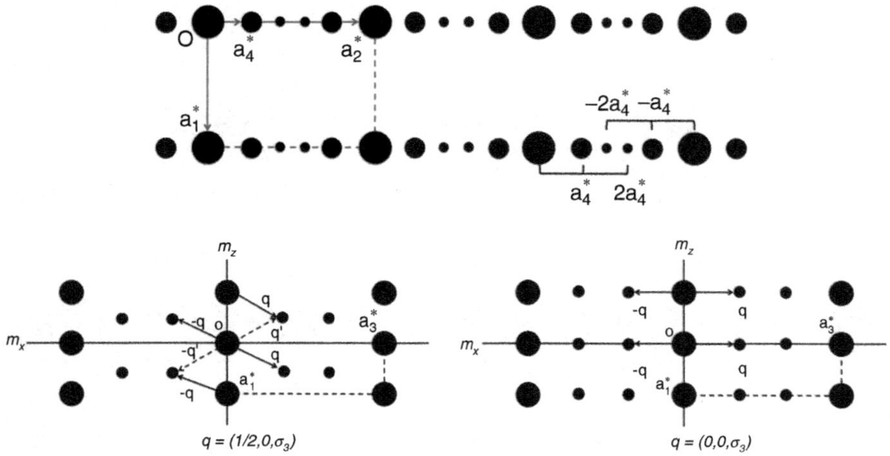

Fig. 2.2 Examples of diffraction patterns of modulated structures: periodic arrays of stronger reflections with weaker satellites [4]

peaks can be described with coefficient h_i. The conventional periodic crystals are a special, though very large, class for which $n = 3$. Crystals for which $n > 3$ are termed aperiodic crystals.

Two fundamentally different types of the aperiodic crystals are known: *incommensurately modulated phases* and *quasicrystals*. The first type relates to periodic crystals: one can find the "main" structural motif, which is periodic, and impose a periodic modulation on this motif, such that the ratio of the two periods is an irrational number. A diffraction pattern in this case will look like a periodic array of stronger reflections with weaker satellites (Fig. 2.2) [4]. The positions of all the reflections can be described as:

$$\mathbf{H} = \mathrm{h}_1\mathbf{a}_1{}^* + \mathrm{h}_2\mathbf{a}_2{}^* + \mathrm{h}_3\mathbf{a}_3{}^* + \mathrm{h}_4\mathbf{a}_4{}^* = \mathrm{h}_1\mathbf{a}_1{}^* + \mathrm{h}_2\mathbf{a}_2{}^* + \mathrm{h}_3\mathbf{a}_3{}^* + \mathrm{m}\mathbf{q}, \quad (2.4)$$

where the first three terms correspond to the positions of "main" reflections, and the modulation vector **q** defines the position of the satellites.

$$\mathbf{q} = \mathbf{a}_4{}^* = \sigma_1\mathbf{a}_1{}^* + \sigma_2\mathbf{a}_2{}^* + \sigma_3\mathbf{a}_3{}^* \quad (2.5)$$

Modulation of the periodicity can be due to a variety of physical phenomena. In some cases modulation arises from variation in the population of positions, conformational and/or orientational variability of molecules as a whole and/or molecular fragments, the rotation of spin, magnetic, or dipole moments, or the incompatibility of the translation periods of different sub-lattices. Incompatibility of sublattices can often arise in the cases where surface layers were grown on a support, or in host-guest compounds, including those where the "host" and "guest" are the same chemical species (Fig. 2.3). There are similarities between modulated structures and structures that contain multiple chemically identical species in the

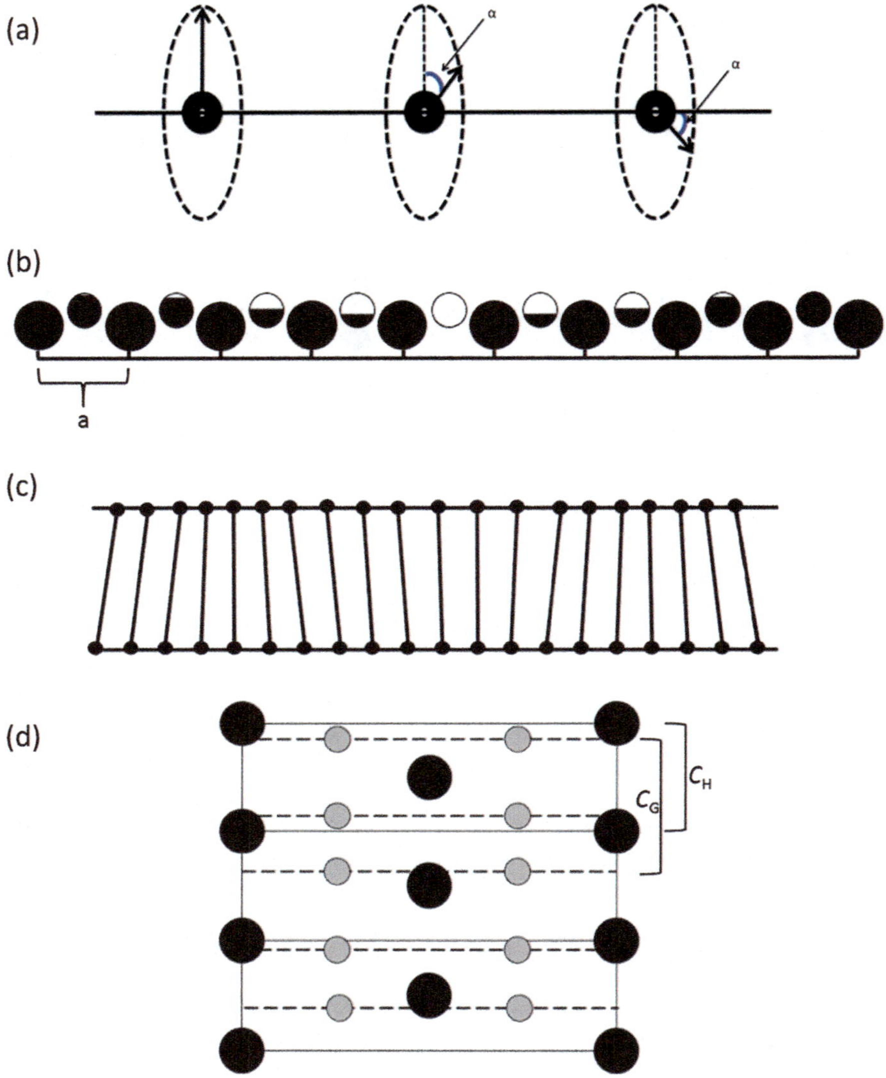

Fig. 2.3 Examples of modulated structures. (**a**) Rotation of fragments, (**b**) modulation of the site occupancies, (**c**) modulation of displacements of species from periodic positions, (**d**) incommensurate translation periods of the sublattices

same unit cell (crystal structures with $z' > 1$) (Fig. 2.4) [2]. Precise diffraction data and skilled data analysis are needed to reliably distinguish between incommensurately modulated structures, structures with disorder, and ordered structures with multiple species in the same unit cell [5].

Aperiodic crystals of this type are fundamentally different from both periodic crystals and from incommensurately modulated phases. Their diffraction patterns

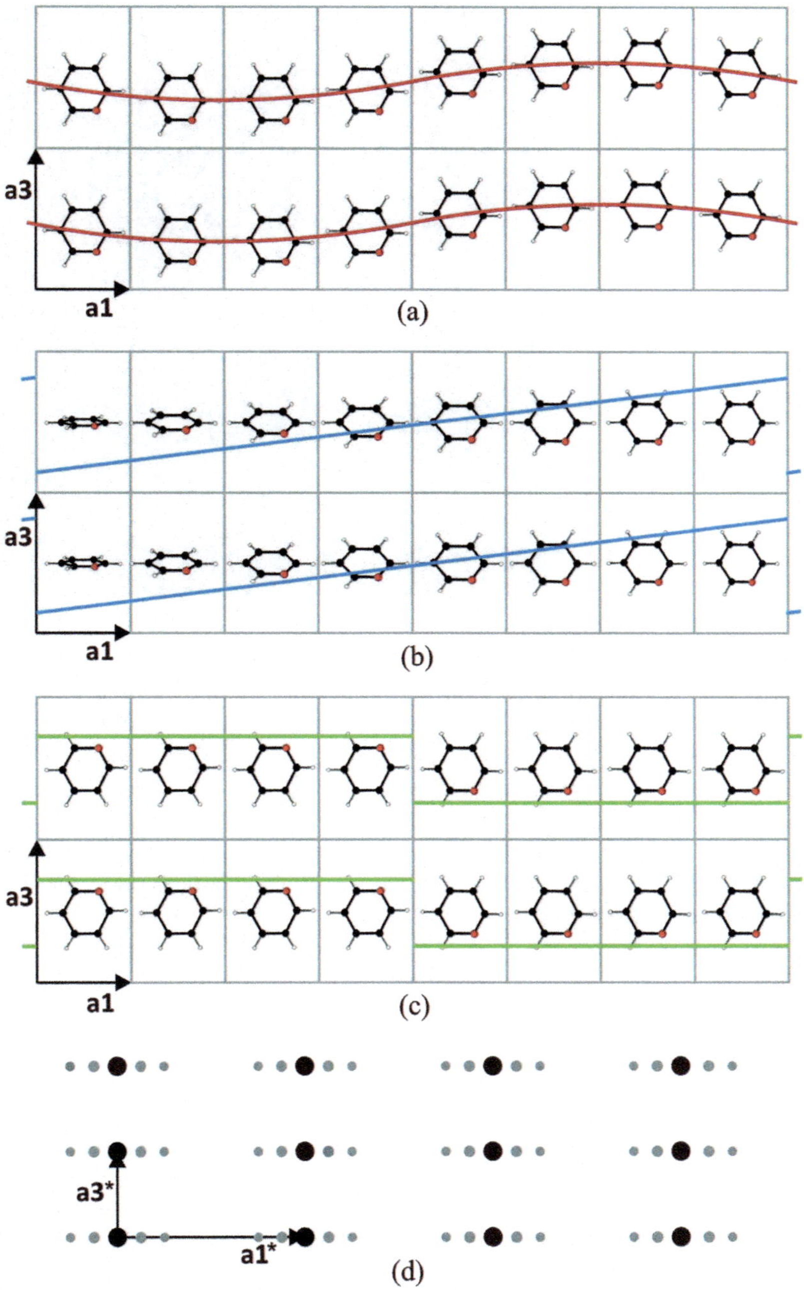

Fig. 2.4 Schematic representation of three variants of a modulated structure with lost translational symmetry along the a_1 axis. All three drawings are derived from the periodic structure shown in the Fig. 2.1 by shifting or rotating the molecules. The atomic modulation functions which can describe the atomic positions are shown as an overlay: (**a**) the molecules are shifted up and down parallel to

Fig. 2.5 A sample diffraction pattern from a quasicrystal [6, 7]

are characterized by sharp intensity maxima and symmetry that is incompatible with lattice translations (Fig. 2.5) [6]. Such phases were first discovered for some Al-containing intermetallics. Today, examples of the quasicrystalline structures have been reported for various classes of compounds, including organic molecular crystals, polymers and even biomolecules [6–17]. Though very different from periodic crystals, quasicrystals are highly ordered: for any site at some distance from another, the structure is unambiguously defined. Penrose tilings play the same role for describing quasicrystalline structures as Bravais lattices do for describing periodic structures (Fig. 2.6).

Fig. 2.4 (continued) a_2 in a continuous harmonic (sinusoidal) way (**red** curve); (**b**) the molecules are rotated around an axis parallel to a_1, the rotation angle can be described using a sawtooth function (**blue**) with a discontinuity between molecules 8 and 1; (**c**) the molecule adopts two different orientations which can be described by a step-like crenel function (**green**); (**d**) schematic diffraction pattern with satellite reflections (**grey** circles) along a_1*. The modulation proceeds only along a, the c direction is not affected. The number of satellite reflections and their intensity distribution depend on the strength and nature of the modulation. For simplicity, only one diffraction scheme was drawn [2]. Reproduced with permission of the International Union of Crystallography (http://journals.iucr.org/)

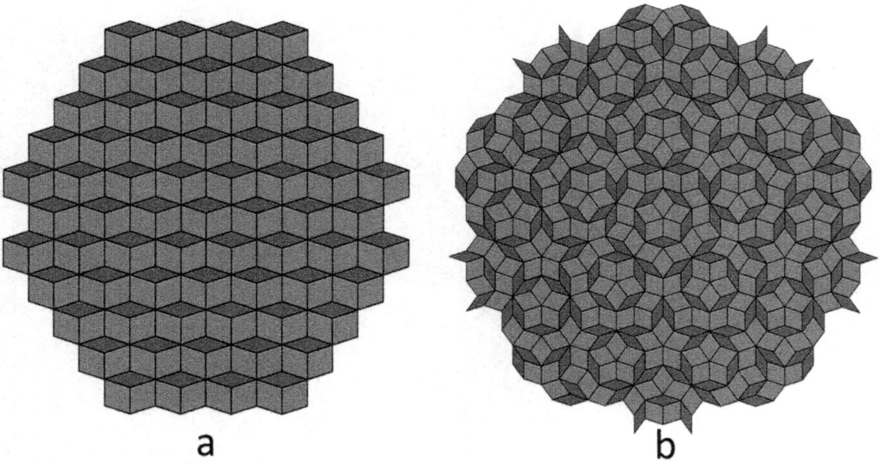

Fig. 2.6 Penrose tilings used to describe periodic (**a**) and aperiodic structures (**b**). In case (**a**) the pattern can be described by a Bravais lattice

2.2 Perfect Disorder (Amorphous Materials)

The structures of both periodic and aperiodic crystals have a common feature: a general law unambiguously defines the structure at any point, i.e. *long-range order* exists. In this respect crystalline structures differ radically from amorphous ones, in which long-range order is absent. This is immediately seen from an X-ray diffraction pattern where no bright diffraction maxima are seen, but instead an "amorphous halo" is present (Fig. 2.7). However, this does not mean that an amorphous structure has no order at all. On the contrary, the structures of amorphous solids are built following certain common, basic principles and can be characterized both qualitatively and quantitatively. In fact, there is often considerable structural order in amorphous solids over length-scales of many Å. The short-range order in amorphous and crystalline solids can be very similar in some cases. This is particularly true in systems such as inorganic oxides [18–21]. However, in other cases, the short-range order differs drastically, e.g. for some organic compounds, where even the molecular structure (conformation) can differ between the crystalline and amorphous phases [22, 23]. The presence of structural disorder in glasses requires statistical structural parameters to provide a spherically averaged description of atomic structure. It is such parameters that are usually measured macroscopically [21].

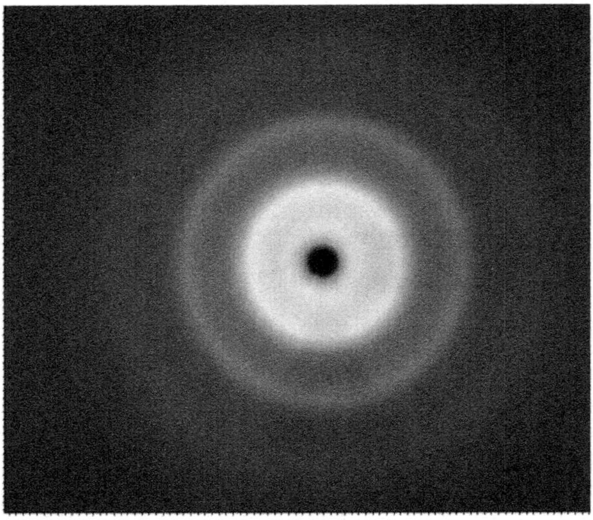

Fig. 2.7 A sample diffraction pattern from an amorphous sample

2.2.1 *Radial Distribution Function*

The information on local structure, intermediate- and long-range order is contained in the pair distribution function (PDF) g($\boldsymbol{r}$), also termed pair correlation function (PCF), or radial distribution function (RDF). The RDF in a system of particles (atoms, molecules, colloids, *etc.*) describes the variation in density as a function of distance from a reference particle. In the simplest terms, the RDF is a measure of the probability of finding a particle at distance $\boldsymbol{r}$ from a given reference particle. The general algorithm involves determining how many particles are within a distance of $\boldsymbol{r}$ and $\boldsymbol{r} + \mathrm{d}\boldsymbol{r}$ from a particle (Fig. 2.8). The PDF is usually determined by calculating the distance between all particle pairs and binning them into a histogram. The histogram is then normalized with respect to the case where histograms are completely uncorrelated. For three dimensions, this normalization is the number density of the system, multiplied by the volume of the spherical shell. Mathematically, this can be expressed as:

$$\mathrm{g}(\boldsymbol{r}) = 4\pi \boldsymbol{r}^2 \rho \mathrm{d}\boldsymbol{r}, \tag{2.6}$$

where ρ is the number density. The value of g($\boldsymbol{r}$) is often plotted as T($\boldsymbol{r}$) = g($\boldsymbol{r}$)/*r*, as in Fig. 2.9.

It is clear from a PDF, that an amorphous structure is not "chaotic", but is characterized by short-range order. This order manifests itself in a series of maxima and minima of the PDF at selected distances. In general, the relative intensities and positions of the maxima of a PDF are characteristic for an amorphous structure. It contains considerable detail about the structural order surrounding each type of atom: positions of peaks give the radii of successive shells of atoms surrounding the average atom and the areas of the Gaussian peaks yield the number of atoms in each

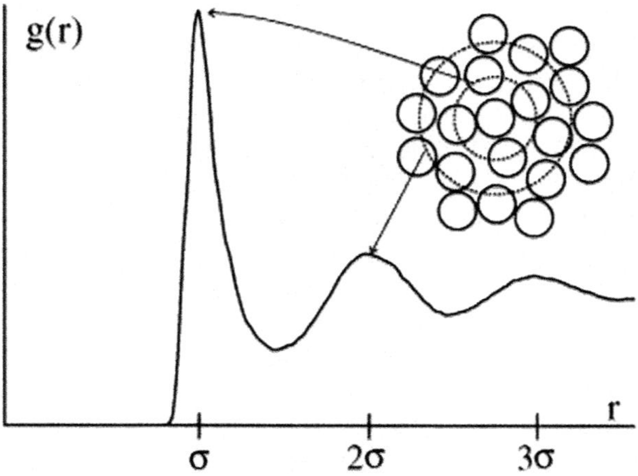

Fig. 2.8 A schematic illustration of the general algorithm of calculating a RDF (the number of particles within a distance of r away from a selected particle) [24]

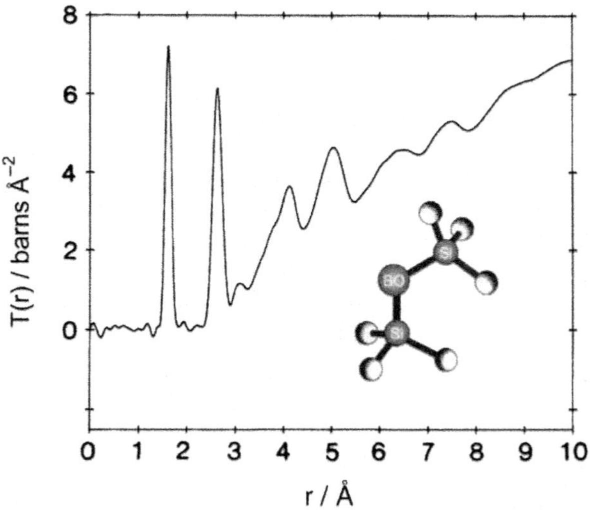

Fig. 2.9 An example of the radial distribution function: the normalized correlation function T(r) = g(r)/r characterizing a silica glass (obtained by neutron scattering). The first three peaks correspond to Si–O, O–O and Si–Si correlations, respectively [20]

of these shells [21]. For example, the first three peaks of the PDF for silica glass (Fig. 2.9) correspond to Si–O, O–O and Si–Si correlations.

One can determine g(***r***) indirectly using neutron scattering or X-ray scattering data [21–23, 25–43]. It is derived from the measured scattered intensity, I(Q), by Fourier transforming the normalised X-ray or neutron structure factor S(Q). Here, Q is the scattering vector $4\pi\sin\Theta/\lambda$, 2Θ is the scattering angle and λ is the X-ray or neutron wavelength. The technique can be used to probe structure at very short length scales (down to the atomic level), but involves significant space and time averaging (over the sample size and the acquisition time, respectively). In this way, the radial distribution function has been determined for a wide variety of systems, ranging from liquid metals to charged colloids [44–49]. It should be noted that going from the experimental data to g(r) is not straightforward and the analysis can be quite complicated. However, modern techniques make it possible to not only solve complex disordered and truly amorphous structures, but also to quantify the ratio of amorphous to crystalline phases in composite multi-phase samples [50].

While the PDF technique has so far been limited to experiments at large scale X-ray and neutron facilities, recent developments in laboratory-based equipment offers the potential for diffraction experiments that are available year round. Another new development is the use of electron diffraction, which could offer new access to instrumentation for the analysis of the local structure of complex materials. At the same time, development at large scale facilities has not stopped, and measurement times have dropped from some 24 hours to fractions of a second. This now allows for the study of kinetics, and the analysis of dynamics in disordered materials. Finally, complementary experiments with X-ray absorption techniques have been long used to study the local structure of materials with sensitivity to the chemical nature of the individual atoms. Traditionally, PDF has been blind to the identity of atoms, unless X-ray and neutron data were combined. This combination gives at least partial insight into atomic identity However, with the use of anomalous powder diffraction methods, this "blindness" can be cured [21].

It is also possible to calculate g(r) directly by extracting particle positions from traditional or confocal microscopy [44–49]. This technique is limited to particles large enough for optical detection (in the micrometer range). While this does limit the application of such techniques, it has the advantage of offering time-resolved experiments so that, aside from statical information, it also gives access to dynamical parameters (*e.g.* diffusion constants [45]) and is also space-resolved (to the level of the individual particle). This spatial-resolution allows for optical techniques to reveal the morphology and dynamics of local structures in colloidal crystals, glasses, and gels [21, 44–49]. Interferometric scattering microscopy enables a researcher to go down to nanometer range.

The measurement of the distribution of pairwise atomic distances is insensitive to orientational order. It cannot therefore provide a complete picture of such phenomena, as supercooling, or the glass transition. Fluctuation scattering with electrons and X-rays can provide this orientational sensitivity. However, it is difficult to interpret the fluctuation data. The fluctuation diffraction data are simpler to interpret if converted into a real-space angular distribution function. The extracted angular distributions contain rich information about orientational order and bond angles. The diffraction fluctuations are potentially measurable with electron sources and also with the brightest X-ray sources, like X-ray free-electron lasers [51, 52].

By definition, the structure of a glass is non-periodic. That said, if one analyses the diffraction pattern (either X-ray or neutron) of such a material, they will observe a sharp first diffraction peak (SFDP). This is indicative of partial order (or remnants of periodicity) within the structure, and makes negligible contribution to the local order of the RDF.

The spacing that corresponds to this 'quasi-periodicity' is given by 2π/QSFDP, and thus relates to the position of the FSDP. The length-scale over which this periodicity is maintained is related to the width of the SFDP, with a sharper peak indicating longer length-scale correlation [53–55]. Experimental data for silica, with a SFDP at 1.52 Å and width of 0.4 Å, suggests that spacing of quasi-periodicity is approximately 4.1 Å, and continues over lengths of 16 Å. It is therefore clear that the correlation length over which this quasi-periodicity spans, determines in part the nature of the long-range order of the system. Instead, the spacing of the quasi-periodic order (here determined by the orientation of corner-sharing SiO_4 units) [54], determines the intermediate-range order.

As with all diffraction, the intensity of the diffraction peak corresponds to the quantity of diffraction electronic density. Thus, by studying the intensity of the SFDP, one can obtain information regarding the distribution of voids associated with corner-sharing polyhedra. [54, 55]. This intermediate-range order resembles the liquid structure [21]. It follows that such information therefore leads to insight into the density of the material. For example, when studying silicas, the SFDP intensity is found to increase with temperature. In contrast, this intensity decreases with applied pressure [54]. In densified silicas, the position of the FSDP moves to larger wave vectors; its strength decreases with increasing density [56].

2.2.2 Voronoi-Dirichlet Polyhedra (VDP)

The Bravais lattice is a universal tool used to describe the translational symmetry of a periodic structure. The three non-coplanar unit translation vectors define a unit cell, which is the same throughout the whole sample (Fig. 2.1). For amorphous structures this approach is not applicable. There is instead another method which can be used to describe the full range of periodic and aperiodic structures: Voronoi-Dirichlet polyhedra (VDP) [57, 58]. A VDP defines the part of space which is closer to a selected point than to any other point. It can be calculated by connecting a given lattice point with all other points by lines. These lines are then bisected by planes, and the polyhedron that results from the intercrossing planes becomes the VDP (Fig. 2.10). For a periodic lattice, the VDP are identical for any lattice point. Instead, for an amorphous lattice a set of VDP will be obtained. These can be characterized by an average, minimum, and maximum size of the VDP, and the width of their size distribution. The number of faces—corresponding to the first coordination number—is another quantitative characteristics that contains information on the amorphous structure and chemical interactions within it. For simple monocomponent glasses the calculation of the VDP is rather straightforward: it is sufficient to connect the centroids of the chemically identical species. For multicomponent systems the calculation of VDP is no longer unambiguous since one can divide a distance between the different species in different ways. One approach in this case is to calculate VDP for each of the components and compare the sub-lattices.

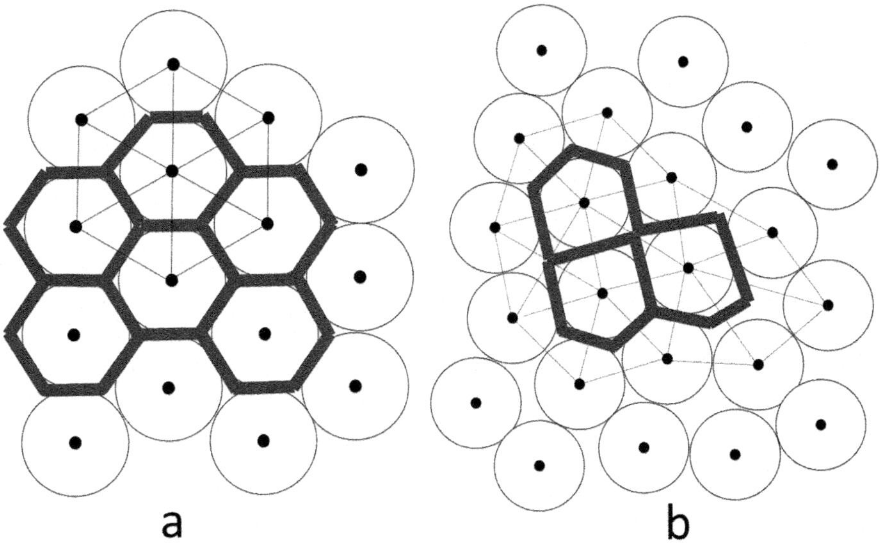

Fig. 2.10 Voronoi tessellation: (**a**) for a periodic lattice, (**b**) for an amorphous structure

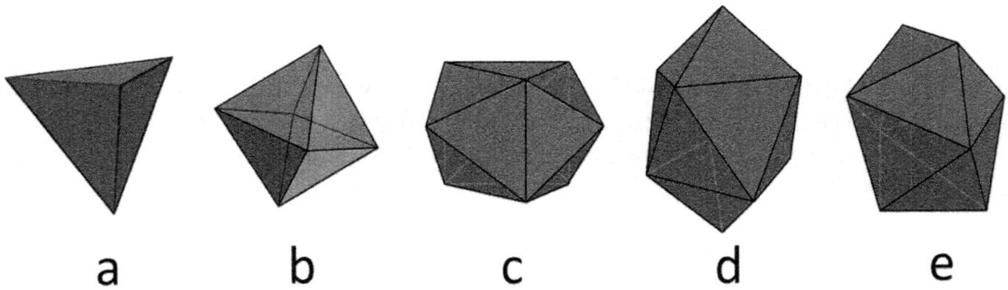

Fig. 2.11 Canonical Bernal polyhedra present in RCP: (**a**) tetrahedron, (**b**) octahedron, (**c**) trigonal prism, (**d**) Archimedian antiprism, (**e**) tetragonal dodecahedron

2.2.3 Random Close Packing (RCP)

Amorphous structures can be described as a random close packing (RCP) of spheres [59–61]. The simplest case, analogous to a model of monoatomic glasses or glasses formed by spherical species, is a RCP of equal spheres. The dominant coordination polyhedron in such a structure is tetrahedron; the polyhedron that corresponds to the densest possible packing if no periodicity condition is imposed. Other canonical Bernal polyhedra that can be found in RCP structures include (semi-)octahedra, trigonal prisms, tetragonal dodecahedra, and Archimedean antiprisms (Fig. 2.11). In this respect, the random close packing is principally different from a periodic close packing, in which only the tetrahedra and octahedra are present as

Table 2.1 Relative proportion of different coordination polyhedra in RCP[a]

Type of polyhedron	Relative proportion/% By number	By volume
Tetrahedron	73.0	48.4
Truncated octahedron	20.3	26.9
Trigonal prism	3.2	7.8
Antiprism	0.4	2.1
Tetragonal dodecahedron	3.1	14.8

[a]For comparison, in a periodic close packing only tetrahedra and octahedra with number ratio 2:1 exist. Average space filling coefficients are 74% and 64% for regular and random close packing, respectively. However, while close packing is uniform in periodic structures, the denser regions (87%) in RCP alternate with looser regions

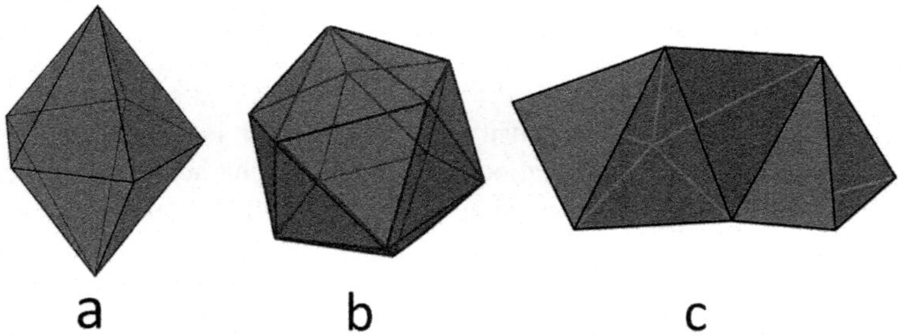

Fig. 2.12 Coordination polyhedral corresponding to clusters of higher density in amorphous structures: (**a**) pentagonal pyramid; (**b**) icosahedron; (**c**) chain of tetrahedra

coordination polyhedra, the number of tetrahedra being equal to the number of semi-octahedra (Table 2.1).

The basic difference in the structures of RCP and periodic close packings explains the difficulties related to transitions between the two and the existence of such phenomena as supercooling of melts and hindered crystallization of glasses. By sharing faces, tetrahedra can form chains, icosahedra, and pentagonal pyramids (Fig. 2.12). These structural units account for the local regions with higher density than the sample average, and alternate with looser packed regions. This spatial inhomogeneity of the RCP is one of the intrinsic features that makes it very different from periodic close packing, in which the density is uniform throughout the entire sample [59–61]. The model can also be extended to the case of packing of spheres of different radii [62]. Density fluctuations in glasses can be measured using small-angle X-ray scattering, sometimes coupled with inelastic X-ray scattering. Long-range order in glass-formers is governed by the fluctuations in density present in the liquid, which become structurally locked in once temperatures go below the glass transition temperature [21]. A long-lasting discussion on the structure of glasses is related to the interpreting of structural inhomogeneities and density fluctuations [18–23, 25]. One point of view is that many glasses contain nanocrystalline inclusions, embedded into a fluidized matrix. In this case, the inclusions and

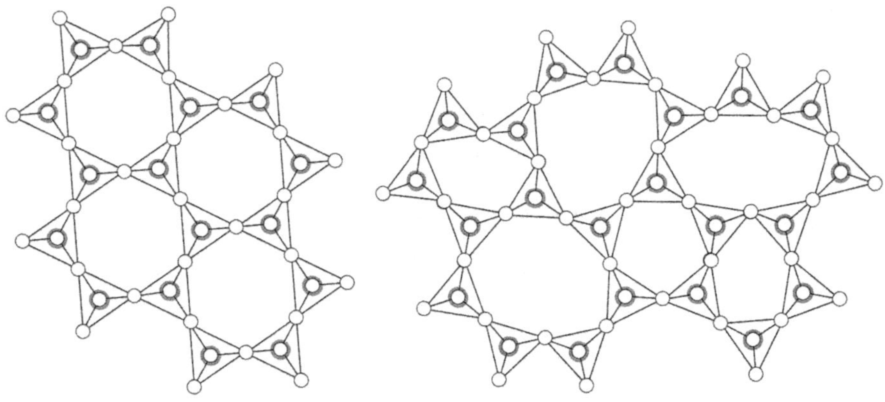

a b

Fig. 2.13 Polyhedral networks: (**a**) regular, (**b**) random

the matrix differ in structure and density. Another opinion is that the whole glass is a random network in which alternations of density are characteristic of the particular network. Recent developments suggest that in fact both variants are possible. Glasses built of "separate particles" (metal atoms, non-hydrogen-bonded organic molecules, macromolecular globules, etc.) are in fact usually heterogeneous and often contain crystalline inclusions. On the contrary, inorganic oxide glasses, which form extended networks of coordination polyhedra, are often single-phase, even if their properties are not uniform throughout the sample.

2.2.4 Polyhedral Networks

For many amorphous solids the structure can be efficiently presented as a network of polyhedra sharing vertices. This is the case of many inorganic oxide glasses [18–21, 63–71]. This way of presenting a structure is similar to the description of ordered periodic structures and is especially practical when a regular structure is compared with the amorphous one (Fig. 2.13).

Many important glasses and melts, like silicates, borates, alumino-silicates, halides and chalcogenides can be described as networks of connected polyhedra. Their structural characteristics can be drawn from the results of spectroscopy and scattering experiments. Nuclear Magnetic Resonance (NMR), X-ray Absorption Fine Structure (XAFS), Neutron Scattering (NS) and both Small- and Wide-angle X-ray Scattering measurements (SAXS/WAXS), which are also often combined with computer simulations, can give detailed information of structure and diffusion in the glassy, as well as the molten, state [21]. All of these techniques reveal a wealth of order, both at the local level of individual coordination polyhedral, and in the intermediate range, governed by the connectivity between adjacent polyhedra.

The more distant, longer-range order, is also present in inorganic glass systems. Its discovery was an exciting development. This distant order includes quasi-periodicity in the network and channels of network modifiers, and has obvious implications for ionic transport. Many tetrahedral glass-formers (oxides, halides or chalcogenides) are chemically and topologically ordered over many nanometres [21]. Over the shorter distances that define intermediate-range order, polyhedra self-assemble to varying extents through quasi-periodic regions. There can be more than one source of quasi-periodicity in the same material, for example, when different polyhedra microsegregate. The persistence of these correlated regions in melts supports the view that the topology of glassy networks can be considered as snapshots of the diffusional dynamics of the liquid state [68]. The dimensionality of the long-range order on the scale of nanometres in modified glasses can range from 3 (in aluminosilicates and boro-silicates) to 1 (in polyphosphates). The presence of the long-range order agrees with the high coefficients of ionic diffusion [64]. The short-range structure of oxide glasses can be extremely well-defined in terms of the coordination polyhedra of the network-forming cations (such as Si, P, B and Ge). The bonding between network-forming cations and oxygen in these glasses is heteropolar. Bond lengths and angles in the first coordination shell of oxygen around these cations are almost the same in all the glasses. Glass-forming cation–oxygen polyhedra like SiO_4 are usually corner-linked through 'bridging' oxygens, and form a three-dimensional extended connected network. Structural disorder in a glass network is partly related to the statistical distribution of angles between the coordination polyhedral (compared to fixed values in regular crystals). This disorder manifests itself also in the distribution of the distances to the next nearest neighbour atoms (e.g. Si–Si in a silicate glass). A random polyhedral network comprises rings of different size. The coordination spheres of atomic neighbours become increasingly broad as the distance from a given atom increases. Nevertheless, the connectivity of a random network is near-perfect, so that the topological and chemical order in network glasses can be deciphered experimentally up to 60 Å from an average atom [63]. The combination of complete connectivity and radial loss of structural order is illustrated in Fig. 2.14 schematically presenting the Continuous Random Network (CRN) structural model for glasses [65].

The so called "network-modifying" cations, such as alkali and alkaline-earth ions, disrupt the connectivity of the oxide network and create 'non-bridging' oxygens that are linked to only one network-forming cation [18–21]. The nearest-neighbour oxygen coordination of these network-modifier polyhedra are usually larger than for network-formers, and the oxygen distance is longer, with a wider distribution of distances, [66]. Likewise, non-bridging oxygen polyhedra can incorporate single network-forming cations, or can incorporate several network modifying cations. The structure of a modified glass can then be pictured as a combination of a partly depolymerized network and an ionically packed modifying oxide. The Modified Random Network (MRN) model is based on the different sizes of network-forming and network-modifying polyhedra that are measured. This model results in networks and modifier components being microsegregated, with the modifier ions forming channels within a depolymerized network structure

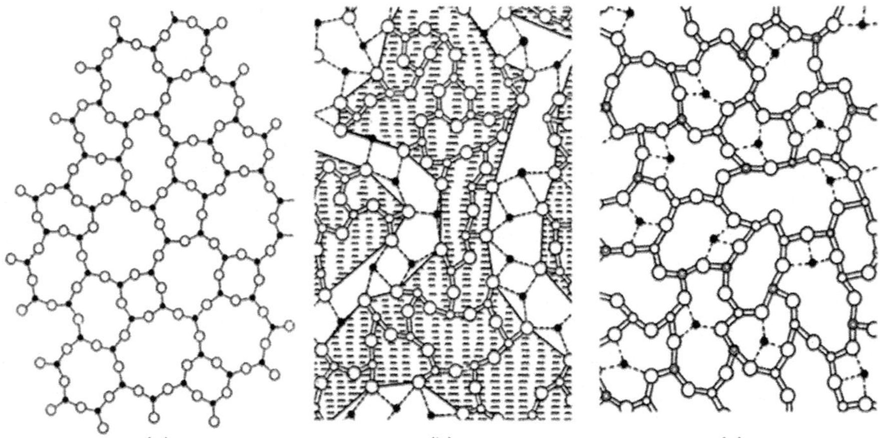

Fig. 2.14 Schematical two-dimensional representations of different random network models for polyhedral glasses (corner-sharing tetrahedral oxides as examples): (**a**) continuous random network model (CRN), SiO_2; (**b**) modified random network (MRN), modified silicate; compensated continuous random network (CCRN), an alumino-silicate glass. In (**b**) the filled circles refer to modifier alkali or alkaline earth cations; bridging O atoms are located within the network, non-bridging O atoms—along the modifier channels. (**c**) [20, 21]

[18, 20, 21, 66] (Fig. 2.14). When networks are modified (network depolymerization or mixing of different glass-formers), some degree of microsegregation is nearly always observed. As a result, the channels are formed in the glass, and the regions formed by different glass-formers segregate. A good example can be found in the separation of silicate regions from borate regions in borosilicate glasses [21].

Structural order beyond the short range of nearest neighbours is usually referred to as intermediate-range order, and involves the important question of how the coordination polyhedra are connected to one another. This is of particular importance when concerning the connectivity of network-forming cations. In glasses like silica, the halides and chalcogenides, a significant amount of thermal and photo-induced structural order–disorder phenomena have been observed. There are also temperature-dependent phenomena in the corresponding liquids prior to quenching, all of which involve changes in the intermediate-range structure. Long-range structural characteristics in glasses (length-scales of 10 Å and above) reflect the degree of homogeneity and dimensionality of the spatial distribution of constituent atoms that have been frozen in from the molten state.

Random networks can be described using statistical approach: characterizing bond angle and bond length distributions. For example, in silica glasses, three-dimensional CRNs provide a direct link with the multiplicity of ring sizes of corner-sharing SiO_4 tetrahedra that are facilitated through variations in the bond angle and the dihedral angle. Topologically variable ring sizes hinder periodic order, but are tolerated in random networks. Ring sizes are generally determined by the shortest path [67]. Computer simulations for silica indicate that five-, six-, seven- and eight-membered rings are the most abundant sizes [68–70]. Three- and four-membered rings are also found, but in smaller numbers.

Silicate oxide glasses encompass fully polymeric structures of glass-formers like SiO_2 together with modified glasses, where structures are depolymerized to varying degrees with commensurate increases in the fragility of the originating glass-forming liquid and often a lowering of the glass transition temperature, Tg. Where correlations between adjacent polyhedra promote pseudo-periodic and microsegregated regions in the long-range, density fluctuations tend to limit the pseudo-periodic correlation lengths. Accordingly, those glasses where short, intermediate- and long-range order is most uniformly connected are the glasses most optically transmitting but also derive from the most fragile liquids [20, 21].

2.2.5 Molecular Glasses and Hybrid Materials

The structures of molecular amorphous solids have much in common with those of metal or inorganic glasses, although they do exhibit some peculiar features. If molecules can form hydrogen bonds, then networks of connected molecules are likely to be formed. The structures then resemble those of CRN inorganic glasses. For example, water is a well-known "tetrahedral liquid" that shows an increased density with increasing temperature [71], similar to silica glass [72], although at notably different temperatures. Glasses formed by molecules that do not form networks via hydrogen bonds or other types of directional intermolecular interactions resemble RCPs of metallic glasses [21]. Like simpler, atomic fluids, these systems exhibit a limiting bulk density below which minimum energy configurations are no longer spatially homogeneous, but consist instead of a locally dense fraction and large, system-spanning voids [23].

At the same time, organic molecular glasses have more potential variables as compared to inorganic glasses [22, 23, 25, 73]. In particular, non-spherical molecules can rotate, and this can account for the orientational disorder. Most organic molecules can change their conformations, provoquing conformational disordering. Many molecules can undergo zwitter-ionic intramolecular isomerization, or a proton can migrate between different molecules, forming a salt in the two-component system. Molecules in different amorphous states can form different clusters (Fig. 2.15), which can potentially serve as nucleation centres to give different crystalline polymorphs on storage of the amorphous samples. Crystallisation nucleated by these clusters can also occur on heating or on compression of the amorphous state (Fig. 2.16) [74–78]. For many organic compounds it has been documented that a so-called "glacial state" is not in fact a homogeneous amorphous phase, but a mixed supercooled liquid/nanocrystallites state [23, 79]. The latter is interpreted as a heavily nucleated state composed of nanocrystals of the stable crystalline phase, embedded in the matrix of the non-transformed supercooled liquid. The origin of the intriguing relative stability

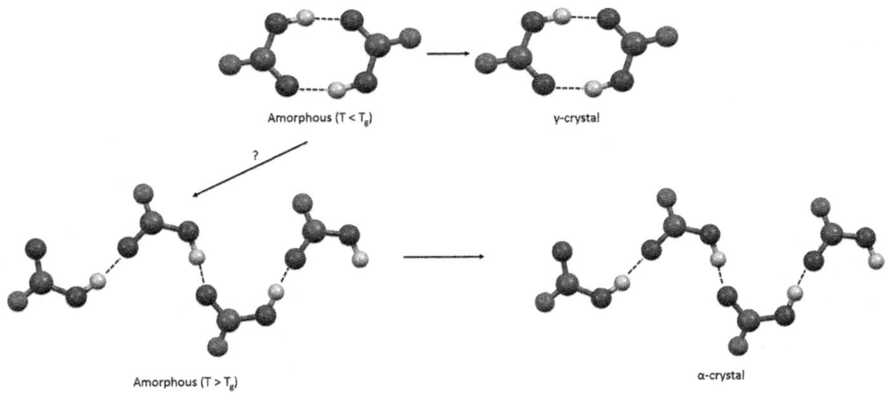

Fig. 2.15 Different molecular clusters in the two different amorphous states of indomethacin giving rise to different polymorphs on crystallization [74]

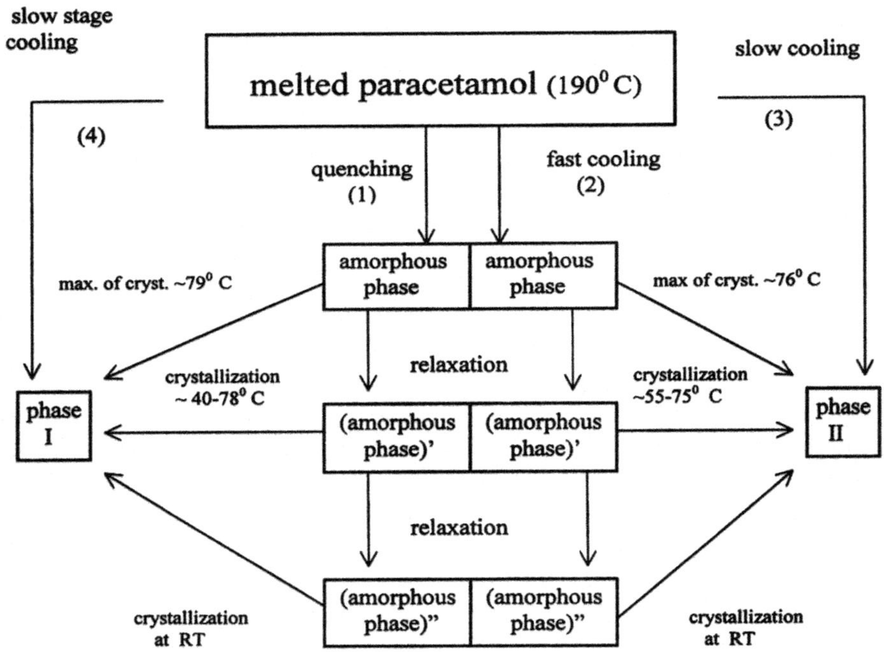

Fig. 2.16 Schematic illustration of the formation of different polyamorphs and polymorphs of paracetamol depending on the conditions of cooling the melt [75]

of this state could result in the combination of two phenomena: the time lag required to attain steady-state values of the nucleation rate (connected to the high viscosity of the liquid), and a high nucleation rate in the temperature range where the growth rate is low [24].

Organic components can also be introduced into inorganic glass structures, such as by sol-gel techniques, to give hybrid materials. The process by which inorganic sol-gel glasses form includes sol formation, gel formation, ageing, drying, and densification by heat at temperatures close to Tg, except for very thin films. The densification step becomes necessary as, due to the three-dimensional network formed by condensation, a porous microstructure is formed. Temperatures around the network softening point are required for further densification. In the case of organics being included in the gel, the inorganic network has to either be widened, the pores have to be smaller, or the pores must simply be absent. In the latter case, shrinkage should decrease substantially or, in other words, the materials should reach its final density at lower temperatures as compared to the "inorganic" T_g. Incorporation of organics leads to a free volume that is more typical for organic polymers than for glasses. The thermoplastic behavior is very similar to that of glasses, but is only shifted to lower temperature [80].

2.2.6 Polyamorphism

Amorphous phases with identical composition can have different structures with different properties, such as different densities, relating to different local order. The different states can be formed under different conditions, such as varying cooling rates of the melt, changes in the compression protocol, or different types and conditions of mechanical treatment. Variations in temperature and pressure can provoque transitions between different amorphous phases [21–25, 73–75, 78, 81–115]. Termed 'polyamorphism', such phenomenon is similar to polymorphism in crystalline materials, where structurally distinct phases of the same composition can be formed and undergo first- or higher-order transitions as a function of pressure and temperature. It is important not to confuse polyamorphism with the chemical phase separation, which is often found in more conventional glass-forming systems like alkali silicates. In this case unmixing occurs through the creation of regions of different composition [20, 21].

In order to distinguish different amorphous states from each other, one can use all the "tools" of characterising their structures that have been mentioned above: pair distribution function, Voronoi-Dirichlet tessellation, parameters of random close packing, or of random network. Depending on the chemical nature of an amorphous sample, the differences between the amorphous phases may be related to either short distances and intramolecular conformations (organic glasses), or to medium- and even long-range structure. The coordination polyhedra formed by immediate neighbours are preserved (inorganic oxide glasses). Different polyamorphs of organic compounds, on the contrary, can differ even in the short-distance range. This is particularly common with changes in molecular conformation, the type and structure of molecular clusters

Table 2.2 The main types of defects in crystals

Type of defects	Examples
Point defects (0D)	Vacancies, impurity atoms, interstitial atoms
Linear defects (1D)	Dislocations, disclinations
Planar defects (2D)	Surface, interface, a disordered plane, a planar intergrowth
Bulk defects (3D)	Clusters of defects, inclusions, three-dimensional intergrowths

linked by hydrogen bonds, and even by the electronic structure of molecules (zwitter-ions or neutral molecules).

Variation of temperature and pressure can trigger transformations between polyamorphs, as well as their crystallization into different polymorphs (Figs. 2.15 and 2.16 as examples). Differences in the structure of polyamorphs can account for differences in their physical and chemical properties, including their relative stability with respect to crystallization, chemical transformations, and, in the case of biologically active compounds, bioactivity.

2.3 Between Order and Disorder

The structures of a completely ordered crystal or completely disordered amorphous solid discussed above are only models, and can hardly exist in reality. The true situation rests with objects that are either imperfect crystals, containing a variety of types of defects, or amorphous solids that contain elements of order. Importantly, there is no sharp or definitive boundary between complete order and complete disorder.

In any real crystal there are deviations from the perfect order described by an idealized structural model. These deviations are termed *defects*. One distinguishes between 0D (point), 1D (linear), 2D (planar) and 3D (bulk) defects (Table 2.2). The defects can be distributed randomly throughout the bulk, or, on the contrary, the structure can be almost perfect in some directions, but strongly defect/disordered in the other. The presence of defects manifests itself in the broadening of the diffraction maxima. This broadening is rather uniform if the defects are random, or selective for certain groups of reflections, if disorder in certain directions dominates [116–125].

The extent of disorder in amorphous solids can be different [24]. A wide spectrum of partially disordered structures makes a continuous bridge between absolutely ordered and absolutely disordered solids [22–24, 126, 127]. Crystalline mesophases are commonly classified according to their translational, orientational, and conformational order as liquid crystals, plastic crystals, and conformationally disordered crystals (condis phases) [24, 126–128]. Crystalline mesophases can be considered as an intermediate state between crystalline and amorphous materials. They resemble amorphous materials in relation to their molecular mobility, and by undergoing glass transitions. At the same time, the crystalline mesophases possess a certain degree of translational periodicity (with the exception of nematic phases). Therefore they show signs of narrow peaks in X-ray diffraction patterns. Plastic crystals, which can be formed both by near-spherical molecules and molecules of lower symmetry (e.g. planar or chain molecules) can serve as an example: they have both extremely sharp X-ray diffraction lines and exhibit a glass transition [24]. In many cases, it is not straightforward to distinguish crystalline mesophases from either crystalline or amorphous states, especially if experimental characterization is limited to a single method. For example, X-ray powder diffraction (XRPD) would not allow one to distinguish between a regular crystal and plastic crystal. The observation of the glass transition by DSC would not necessarily be a unique signature of an amorphous material, because it is an inherent property of crystal mesophases as well [24, 126, 127, 129]. It is also possible that what is considered to be an amorphous state is in fact a plastic crystal composed of nanocrystals (a polyamorphic form of triphenyl phosphite, so called "glacial phase", can be mentioned as an example [79]). An important fundamental feature of mesophases, which distinguishes them from amorphous glasses, is that the glass transition can take place in a thermodynamically stable crystalline state [24].

Mesophases can be converted to a low temperature (and thermodynamically stable) crystalline phase, although such transition can be kinetically hindered under typical experimental conditions. In this respect they are similar to amorphous materials. If sufficiently undercooled ("supercooled"), molecular motions in the mesophases "freeze", and this phenomenon is similar to the liquid-to-glass transition. Correspondingly, the mesophases can be in a state with either dynamic disorder (above the calorimetric glass transition temperature, Tg), or static-frozen disorder (below the Tg) [24]. For example, if of a metastable disordered crystalline phase of ethanol is cooled continuously, then the disorder is frozen at temperatures below Tg, which represents a monotropic situation with the ordered stable phase [24].

Different completely ordered (crystalline), disordered (amorphous) and partially disordered (mesophases) states of condensed matter are presented schematically in Fig. 2.17 [24]. Translationally and rotationally ordered, but also having partial or complete conformational disorder condis crystals are structurally the closest to common crystalline materials [130]. They are typically formed by macromolecules, such as synthetic polymers, with two or more conformers of similar overall molecular shape, though conformational disorder is also known for small-molecule crystals. A condis phase can potentially be converted to either a plastic crystal or a liquid crystal, depending on the specific type of order it loses during the conversion [24].

Liquid crystals are usually formed by either rod- or disc-shaped molecules. They have orientational order but conformational and three-dimensional translational disorder. Depending on the shape and flexibility of a particular molecule, liquid crystals can form nematic, smectic, or hexagonal structures [131, 132]. Liquid crystals can form different mesostructures depending on variations in either temperature (thermotropic transitions) or in solvent content (lyotropic transitions).

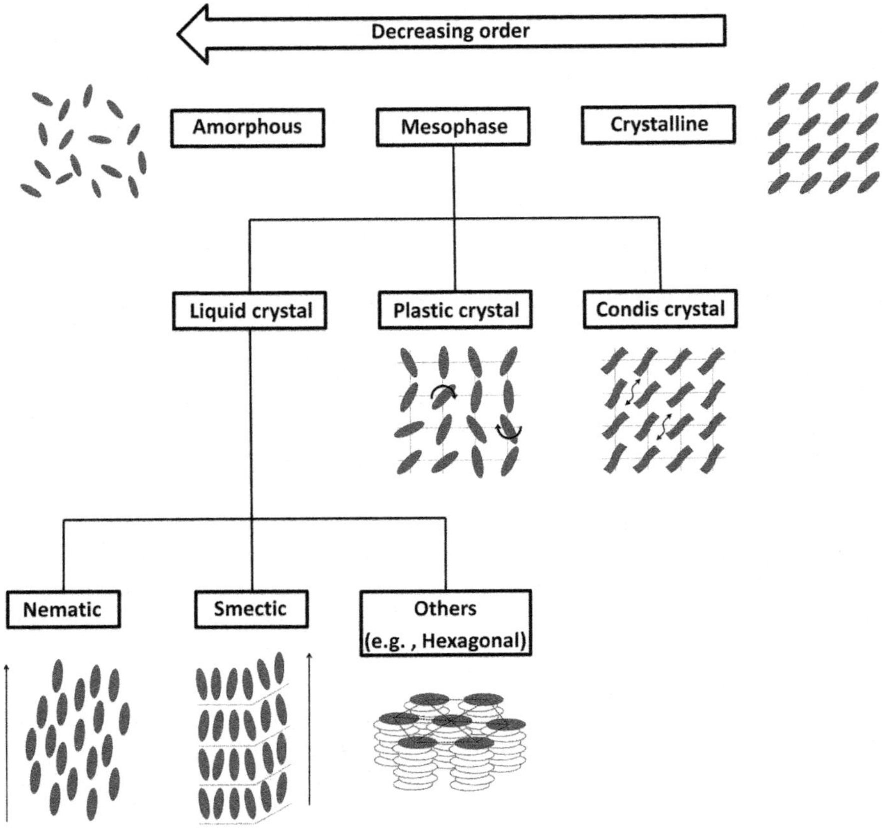

Fig. 2.17 Hierarchy of different ordered and disordered states of condensed matter [24]

Nematic liquid crystals have only orientational order. These phases are the most similar to liquids. Smectic or hexagonal phases have both orientational order and partial positional order in one or two dimensions. They can undergo either thermotropic, or lyotropic transitions. In addition, liquid crystals can often be differentiated based on molecular geometry (calamitic vs. discotic) or type of bond (molecular vs. ionic).

Plastic crystals (also called orientationally disordered crystals, ODIC phases) have three-dimensional translational order. They are usually formed by near-spherical molecules, although plastic crystals of less symmetrical molecules (e.g. planar or chain molecules), are also known. They commonly have a simple crystal structure with high symmetry, such as face centered cubic or hexagonal. Molecular plastic crystals can be easily deformed by a modest mechanical force. They even flow under their own weight (this is why they are termed "plastic crystals") [133, 134]. Ionic ODIC phases are often more rigid, than the molecular ones [24].

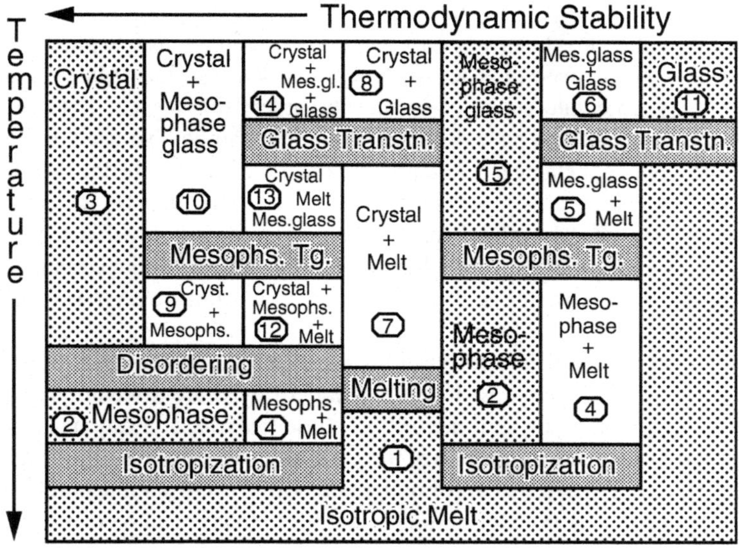

Fig. 2.18 An oversimplified scheme of possible transitions between the phases with different degree of order vs. disorder [126]

Some phases can be considered as belonging to several different categories at the same time. For example, the common form of crystalline ice, hexagonal ice, can be simultaneously classified as a conformationally and an orientationally disordered crystal; the same holds for the plastic crystal form of succinonitrile [24].

Tg in mesophase glasses depends on both the chemical nature of the molecule and the presence and concentration of water. In this respect the mesophase glasses are similar to isotropic (i.e., amorphous) glasses. Another important parameter is the shape of the molecule itself (e.g. rod like, banana type, spherical). It has a strong impact on the interactions and consequently on the dynamics of molecules. Tg can also be different for different (e.g. inverted hexagonal vs. lamellar) mesostructures formed by the same molecules [24]. Similarly to the relaxation of amorphous glasses, the relaxation of mesophase glasses during annealing is characterized by a broad distribution of relaxation times [24]. Intuitively, the free volume in the more disordered (amorphous) glass could be expected to be higher, and the Tg lower, than that in a more ordered, liquid crystalline, glass formed by the same molecules. This in fact holds for some compounds, e.g. for polyacryloxybenzoic acid [126]. However, the situation appears to be more complex. Ethanol provides an example, when the Tg of an amorphous state can be higher than that of a mesophase system of an identical chemical composition [24].

An oversimplified scheme of possible transitions between the phases with different degree of order vs. disorder has been proposed by Wunderlich [126] (Fig. 2.18).

The thermodynamic stability increases from the right side of the diagram to its left side, i.e. in parallel with an increase in the extent of ordering. At a fixed low temperature, for example, both the extent of order and the thermodynamic stability increases from the glass to a two-phase system consisting of a mesophase glass + an isotropic glass. It increases further to a single-phase mesophase glass. Finally (via three more intermediate states of increasing order) this system transforms to a highly ordered crystalline state. The same diagram can be also used to follow the temperature-induced changes. A common example of this is the heating of a two-phase system of crystal + mesophase glass above the Tg to give a crystal + mesophase mixture, after what first a single-phase mesophase system, and finally an isotropic melt form. This diagram is applicable to all three main classes of mesophases, i.e., conformationally disordered crystals, plastic crystals (ODIC), and liquid crystals [126].

In systems with an ODIC phase heating most commonly results in the first-order phase transitions (associated with "unfreezing" of dynamic orientational disorder) from a low-temperature crystalline (brittle) phase to a rotator mesophase, before an isotropic melt has been formed. In some cases, e.g. in succinonitrile, such transitions can be also assisted by conformational flexibility and a solid state isomerization reaction [24]. The transformation to the plastic phase can be considered as an "orientational melting". If just the translational degrees of freedom are unlocked, the melting event temperature is higher, and the melting enthalpy is low. Exceptions do exist, ice being one of then [24]. The behavior of the mesophase-forming systems is often even more diverse and complex.

References

1. Janssen T, Janner A, Looijenga-Vos A, de Wolff PM (2006) Incommensurate and commensurate modulated structures. Int Tables Crystallogr C:907–955. https://doi.org/10.1107/97809553602060000624
2. Wagner T, Schoenleber A (2009) A non-mathematical introduction to the superspace description of modulated structures. Acta Crystallogr B 65(3):249–268. https://doi.org/10.1107/S0108768109015614
3. Souvignier B, Wondratschek H, Aroyo MI, Chapuis G, Glazer AM (2016) Space groups and their descriptions. Int Tables Crystallogr A:42–74. https://doi.org/10.1107/97809553602060000922
4. Smaalen SV (2007) Incommensurate crystallography. Oxford University Press, New York
5. Arakcheeva A, Bykov M, Bykova E, Dubrovinsky L, Pattison P, Dmitriev V, Chapuis G (2017) Incommensurate atomic density waves in the high-pressure IVb phase of barium. IUCrJ 4(2):152–157. https://doi.org/10.1107/S2052252517000264
6. Shechtman D, Blech I, Gratias D, Cahn JW (1984) Metallic phase with long-range orientational order and no translational symmetry. Phys Rev Lett 53(20):1951–1953. https://doi.org/10.1103/PhysRevLett.53.1951
7. Levine D, Steinhardt PJ (1984) Quasicrystals: a new class of ordered structures. Phys Rev Lett 53(26):2477–2480. https://doi.org/10.1103/PhysRevLett.53.2477
8. Zong X, Ungar G, Liu Y, Percec V, Dulcey E, Hobbs JK (2004) Supramolecular dendritic liquid quasicrystals. Nature 428(11):157–160. https://doi.org/10.1038/nature02368
9. Hayashida K, Dotera T, Takano A, Matsushita Y (2007) Polymeric quasicrystal: mesoscopic quasicrystalline tilling in ABC star polymers. PRL 98:195502-1–195502-4. https://doi.org/10.1103/PhysRevLett.98.195502

10. Ivanov EY, Konstanchuk IG, Bokhonov BD, Boldyrev VV (1989) Mechanochemical synthesis of icosahedral phases in Mg-Zn-Al and Mg-Cu-Al alloys. React Solids 7(2):167–172. https://doi.org/10.1016/0168-7336(89)80026-9
11. Bokhonov B, Konstanchuk I, Ivanov E, Boldyrev V (1992) Stage formation of quasi-crystals during mechanical treatment of the cubic Frank-Kasper phase $Mg_{32}(Zn, Al)_{49}$. J Alloys Compd 187(1):207–214. https://doi.org/10.1016/0925-8388(92)90534-G
12. Janot C (1992) Quasicrystals A. Primer. Monographs on the physics and chemistry of materials. Oxford University Press, Oxford
13. Bokhonov B, Konstanchuk I, Boldyrev V, Ivanov E (1993) HRTEM study of milling induced phase transition and quasicrystalline formation in $Mg_{32}(Al, Zn)_{49}$ cubic Frank-Kasper phase. J Non Cryst Solids 153:606–610. https://doi.org/10.1016/0022-3093(93)90424-V
14. Twarock R (2004) A tiling approach to virus capsid assembly explaining a structural puzzle in virology. J Theor Biol 226(4):477–482. https://doi.org/10.1016/j.jtbi.2003.10.006
15. Bokhonov BB (2008) Mechanical alloying and self-propagating high-temperature synthesis of stable icosahedral quasicrystals. J Alloys Compd 461(1):150–153. https://doi.org/10.1016/j.jallcom.2007.07.015
16. Steurer W, Deloudi S (2008) Fascinating quasicrystals. Acta Cryst A 64(1):1–11. https://doi.org/10.1107/S0108767307038627
17. Jaric M (2012) Introduction to quasicrystals. Elsevier, Burlington
18. Kobeko PP (1952) Amorphous compounds. USSR Academy of Sciences Publishing House, Moscow
19. Porai-Koshits EA (1958) The possibilities and results of x-ray methods for investigation of glassy substances. In: Lebedev AA (ed) The structure of glass. Consultants Bureau, New York, pp 25–35
20. Greaves GN, Sen S (2007) Inorganic glasses, glass-forming liquids and amorphising solids. Adv Phys 56(1):1–166. https://doi.org/10.1080/00018730601147426
21. Henderson GS (2005) The structure of silicate melts: a glass perspective. Can Miner 43 (6):1921–1958. https://doi.org/10.2113/gscanmin.43.6.1921
22. Descamps M (ed) (2016) Disordered pharmaceutical materials. Wiley, Weinheim
23. Hancock BC, Shalaev EY, Shamblin SL (2002) Polyamorphism: a pharmaceutical science perspective. J Pharm Pharmacol 54(8):1151–1152. https://doi.org/10.1211/002235702320266343
24. Shalaev E, Wu K, Shamblin S, Krzyzaniak JF, Descamps M (2016) Crystalline mesophases: structure, mobility, and pharmaceutical properties. Adv Drug Deliv Rev 100:194–211. https://doi.org/10.1016/j.addr.2016.04.002
25. Bates S, Zografi G, Engers D, Morris K, Crowley K, Newman A (2006) Analysis of amorphous and nanocrystalline solids from their x-ray diffraction patterns. Pharm Res 23 (10):2333–2349. https://doi.org/10.1107/s11095-006-9086-2
26. Doi K (1976) Profile analysis of amorphous haloes by means of a Fourier transformation, with special reference to the structures of amorphous Pt-C and Ni-P. J Appl Cryst 9:382–390. https://doi.org/10.1107/S0021889876011679
27. Stetsko YP, Shanahan N, Deford H, Zayed A (2017) Quantification of supplementary cementitious content in blended Portland cement using an iterative Rietveld–PONKCS technique. J Appl Cryst 50:498–507. https://doi.org/10.1107/S1600576717002965
28. Bergese P, Colombo I, Gervasoni D, Depero LE (2003) Assessment of the x-ray diffraction-absorption method for quantitative analysis of largely amorphous pharmaceutical composites. J Appl Cryst 36:74–79. https://doi.org/10.1107/S002188980201926X
29. Proffen T, Neder RB (2012) Analysis of complex materials through the application and analysis of the pair distribution function (PDF). Z Kristallogr Cryst Mater 227(5). https://doi.org/10.1524/zkri.2012.0002
30. Wright AC, Hulme RA, Grimley D, Sinclair RN, Martin SW, Price DL, Galeener FL (1991) The structure of some simple amorphous network solids revisited. J Non Cryst Solids 129:213–232. https://doi.org/10.1016/0022-3093(91)90098-Q

31. Yarnell J, Katz M, Wenzel R, Koenig S (1973) Structure factor and radial distribution function for liquid argon at 85°K. Phys Rev A 7(6):2130–2144. https://doi.org/10.1103/PhysRevA.7.2130
32. Gingrich NS, Heaton L (1961) Structure of alkali metals in the liquid state. J Chem Phys 34 (3):873–878. https://doi.org/10.1063/1.1731688
33. Sirota E, Ou-Yang H, Sinha S, Chaikin P, Axe J, Fujii Y (1989) Complete phase diagram of a charged colloidal system: a synchrotron X-ray scattering study. Phys Rev Lett 62 (13):1524–1527. https://doi.org/10.1103/PhysRevLett.62.1524
34. Pedersen JS (1997) Analysis of small-angle scattering data from colloids and polymer solutions: modeling and least-squares fitting. Adv Colloid Interface Sci 70:171–210. https://doi.org/10.1016/S0001-8686(97)00312-6
35. Toby BH, Egami T (1992) Accuracy of pair distribution function analysis applied to crystalline and non-crystalline materials. Acta Cryst A 48:336–346. https://doi.org/10.1107/S0108767391011327
36. Prill D, Juhás P, Billinge SJL, Schmidt MU (2016) Towards solution and refinement of organic crystal structures by fitting to the atomic pair distribution function. Acta Cryst A 72:62–72. https://doi.org/10.1107/S2053273315022457
37. Granlund L, Billinge SJL, Duxbury PM (2015) Algorithm for systematic peak extraction from atomic pair distribution functions. Acta Cryst A 71:392–409. https://doi.org/10.1107/S2053273315005276
38. Chapman KW, Lapidus SH, Chupas PJ (2015) Applications of principal component analysis to pair distribution function data. J Appl Cryst 48:1619–1626. https://doi.org/10.1107/S1600576715016532
39. Prill D, Juhás P, Schmidt MU, Billinge SJL (2015) Modelling pair distribution functions (PDFs) of organic compounds: describing both intra- and intermolecular correlation functions in calculated PDFs. J Appl Cryst 48:171–178. https://doi.org/10.1107/S1600576714026454
40. Peterson PF, Bozin ES, Proffen T, Billinge SJL (2003) Improved measures of quality for the atomic pair distribution function. J Appl Cryst 36:53–64. https://doi.org/10.1107/S0021889802018708
41. Mu X, Neelamraju S, Sigle W, Koch CT, Totò N, Schön JC, Bach A, Fischer D, Jansen M, van Aken PA (2013) Evolution of order in amorphous-to-crystalline phase transformation of MgF_2. J Appl Cryst 46:1105–1116. https://doi.org/10.1107/S0021889813011345
42. Tran DT, Svensson G, Tai CW (2017) SUePDF: a program to obtain quantitative pair distribution functions from electron diffraction data. J Appl Cryst 50:304–312. https://doi.org/10.1107/S160057671601863X
43. Masson O, Thomas P (2013) Exact and explicit expression of the atomic pair distribution function as obtained from x-ray total scattering experiments. J Appl Cryst 46:461–465. https://doi.org/10.1107/S0021889812051357
44. Crocker JC, Grier DG (1996) Methods of digital video microscopy for colloidal studies. J Colloid Interface Sci 179:298–310. https://doi.org/10.1006/jcis.1996.0217
45. Nakroshis P, Amoroso M, Legere J, Smith C (2003) Measuring Boltzmann's constant using video microscopy of Brownian motion. Am J Phys 71(6):568–573. https://doi.org/10.1119/1.1542619
46. Gasser U, Weeks ER, Schofield A, Pusey PN, Weitz DA (2001) Real-space imaging of nucleation and growth in colloidal crystallization. Science 292(5515):258–262. http://www.jstor.org/stable/3082728
47. Weeks ER, Crocker JC, Levitt AC, Schofield A, Weitz DA (2000) Three-dimensional direct imaging of structural relaxation near the colloidal glass transition. Science 287 (5453):627–631. http://www.sciencemag.org
48. Cipelletti L, Manley S, Ball RC, Weitz DA (2000) Universal aging features in the restructuring of fractal colloidal gels. Phys Rev Lett 84(10):2275–2278. https://doi.org/10.1103/PhysRevLett.84.2275

49. Varadan P, Solomon MJ (2003) Direct visualization of long-range heterogeneous structure in dense colloidal gels. Langmuir 19(3):509–512. https://doi.org/10.1021/la026303jCCC
50. Peterson J, TenCate J, Proffen T, Darling T, Nakotte H, Page K (2013) Quantifying amorphous and crystalline phase content with the atomic pair distribution function. J Appl Cryst 46:332–336. https://doi.org/10.1107/S0021889812050595
51. Martin AV (2017) Orientational order of liquids and glasses via fluctuation diffraction. IUCrJ 4(1):24–36. https://doi.org/10.1107/S2052252516016730
52. Zobel M (2016) Observing structural reorientations at solvent-nanoparticle interfaces by x-ray diffraction – putting water in the spotlight. Acta Cryst A72:621–631. http://doi.org.ololo.sci-hub.bz/10.1107/S2053273316013516
53. Salmon PS (1994) Real space manifestation of the first sharp diffraction peak in the structure factor of liquid and glassy materials. Proc R Soc Lond A 445(1924):351–365. http://www.jstor.org/stable/52602
54. Elliott SR (1991) Origin of the first sharp diffraction peak in the structure factor of covalent glasses. Phys Rev Lett 67:711. https://doi.org/10.1103/PhysRevLett.67.711
55. Elliott SR (1995) Extended-range order, interstitial voids and the first sharp diffraction peak of network glasses. J Non Cryst Solids 182(1–2):40–48. https://doi.org/10.1016/0022-3093(94)00539-7
56. Inamura Y, Arai M, Nakamura M, Otomo T, Kitamura N, Bennington SM, Hannon AC, Buchenau U (2001) Intermediate range structure and low-energy dynamics of densified vitreous silica. J Non Cryst Solids 283–295:389–393. https://doi.org/10.1016/S0022-3093(01)00824-9
57. Voronoi G (1908) Nouvelles applications des paramètres continus à la théorie des formes quadratiques. J die Reine und Angew Mathematik 133(133):97–178. https://doi.org/10.1515/crll.1908.134.198
58. Brumberger H, Goodisman J (1983) Voronoi cells: an interesting and potentially useful cell model for interpreting the small-angle scattering of catalysts. J Appl Cryst 16:83–88. https://doi.org/10.1107/S002188988300998X
59. Bernal JD (1960) Geometry of the structure of monatomic liquids. Nature 185(4706):68–70. https://doi.org/10.1038/185068a0
60. Bernal JD, Mason J, Knight KR (1962) Radial distribution of the random close packing of equal spheres. Nature 194(4832):957–958. https://doi.org/10.1038/194957a0
61. Finney JL, Woodcock LV (2014) Renaissance of Bernal's random close packing and hypercritical line in the theory of liquids. J Phys Condens Matter 26(46):463102(19pp). https://doi.org/10.1088/0953-8984/26/46/463102
62. Li F, Liu XJ, Lu ZP (2014) Atomic structural evolution during glass formation of a Cu–Zr binary metallic glass. Comput Mater Sci 85:147–153. https://doi.org/10.1016/j.commatsci.2013.12.058
63. Salmon PS, Martin RA, Mason PE, Cuello GJ (2005) Topological versus chemical ordering in network glasses at intermediate and extended length scales. Nature 435(5):75–78. http://www.nature.com/nature
64. Greaves GN, Ngai KL (1995) Reconciling ionic-transport properties with atomic structure in oxide glasses. Phys Rev B52(9):6358–6380. https://doi.org/10.1103/PhysRevB.52.6358
65. Zachariasen WH (1932) The atomic arrangement in glass. J Am Chem Soc 54(10):3841–3851. https://doi.org/10.1021/ja01349a006
66. Greaves GN (1985) EXAFS and the structure of glass. J Non Cryst Solids 71(1–3):203–217. https://doi.org/10.1016/0022-3093(85)90289-3
67. Franzblau DS (1991) Computation of ring statistics for network models of solids. Phys Rev B 44(10):4925–4930. https://doi.org/10.1103/PhysRevB.44.4925
68. Gladden LF (1990) Medium-range order in v-SiO_2. J Non Cryst Solids 119(3):318–330. https://doi.org/10.1016/0022-3093(90)90305-6

69. Balducci R, Pearlman RS (1994) Efficient exact solution of the ring perception problem. J Chem Inf Comput Sci 34(4):822–831. http://pubs.acs.org/doi/abs/10.1021/ci00020a016?journalCode=jcics1
70. Yuan X, Cormack AN (1997) MD simulated structures of soda-lime-silica glass and its surface. Ceram Trans 82:281–286
71. Angell CA, Kanno H (1976) Density maxima in high-pressure supercooled water and liquid silicon dioxide. Science 193(4258):1121–1122. https://doi.org/10.1126/science.193.4258.1121
72. Sen S, Andrus RL, Baker DE, Murtagh MT (2004) Observation of an anomalous density minimum in vitreous silica. Phys Rev Lett 93(12):125902-1–125902-4. https://doi.org/10.1103/PhysRevLett.93.125902
73. Angell CA (1995) Formation of glasses from liquids and biopolymers. Science 267 (5206):1924–1935. https://doi.org/10.1126/science.267.5206.1924
74. Andronis V, Zografi G (2000) Crystal nucleation and growth of indomethacin polymorphs from the amorphous state. J Non Cryst Solids 271(3):236–248. https://doi.org/10.1016/S0022-3093(00)00107-1
75. Politov AA, Kostrovskii VG, Boldyrev VV (2001) Conditions of preparation and crystallization of amorphous paracetamol. Russ J Phys Chem A 75(11):1903–1911. http://cat.inist.fr/?aModele=afficheN&cpsidt=13620211
76. Taylor LS, Zografi G (1997) Spectroscopic characterization of interactions between PVP and indomethacin in amorphous molecular dispersions. Pharm Res 14(12):1691–1698. https://doi.org/10.1023/A:1012167410376
77. Andronis V, Zografi G (1997) Molecular mobility of supercooled amorphous indomethacin, determined by dynamic mechanical analysis. Pharm Res 14(4):410–414. https://doi.org/10.1023/A:1012026911459
78. Turnbull D (1976) Relation of crystallization behavior to structure in amorphous systems. Ann NY Acad Sci 279:185. https://doi.org/10.1111/j.1749-6632.1976.tb39706.x
79. Demirjian BG, Dosseh G, Chauty A, Ferrer ML, Morineau D, Lawrence C, Takeda K, Kivelson D, Brown S (2001) Metastable solid phase at the crystalline amorphous border: the glacial phase of triphenyl phosphite. J Phys Chem B 105:2107–2116. https://doi.org/10.1021/jp000765t
80. Schmidt H (1989) Organic modification of glass structure. New glasses or new polymers? J Non Cryst Solids 12:419–423. https://doi.org/10.1016/0022-3093(89)90565-6
81. Mishima O, Calvert LD, Whalley E (1984) Melting ice' I at 77 K and 10 kbar: a new method of making amorphous solids. Nature 310:393–395. https://doi.org/10.1038/310393a0
82. Grimsditch M (1984) Polymorhism in amorphous SiO_2. Phys Rev Lett 52:2379–2381
83. Mitius AC, Patashinskii AZ, Shimilo BI (1985) The liquid-liquid phase transition. Phys Lett A 113:41–44. https://doi.org/10.1016/0375-9601(85)90602-4
84. Thiel MV, Ree FH (1993) High-pressure liquid-liquid phase change in carbon. Phys Rev B 48 (6):3591–3599. https://doi.org/10.1103/PhysRevB.48.3591
85. Riebling EF (1970) Relationships between phase diagrams and the structure of glass-forming oxide melts. Phase diagrams. Materials science and technology, vol III. Academic, New York, pp 253–270. https://doi.org/10.1016/B978-0-12-053203-2.50015-6
86. Tsukumi I, Yamamuro O, Suga H (1994) Heat capacities and glass transitions of ground amorphous solid and liquid-quenched glass of tri-O-methyl-ß-cyclodextrin. J Non Cryst Solids 175:187–194. https://doi.org/10.1016/0022-3093(94)90010-8
87. Aasland S, McMillan PF (1994) Density-driven liquid-liquid phase separation in the system Al_2O_3-Y_2O_3. Nature 369:633–636. https://doi.org/10.1038/369633a0
88. Flory PJ (1973) Macromolecular chemistry-8. International symposium macromolecules held in Helsinki, Molecular configuration in bulk polymers, Finland, pp 1–15, 2–7 July 1972. https://doi.org/10.1016/B978-0-408-70516-5.50004-0
89. Franzese G, Malescio G, Skibinsky A, Buldyrev SV, Stanley HE (2001) Generic mechanism for generating a liquid-liquid phase transition. Nature 409(6821):692–695. https://doi.org/10.1038/35055514

90. Boyer RF (1976) General reflections on the symposium on physical structure of the amorphous state. J Macromol Sci Phys B 12(2):253–301. https://doi.org/10.1080/00222347608212774
91. Ishii K, Nakayama H, Koyama K, Yokoyama Y, Ohashi Y (1997) Molecular conformation of butanenitrile in gas, liquid, glass, and crystalline states: in relation to the stability of the glass state. Bull Chem Soc Jpn 70(9):2085–2091. https://doi.org/10.1246/bcsj.70.2085
92. Jarmelo S, Maria TMR, Leitao MLP, Fausto R (2001) The low temperature crystalline and glassy states of methyl α-hydroxy-isobutyrate. Phys Chem Chem Phys 3:387–392. https://doi.org/10.1039/B007722O
93. Ganguli D (2009) Polyamorphism in liquids and amorphous substances: an analogue of polymorphism in crystalline solids. Trans Indian Ceram Soc 68(2):65–80. https://doi.org/10.1080/0371750X.2009.11082161
94. Mishima O, Calvert LD, Whalley E (1985) An apparently first-order transition between two amorphous phases of ice induced by pressure. Nature 314(6006):76–78. https://doi.org/10.1038/314076a0
95. Poole PH, Grande T, Angell CA, McMillan PF (1997) Polymorphic phase transitions in liquids and glasses. Science 275(5298):322–323. https://doi.org/10.1126/science.275.5298.322
96. Poole PH, Sciortino F, Essmann U, Stanley HE (1992) Phase behaviour of metastable water. Nature 360(6402):324–328. https://doi.org/10.1038/360324a0
97. Debenedetti PG (2003) Supercooled and glassy water. J Phys Condens Matter 15(45):R1669–R1726. https://doi.org/10.1088/0953-8984/15/45/R01
98. Sciortino F, Essmann U, Stanley HE, Hemmati M, Shao J, Wolf GH, Angell CA (1995) Crystal stability limits at positive and negative pressures, and crystal-to-glass transitions. Phys Rev E 52(6):6484–6491. https://doi.org/10.1103/PhysRevE.52.6484
99. Rapoport E (1967) Model for melting-curve maxima at high pressure. J Chem Phys 46 (8):2891–2896. https://doi.org/10.1063/1.1841150
100. Rapoport E (1967) Melting-curve maxima at high pressure. II. Liquid cesium. Resistivity, Hall Effect, and composition of molten tellurium. J Chem Phys 48(4):1433–1438. https://doi.org/10.1063/1.1668858
101. Sastry S, Debenedetti PG, Sciortino F, Stanley HE (1996) Singularity-free interpretation of the thermodynamics of supercooled water. Phys Rev E 53(6):6144–6154. https://doi.org/10.1103/PhysRevE.53.6144
102. Deb SK, Wilding M, Somayazulu M, McMillan PF (2001) Pressure-induced amorphization and an amorphous–amorphous transition in densified porous silicon. Nature 414 (6863):528–530. https://doi.org/10.1038/35107036
103. Monahan AR, Kuder JE (1972) Spectroscopic differences between crystalline and amorphous phases of indigo. J Org Chem 37(25):4182–4184. https://doi.org/10.1021/jo00798a048
104. Hédoux A, Paccou L, Guinet Y, Willart JF, Descamps M (2009) Using the low-frequency Raman spectroscopy to analyze the crystallization of amorphous indomethacin. Eur J Pharm Sci 38(2):156–164. https://doi.org/10.1016/j.ejps.2009.06.007
105. Wang B, Pikal MJ (2010) The impact of thermal treatment on the stability of freeze dried amorphous pharmaceuticals: I. Dimer formation in sodium ethacrynate. J Pharm Sci 99 (2):663–682. https://doi.org/10.1002/jps.21959
106. Sastry S, Angell CA (2003) Liquid–liquid phase transition in supercooled silicon. Nat Mater 2(11):739–743. https://doi.org/10.1038/nmat994
107. Hedler A, Klaumunzer SL, Wesch W (2004) Amorphous silicon exhibits a glass transition. Nat Mater 3(11):804–809. https://doi.org/10.1038/nmat1241
108. McMillan PF (2000) Phase transitions: Jumping between liquid states. Nature 403 (6667):151–152. https://doi.org/10.1038/35003088
109. Wilding MC, McMillan PF (2001) Polyamorphic transitions in yttria–alumina liquids. J Non Cryst Solids 293–295:357–365. https://doi.org/10.1016/S0022-3093(01)00686-X
110. Wilding MC, Wilson M, McMillan PF (2005) X–ray and neutron diffraction studies and MD simulation of atomic configurations in polyamorphic Y_2O_3-Al_2O_3 systems. Proc R Soc London, Ser A 363:589–607. https://doi.org/10.1098/rsta.2004.1510

111. Katayama Y, Mizutani T, Utsumi W, Shimomura O, Yamkata M, Funakoshi K-I (2000) A first-order liquid–liquid phase transition in phosphorus. Nature 403(6766):170–173. https://doi.org/10.1038/35003143
112. Senda Y, Shimojo F, Hoshimo K (2002) The liquid–liquid phase transition of liquid phosphorus studied by ab initio molecular-dynamics simulations. J Non Cryst Solids 312–314:80–84. https://doi.org/10.1016/S0022-3093(02)01653-8
113. Katayama Y, Inamura Y, Mizutani T, Yamakata M, Utsumi W, Shimomura O (2004) Macroscopic separation of dense fluid phase and liquid phase of phosphorus. Science 306 (5697):848–851. https://doi.org/10.1126/science.1102735
114. Brazhkin VV, Popova SV, Voloshin RN (1999) Pressure–temperature phase diagram of molten elements: selenium, sulfur and iodine. Phys B Condens Matter 265(1–4):64–71. https://doi.org/10.1016/S0921-4526(98)01318-0
115. Monaco G, Falconi S, Crichton WA, Mezouar M (2003) Nature of the first-order phase transition in fluid phosphorus at high temperature and pressure. Phys Rev Lett 90(25):255701. https://doi.org/10.1103/PhysRevLett.90.255701
116. Tsybulya SV, Kryukova GN (2008) Nanocrystalline transition aluminas: nanostructure and features of x-ray powder diffraction patterns of low-temperature Al_2O_3 polymorphs. Phys Rev B 77(2):024112. https://doi.org/10.1103/PhysRevB.77.024112
117. Isupova LA, Alikina GM, Tsybulya SV, Boldyreva NN, Kryukova GN, Yakovleva IS, Isupov VP, Sadykov VA (2001) Real structure and catalytic activity of $La_{1-x}Sr_xCoO_3$ perovskites. Int J Inorg Mater 3(6):559–562. https://doi.org/10.1016/S1466-6049(01)00062-9
118. Nikulina O, Yatsenko D, Bulavchenko O, Zenkovets G, Tsybulya S (2016) Debye function analysis of nanocrystalline gallium oxide γ-Ga_2O_3. Z Kristallogr Cryst Mater 231 (5):261–266. https://doi.org/10.1515/zkri-2015-1895
119. Kryukova GN, Klenov DO, Ivanova AS, Tsybulya SV (2000) Vacancy ordering in the structure of γ-Al_2O_3. J Eur Ceram Soc 20(8):1187–1189. https://doi.org/10.1016/S0955-2219(99)00278-2
120. Cherepanova SV, Tsybulya SV (2000) Simulation of X-ray powder diffraction patterns for low-ordered materials. J Mol Catal A Chem 158(1):263–266. https://doi.org/10.1016/S1381-1169(00)00087-X
121. Sadykov VA, Isupova LA, Tsybulya SV, Cherepanova SV, Litvak GS, Burgina EB, Kustova GN, Kolomiichuk VN, Ivanov VP, Paukshtis EA, Golovin AV, Avvakumov EG (1996) Effect of mechanical activation on the real structure and reactivity of iron (III) oxide with corundum-type structure. J Solid State Chem 123(2):191–202. https://doi.org/10.1006/jssc.1996.0168
122. Isupova LA, Sadykov VA, Tsybulya SV, Kryukova GN, Ivanov VP, Petrov AN, Kononchuk OF (1997) Effect of structural disorder on the catalytic activity of mixed La–Sr–Co–Fr–O perovskites. React Kinet Catal Lett 62(1):129–135. https://doi.org/10.1007/BF02475723
123. Cherepanova SV, Tsybulya SV (2004) Simulation of X-ray powder diffraction patterns for one-dimensionally disordered crystals. Mater Sci Forum 443:87–90. https://doi.org/10.4028/www.scientific.net/MSF.443-444.87
124. Cherepanova SV, Tsybulya SV (2006) Influence of coherent connection of crystalline blocks on the diffraction pattern of nanostructured materials. Z Kristallogr Suppl 23:155–160. https://doi.org/10.1524/zksu.2006.suppl_23.155
125. Kryukova GN, Tsybulya SV, Solovyeva LP, Sadykov VA, Litvak GS, Andrianova MP (1991) Effect of heat treatment on microstructure evolution of haematite derived from synthetic goethite. Mater Sci Eng A 149(1):121–127. https://doi.org/10.1016/0921-5093(91)90793-M
126. Wunderlich B (1999) A classification of molecules, phases, and transitions as recognized by thermal analysis. Thermochim Acta 340–341:37–52. https://doi.org/10.1016/S0040-6031(99)00252-X

127. Levine H, Shalaev E, Zografi G (2002) The concept of 'structure' in amorphous solids from the perspective of the pharmaceutical sxciences. In: Levine H (ed) Amorphous food and pharmaceutical systems. Royal Society of Chemistry, Cambridge, pp 11–30. https://doi.org/10.1039/9781847550118-00011
128. Cui Y (2007) A material science perspective of pharmaceutical solids. Int J Pharm 339:3–18. https://doi.org/10.1016/j.ijpharm.2007.04.021
129. Seyer JJ, Luner PE, Kemper MS (2000) Application of diffuse reflectance near-infrared spectroscopy for determination of crystallinity. J Pharm Sci 89(10):1305–1316. https://doi.org/10.1002/1520-6017(200010)89:10<1305::AID-JPS8>3.0.CO;2-Q
130. Bernhard W, Wei C (1996) The difference between liquid crystals and conformationally disordered crystals. In: Isayev AI, Kyu T, Cheng SZD (eds) Liquid-crystalline polymer systems. ACS symposium series, vol 632. American Chemical Society, Washington, DC, pp 232–248. https://doi.org/10.1021/bk-1996-0632.ch015
131. Singh S (2002) Liquid crystals: fundamentals. World Scientific, Singapore
132. Kumar S (2001) Liquid crystals: experimental study of physical properties and phase transitions. Cambridge University Press, Cambridge
133. Sherwood JN (ed) (1978) The plastically crystalline state. Wiley, Hoboken, NJ
134. Billinge SJL, Thorpe MF (eds) (1998) Local structure from diffraction. Springer, New York. 399 p

Chapter 3
Amorphous Drug Solubility and Absorption Enhancement

3.1 Introduction

The poor oral bioavailability of many active pharmaceutical ingredients (APIs) resulting from low solubility is one of the important challenges in pharmaceutical technology. Over the last two decades the number of relatively insoluble drugs has grown steadily. Nowadays it is estimated that approximately 70% of new drug candidates are characterized by poor solubility. In order to ensure the optimum therapeutic efficacy, the selection of the drug substance and formulation is crucial in drug design. The development and approval of new, innovative and safe drugs is tremendously complex and requires extensive knowledge of materials, current technological processes and regulations. Bearing in mind that the form of drug should be suitable for the administration route and safe to apply, understanding of manufacturing process is of key importance to successful dosage form development.

Among the different dosage forms, oral products have been the most extensively investigated in pharmaceutical technology. Up to now, oral route still remains the favorite route of drug administration in many disease states. Due to high patient compliance, flexibility of oral dosage form design and cost effective manufacturing this route of administration is still the first way investigated in pharmaceutical technology.

The major problem in design of oral drug delivery systems is low or erratic bioavailability, which results from one or more factors. Poor solubility is one of the most important physicochemical parameter affecting drug absorption and therefore delayed or limited dissolution rate of drug substance is a matter of inadequate bioavailability. Other factors related to it are: low permeability through the gut mucosa, instability in gastrointestinal environment as well as biodegradation during first-pass hepatic/intestinal metabolism. Moreover, loss of large portion of drug dose raises the cost of drug therapy. Therefore, it is important to establish the reason of the low bioavailability and determine the improvement methods to avoid the problem of subtherapeutic drug level and unreproducible response. Based on

M. Rams-Baron et al., *Amorphous Drugs*,
https://doi.org/10.1007/978-3-319-72002-9_3

aforementioned arguments, the development and optimization of pharmaceutical formulations with improved clinical performance and safety in applying is of great importance.

The design of oral dosage forms requires deep understanding of physicochemical, physiological, and biopharmaceutical factors to obtain sufficient dissolution rate and bioavailability. Thus, the primary goal of preformulation stage is to gain detailed knowledge about physicochemical properties of the drug substances and excipients to achieve dosage forms of desired quality, because the dissolution rate of API depends not only on drug substance properties but also on dosage form and physiological factors of gastrointestinal tract. Therefore, the properties of critical meaning for drug formulation should be firstly considered.

Solubility of drug is an essential determinant of its dissolution. However, both factors i.e. solubility and dissolution might affect the concentration of the drug substance that may be achieved at the site of absorption. The majority of drug molecules were classified according to biopharmaceutical classification system (BCS) which categorizes drug substances into four groups (Fig. 3.1). These BCS categories are defined based on known data of two parameters as follows: high solubility/high permeability—BCS class I, low solubility/high permeability—BCS class II, high solubility, low permeability—BCS class III, low solubility/low permeability—BCS class IV. These data are useful for formulation design to overcome absorption barriers posed by solubility and permeability changes. According to FDA guideline, the drug substance is considered to be soluble when its highest dose strength is soluble in <250 mL of aqueous media over a pH range 1–7.5 at 37 ± 1 °C, otherwise, the drugs is characterized as poorly soluble. Generally, it was estimated that solubility lower than 100 μg/mL might affect limited absorption [1]. The term "high" permeability is used when the extent of absorption in humans is 90% or more of administered dose. The classification of high/low permeability is based on the rate of mass transfer across human intestinal membrane.

The drug substances with low solubility are categorized in BCS class II or IV. They differ in permeability which is high or low respectively. In the case of BCS class II drugs the dissolution rate depends on saturation solubility, diffusion coefficient, diffusion layer thickness, active surface area and dissolution media volume. With increasing surface area and saturation solubility, the dissolution rate could be increased. Therefore, crystal modifications, particle size reduction and amorphization are considered as effective approaches for dissolution enhancement [2].

BCS class IV drugs with low solubility and permeability need similar approaches, however absorption enhancers are required, because after drug dissolution in physiological fluids the absorption could be limited by intestine permeability. Additionally, the large intra- and intersubject variability in drug absorption poses the challenge in drug formulation. Thus, for these active ingredients the structural modifications or the use of alternative routes of drug administration are considered.

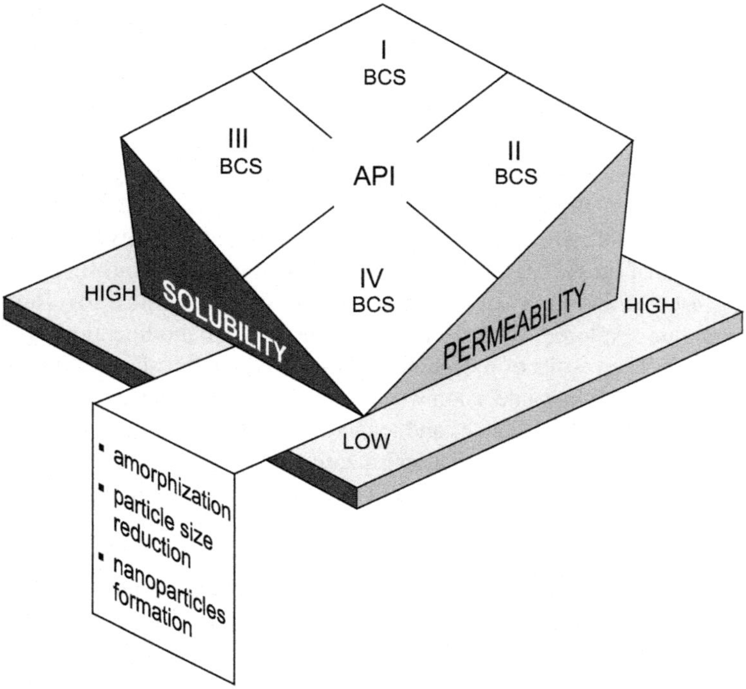

Fig. 3.1 BCS categories

The bioavailability of BCS class III drugs with high solubility and low permeability is rate-limited by gastrointestinal membrane permeability. These drugs are absorbed from intestinal lumen mainly via transcellular passive transport, transcellular carrier-mediated active transport or paracellular transport. As modification of the permeation characteristics is not an easy task, the use of permeation enhancers as excipients in the not membrane damaging amount is suggested.

For BCS class I drugs being highly soluble and permeable there are not dissolution rate-limiting factors. In the context of BCS classification, the rapid dissolution of drugs in gastrointestinal fluids and quick permeability across the membrane are ensured. However, other factors such as luminal degradation or complexation, pre-systemic metabolism and active secretion by efflux mechanism may also reduce the drug absorption.

3.2 The Strategies in Drug Solubility Improvement

The pharmaceutical products need to have adequate chemical and physical stability, pharmacokinetic properties and safe profile. Poor oral drug bioavailability resulting from low aqueous solubility is the matter that causes various concerns during drug development. Due to poor solubility, the BCS class II and IV drugs pose challenges for dosage forms elaboration, particularly if drug in large doses are necessary. The dissolution process can quickly achieve equilibrium solubility in physiological fluids but the gradient of drug concentration may be too low to let molecules cross through passive diffusion. The therapeutic dose cannot dissolve in the available volume of physiological fluids what leads to insufficient pharmacodynamic response and thus, the modulation of drug solubility is more viable option to improve bioavailability of drug than modification of permeability.

In the light of these data, different technological options focusing on improvement of drug dissolution are considered. Physical, chemical or formulation approaches are explored. The decision on the type of modification is made in several stages. The results of morphology, solubility, chemical and physical stability analyses lead to make a decision whether alterations of drug substance alone are sufficient or additional substances and technologies should be implemented in drug design project. It is well known that optimization of the physical form of the drug is essential in dosage form development and makes the processes simpler and faster to draw up.

Solubility problems of the drug can be solved in three ways (Fig. 3.2). First, the reduction of the particles sizes or new solid form such as salts and cocrystals can be attempted. Also the prodrug formulation as bioreversible derivatives is an interesting chemical/biochemical approach. Second, the binary systems with the carrier based on different technological processes can be proposed. In both ways amorphisation of drug obtained by various methods plays an important role. Alternatively to mentioned above, multicompartment dosage forms such as micro- and nanoparticles, microemulsions or self-emulsifying systems can be elaborated.

3.2.1 Amorphous Pharmaceuticals

Crystalline active pharmaceutical ingredients are frequently used in the drug dosage forms formulation. Most of them exist in different polymorphic forms or as hydrates. Each of these forms exhibit a variety of different properties. The molecules of crystalline material are "packed" in a highly ordered fashion in contrast to amorphous materials characterized by lack of distinct intramolecular arrangement and poor thermodynamic stability. Many excipients used in drug formulation such as natural or synthetic polymer and their modified forms are usually amorphous or partially amorphous. Their properties combined with those of APIs i.e. packing, thermodynamic, surface or mechanical properties may have significant impact on the drug formulation. Therefore, selection of the appropriate drug substance and excipients combined with the choice of the technological processes providing formation of desired particle size product and drug dose are crucial for high quality of pharmaceutical product. It was determined that the solubility of an amorphous drug could be 1.1- to 1000-fold higher in comparison to crystalline form [3–5]. The

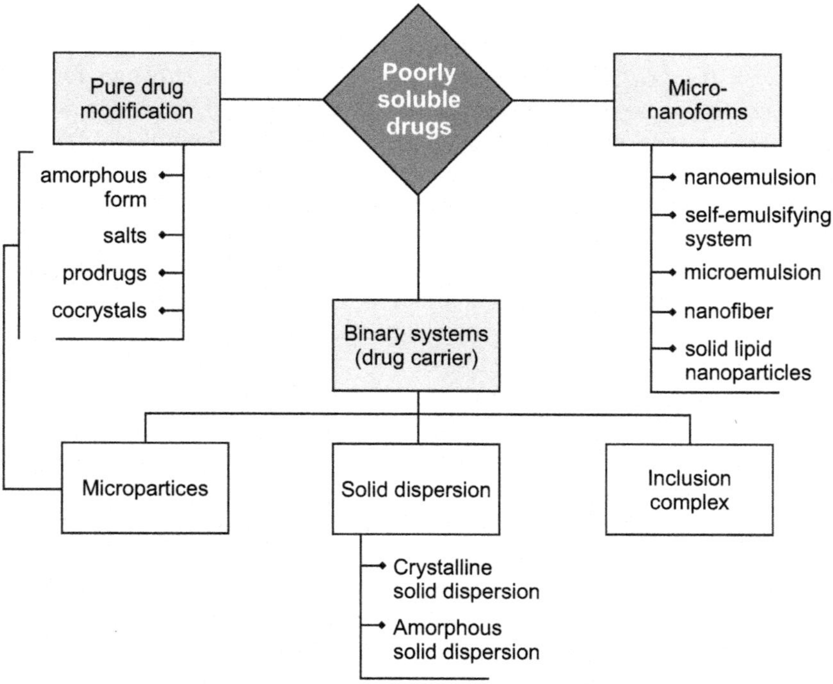

Fig. 3.2 Ways of physiochemical modification of poorly soluble drugs

significant improvement in saturated solubility of amorphous drug may ensure also noticeable improvement of drug bioavailability. It is the reason why technological processes in wide range are intensively explored for selecting the best way of manufacturing to obtain stable amorphous product.

3.2.2 Amorphous Solids

Active pharmaceutical ingredients can exist in crystalline or amorphous forms. Converting crystalline compounds to its amorphous counterpart is one of the promising tools in dosage form technology. Generally, amorphous products are classified in two groups as molecularly pure drug or solid dispersions being binary systems of drug molecules dispersed within carriers. Both types are characterized by solid-state nature without crystalline structure, higher free energy, enthalpy, enthropy and volume in comparison to the crystalline form. Due to thermodynamic properties the amorphous products present higher apparent water solubility and dissolution rate leading to better oral absorption. Moreover, lower energy is needed to enhance dissolution rate of the drug. Despite these advantages, pure amorphous active ingredients are seldom drug candidate in dosage form development because of the difficulties in scale up resulting from thermodynamic instability. Usually pure amorphous drugs are prepared in laboratorial scale by spray-drying or freeze–drying with the fast solvent removal, that prevent crystal structure formation. Stabilization of metastable forms is not easy task but several ways such as the use of appropriate excipients in drug formulation are proposed. Zafirlukast (Accolate®) is one of few examples available commercially in amorphous form [6, 7].

The combination of drug molecules with excipients that stabilize amorphous state and prevent crystallization of drug in gastrointestinal fluids, in form of solid dispersion, seems to be major formulation options, providing the implementation of the product in form of solid dispersion into the market.

3.2.3 Solid Dispersions

Since more than 50 years, the solid dispersions have been one of the most exciting and intensively investigated scientific area and up to now they are still promising forms for pharmaceutical applications.

The solid dispersion has been firstly described by Sekiguchi and Obi [8] as eutectic mixture prepared by melting API and carrier together and cooling in an ice bath. The resultant sulfathiazole-urea eutectic combination exhibited not only faster drug dissolution but also better bioavailability. Ten years later, Chiou and Riegelman [9] defined solid dispersion as a " dispersion of one or more active ingredients in an inert carrier in solid state, prepared by solvent, melting or solvent-melting method". Goldberg et al. [10] emphasized that the drug in solid dispersion can be in crystalline state as well as molecularly dispersed an in inert carrier. Based on molecular components arrangement and physical states, solid dispersions were divided into eutectic mixtures, solid solutions, glass solutions and amorphous precipitation in crystalline or amorphous carriers. Vasconcelos et al. [11] distinguished three generations of solid dispersions taking into account their physical state and carrier properties—crystalline solid dispersions with urea or sugars in the first generation, amorphous solid dispersions with fully synthetic or natural polymeric carriers in second one and amorphous solid dispersions with amorphous carriers having surface activity or amorphous carrier with surfactants in third generation. Recently, the fourth generation is also developed with swellable or water insoluble polymers [12]. The development of solid dispersion generations is schematically illustrated in Fig. 3.3.

Solid dispersion can exist as the one-phase system (solid solution, glass solution) or two phases (eutectic mixture, glass suspension), in which API is present in amorphous or crystalline form [13]. According to the miscibility of both components, solid solutions are classified as a continuous, if both components are miscible in all proportions and discontinuous when the one component is soluble in other one in a specific compositional region. One-phase continuous solid dispersion system is formed when the Tg parameters has sufficient high value whereas, discontinuous solid solution is two-phase form. Taking into account the distribution of solute

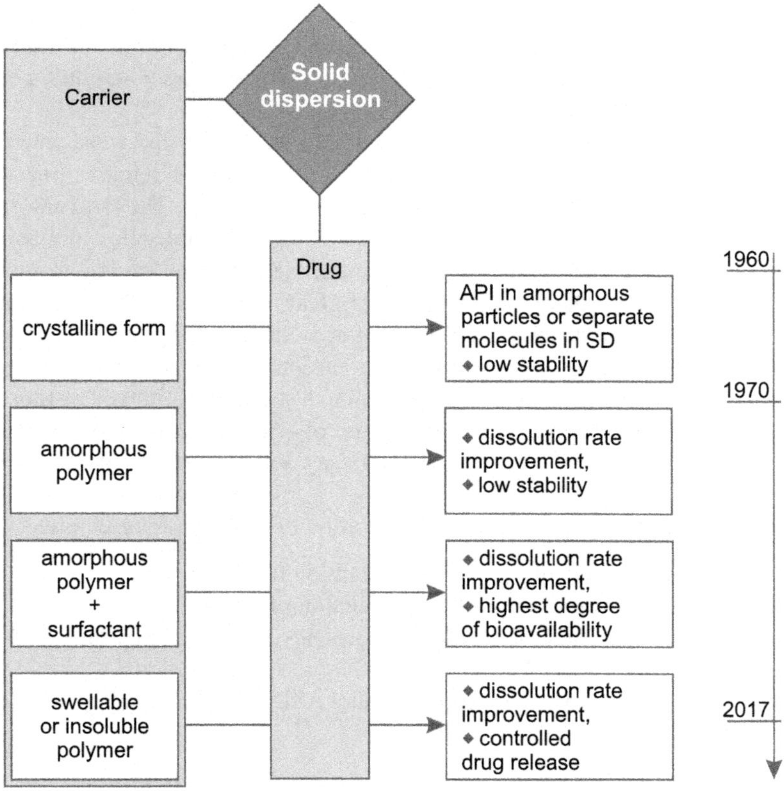

Fig. 3.3 The evaluation steps of solid dispersion development

molecules in solvent, substitutional, interstitial and amorphous solid solutions are being singled out. Amorphous solid solutions are characterized as molecular dispersion of drug molecules (solute phase) in inert carriers (polymeric continuous phase—amorphous solvent) but in irregular way while substitutional and interstitial solid solutions have a crystalline structure. In the case of first one, the differences in size of both types of molecules are less than 15% [14], in second, the volume of solute molecules should be less than 20%. In contrast, solid suspensions are characterized as dispersion of amorphous drug particles within polymeric matrix and they are distinguished by two Tg values. In amorphous solid solutions, the drug molecules are dispersed molecularly within amorphous polymer as a currier in irregular way.

Currently, solid dispersions are recognized as a valuable formulation strategy to improve dissolution profile and apparent solubility of poorly soluble APIs [7, 15–17]. They consist of at least two components, hydrophobic APIs and hydrophilic matrix, in which API as amorphous or crystalline particles are dispersed in amorphous or crystalline matrices of hydrophilic carriers. Numerous studies have shown the advantages of these formulations. Particle size reduction even to molecular level provides increasing wettability and dispersibility of drugs while changing the crystalline state into amorphous one prevents the molecules aggregation due to drug-carrier interaction as well as precipitation upon the contact with gastrointestinal medium.

Nowadays, amorphous solid dispersions (ASDs) depict the most interesting approach because the improvement of dissolution behavior remains one of the most challenging tasks in drug delivery system development. The term amorphous solid dispersion is determined as the dispersion of API molecules in stabilizing hydrophilic carriers. From the industrial point of view the way leading to show the ability of more potent drugs production is of great importance. Therefore, there is growing interest in the development of ASDs approaches to improve bioavailability of sparingly soluble drugs. Amorphous drug-carrier molecular dispersions application has received considerable attention over the last decade, followed by more than 1000 papers published in the literature. The use of ASDs affect the physicochemical properties of drug, bioavailability and efficacy in comparison to conventional formulation.

The main goals of pharmaceutical application of ASD are focused on:

- enhancement of drug solubility and dissolution behavior;
- enhancement of drug absorption and reduction of side-effects;
- stabilization of unstable drugs and amorphous state of ASD, and protection against decomposition;
- formulation of oral dosage forms containing ASD providing the improvement of the release of poorly soluble drug.

The observed improvement of dissolution rate of sparingly soluble drug molecules were generally attributed to:

- significantly increased drug surface area in comparison to conventional forms;
- modification of drug molecules surface properties through intimate contact with the carrier;
- increasing solubility of drug molecularly dispersed within polymer matrix.

Recently, different technological processes are developed to produce ASD with appropriate robustness and reproducibility. The impact of ASD solid state properties on developability assessment in drug formulation should be also recognized. The following formulation aspects should be taken into account in drug formulation:

- aqueous solubility and dissolution rate as related to bioavailability;
- chemical stability of drug in relation to ASD and dosage form formulation;
- stability at ambient temperature and humidity as well as photostability;
- relationship between solid state properties and drug dissolution behavior and stability;
- compactibility with carriers and other excipients used in drug formulation;
- technological processes scalability.

The physical forms of ASD depend on manufacturing processes, therefore, they can be developed in different forms such as powders, extrudates, films, foams. Solid state properties can have a profound impact on two major factors that are essential to successful development of ASD formulation: solubility and stability. Their poor properties can lead to significant drug development problems.

ASDs are investigated to develop mainly oral products. The research interests are focused on:

- applying new carriers or well-known with additional excipients such as surfactants, pH modulators or superdisintegrants;
- developing new processes or applying modifications in used technologies;
- elucidating the mechanism of the processes on each stage of drug formulation due to better understanding the thermodynamic properties of prepared ASD.

Several problems are identified in commercial product elaboration containing solid dispersion which mainly concern technological aspects, equipment availability as well as drug-polymer solubility and stability during storage. Physical instability of ASDs may be the main impediment in widespread use. Therefore, the drug recrystallization is emphasized as the most important because it may lead to loss of solubility advantage and also bioavailability decrease.

3.3 Solubility and Dissolution Strategy for Bioavailability Improvement

Oral drug absorption depends on the desired solubility, permeability through the cellular membrane and the required dose. In literature, many cases of solubility-limited absorption have been reported, however the dissolution rate over time needed to drug dissolve in gastrointestinal medium at the absorption site, is also a limiting step.

Only drug substances in solution are available for permeation across gut membrane and undergo subsequent absorption in the systemic circulation. Thus, drug solubility is quality indicator determining its usefulness in many aspects of drug research, especially in drug discovery, dosage form formulation as well as development of analytical methods for drug estimation and dosage forms evaluation. Therefore, solubility is described at early stage of drug research. Based on USP, Ph.Eur. data, the descriptive terms very slightly soluble drug is determined by solubility range 0.1–1.0 mg/mL, while below 0.1 mg/mL drug is defined as practically insoluble. Currently, much research is focused on the drugs with aqueous solubility lower than 100 μg/mL, presenting dissolution-limited absorption.

The aqueous solubility defined as the amount of drug that dissolves in given volume of solvent at which solution and solid phase are in equilibrium at specified temperature, pH and pressure can be expressed by van't Hoff equation. The ideal solubility of a solute can be mathematically expressed:

$$-logX = \frac{\Delta H_f}{2.303R}\left(\frac{T_0 - T}{T_0 T}\right) + log\vartheta \tag{3.1}$$

where X is solubility of solute in mol fraction, ΔH_f is the enthalpy of melting, T_0 corresponds to melting point of the solid solute in absolute degrees, T is absolute temperature of the solution, R is the gas constant and ϑ is activity coefficient. Two parameters, melting point and activity coefficient affect drug solubility. If the drug is characterized by low melting point and high value of log activity coefficient, its molecular structure can limit the solubility [18].

Numerous studies have considered drug amorphisation as an important aspect of drug absorption improvement as metastable amorphous state. Due to higher free energy and higher apparent solubility this form is suitable for slightly soluble drug formulation. Amorphous pharmaceuticals are attractive for solubility enhancement if they demonstrate higher concentration in gastrointestinal fluids and faster dissolution in comparison to their crystalline counterparts [5, 19]. However in reality, the expected dissolution behavior, particularly higher solubility is sometimes difficult to detected, due to problems in determination of the solubility of amorphous solids under equilibrium conditions.

The solubility determination is usually considered as a routine experiments in drug development, however it is important to identify the requirements for thermodynamic equilibrium solubility or kinetic dissolution rate determination at first [20]. Based on literature, "equilibrium solubility" is usually used for formulation with stable equilibrium and refers to the maximum quantity of drug which can be dissolved at given temperature and solvent conditions. Generally, solubility is determined by adding excess of drug to defined volume of aqueous medium and agitation until the equilibrium is achieved. The drug concentration in solvent at which the equilibrium is reached, is defined as thermodynamic solubility equilibrium. Traditionally, phase-solubility technique is used, however the external conditions of performed experiments have a great impact on solubility results. The "apparent or kinetic solubility" refers to the maximum drug concentration in supersaturated state that is in equilibrium with metastable forms of drug and is determined from the maximum of kinetic solubility profile [21].

The supersaturation state is determined by drug concentration in solution above its saturation solubility. In such conditions APIs can either convert from amorphous to crystalline form or change the polymorphic phase. The supersaturation can be expressed by supersaturation ratio S and relative supersaturation index σ:

$$\sigma = S - 1 = \frac{c - c_{eq}}{c_{eq}} = \ln S \tag{3.2}$$

where: c_{eq} is the equilibrium solubility [22]. For supersaturated solution the values relations are following: $S > 1, \sigma > 0$.

High-energy amorphous forms are characterized by higher kinetic solubility values, however when supersaturation is achieved crystal growth, can be initiated by nucleation. Thus, the formation of supersaturated solution requires usually stabilization to prevent precipitation as rapid drug precipitation can hinder its dissolution. Hancock and Parks [5] showed the significant solubility differences between amorphous and crystalline forms. In case of drugs that poor solubility is due to high crystallinity, the noticeable differences between kinetic and equilibrium solubility are noted. This difference is less important if the poor solubility is caused by nucleation high lipophilicity of drug. Huang and Tong [23] stated that the solubility differences between amorphous and crystalline forms can be up to several hundred times, whereas between polymorphs usually less than 10-folds. Lipinski et al. [24] underlined three energy terms that have impact on solubility: crystal packaging, cavitation energy and solvatation energy from which the crystal packaging energy is larger than others. Therefore in ASDs formulation, minimizing the components energy by disrupting the crystal drug lattice is favorable.

In order to prevent crystallization, polymers are used as ASDs matrices and stabilizers. The polymers being the best crystallization inhibitor in solid formulation cannot be effective in aqueous solution and vice versa [25]. It is noted that polymer in ASD increases phase transition temperature (Tg) of amorphous drugs and lowers mobility of drug molecules in the matrix what leads to inhibition of the crystal nuclei formation. It means that choice of polymer should be considered as the integral part of developability assessment. The formulation of ASDs in a form of glass solution with high polymer weight percentage leads usually to increase in kinetic barrier for phase separation and consequently crystallization. Amorphous solids require less energy to dissolve, therefore, the dissolution rate can increase. High rate and the extent of supersaturation generated in gastrointestinal lumen after oral administration of drug in ASD form can result in absorption enhancement of the drug with good permeability. The drug should be maintained in contact with gastrointestinal fluids for sufficiently long period for its absorption before supersaturation-induced precipitation. The higher initial solubility promotes rapid drug absorption.

In ASDs preparation the rational selection of the carriers can be provided by solubility parameters calculations in combination with analysis using thermal methods. The total solubility parameters of drug and polymer as the components of ASDs can be calculated according to Hoftyzer and van Krevelen method [26]. Solubility parameter (δ_t) is determined from the interactions between dispersion forces (δ_d), hydrogen bonding (δ_b) and polar interactions(δ_p) of functional groups in the parent molecule:

$$\delta_t^2 = \delta_d^2 + \delta_b^2 + \delta_p^2 \tag{3.3}$$

Solubility parameter for polymeric carrier is based on average molecular weight. For evaluation of drug miscibility in polymeric carrier thermal analysis by DSC is usually used. A single T_g value that ranges between the T_g of pure components, confirms miscibility both of them [15]. It is valuable factor for predicting the physical nature of solid dispersions as a single or two phases systems that can by expressed by Gordon-Taylor equation:

$$T_g = \frac{W_1 \times T_{g1} + K \times W_2 \times T_{g2}}{W_1 + K + W_2} \quad K = \frac{T_{g1} \times \rho_1}{T_{g2} \times \rho_2} \tag{3.4}$$

where T_{g1} and T_{g2} values are the glass transition temperatures of components, W_1 and W_2 are the weight fractions, K is calculated from the T_g and density values of amorphous components.

It should be taken into account that solubility requirement for the drug is dose dependent. For oral administered drug with the dose of 1 mg/kg, its solubility greater than 65 μg/mL cannot limit fraction of absorbed drug whereas the solubility less than 10 μg/mL can limit drug absorption [27]. Through different mathematical models describing the interplay between solubility and physiological parameters, the useful and simplistic model is maximum absorbable dose (MAD) [23, 28, 29]. The MAD estimates the maximum amount of drug that can be absorbed during its presence in solution in the intestine and is related to the dose required to achieve the therapeutic effect. The maximum absorbable dose is given by Eq. (3.5):

$$MAD = S \times K_a \times SIWV + SITT \tag{3.5}$$

where S is solubility (mg/ml) in intestinal condition (at pH 6.5), K_a is transintestinal absorption rate constant (min^{-1}), SIWV is small intestine water volume (mL), generally 250 mL, SITT is small intestinal transit time, approximately 270 min. The concept of MAD equation assume that solubility and permeability are compensatory. Several version of MAD are proposed using different means of permeability estimation. Other absorption models are also used to estimate the fraction of absorbed drug based on following concepts:

- drug will be solubility-limited if the fluid volume required to dissolve desired dose in gastrointestinal tract is inadequate;
- drug will be dissolution rate-limited if the rate of dissolution is not sufficient to dissolve all drug particles during the transit time to absorption site;
- drug will be permeability-limited if inadequate rate of transit from gastrointestinal tract lumen across the gut wall is identified [28].

As it was mentioned above, the solubility describes an equilibrium state, whereas drug dissolution is the dynamic process characterized by rate, i.e. the amount of drug dissolved per time unit is determined. Poor aqueous solubility of drug has impact on dissolution rate that is particularly important in oral drug elaboration. To maximize drug absorption in intestinal transit time, the drug should be completely dissolved with high enough solubility in gastrointestinal juices. It is problematic if the required dose cannot dissolve in desired volume of physiological fluids. Therefore, the dissolution testing is recognized as a prognostic tool for in vivo drug absorption and should be validated according to the regulatory standards. A control strategy is designed to ensure that ASDs of required quality is formulated consistently. It is challenging to establish an accurate in vitro-in vivo correlation because of complex nature of dissolution behavior of polymer ASDs.

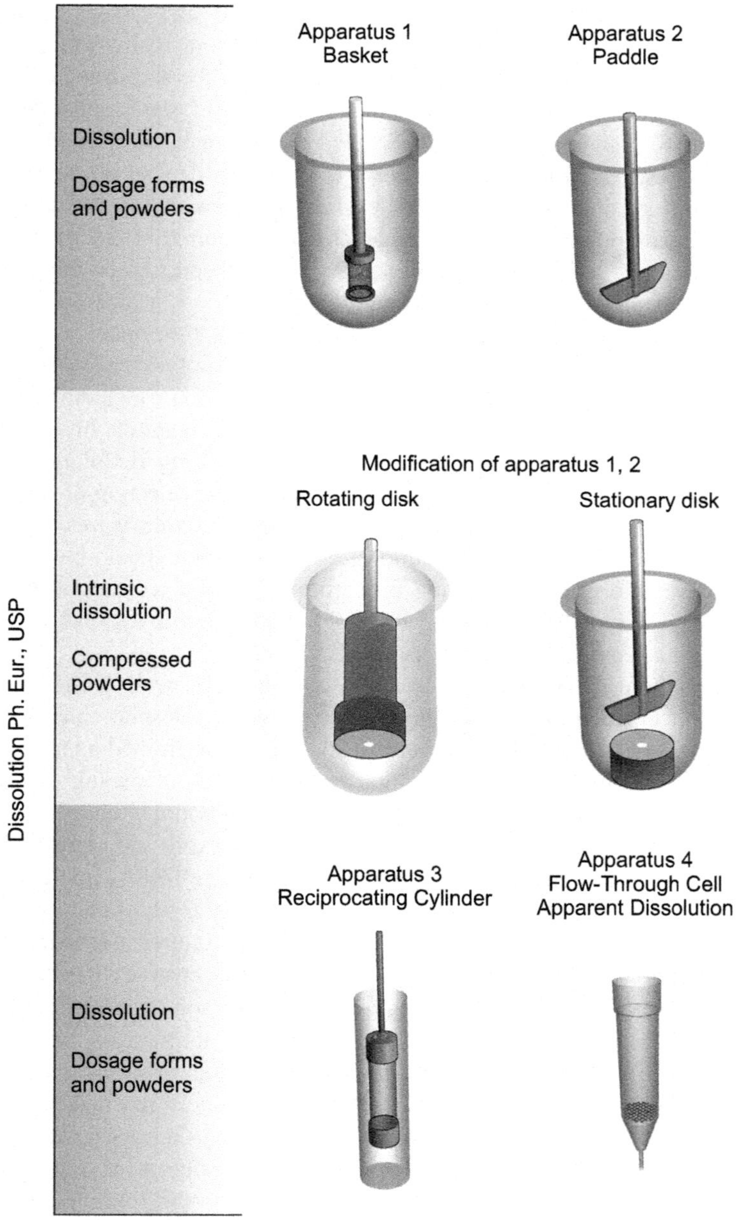

Fig. 3.4 Common apparatus used in dissolution studies

Dissolution methods need to be adequately defined, in order to properly characterize and interpret the experimental results. For dissolution studies of amorphous pharmaceuticals different apparatus for dissolution test may be chosen to determine critical factors. Dependently on formulation properties paddle, basket, reciprocating cylinder and flow-through apparatus are used (Fig. 3.4). As dissolution media, aqueous-based simulated physiological fluids, gastric and intestinal, are used [30]. Dressman et al. [31] promote the concept of using biorelevant dissolution media with bile acids as solubilizers rather than buffers for assessment particularly formulation performance of poorly soluble drugs. Fed and fasted simulated intestinal (FeSSIF, FaSSIF) and gastric fluids (FeSSGF, FaSSGF) make the dissolution studies more biorelevant. The type and volume of the medium depend on dose-solubility rate. Usually, a recommendation is made to sink conditions, defined as the volume of dissolution medium at least three times greater than required in order to form drug saturated solution, however non-sink conditions are accepted when the dissolution medium ensures the supersaturation lasting for long period of time that allows the determination of the amount of dissolved drug. The agitation speed 50 or 100 rpm appropriately for paddle or basket method is suggested, however the lower and higher speed are also applied dependently on formulation properties, including changes in amorphous/crystalline ratio of drug in ASD. In such case both, the dissolution drug profile and ability to form supersaturated state upon drug dissolution must be characterized. Therefore, as it was mentioned before, the selection of polymers as the carriers and stabilizers of ASDs plays an important role in retaining of the supersaturated solution during dissolution test. An attention should be also paid to polymer dissolution and/or swelling properties. Upon the contact with aqueous medium or physiological fluids, the hydrophilic carrier will dissolve and release molecularly dispersed drug into supersaturated solution.

Determination of intrinsic dissolution rate (IDR) from well-defined surface is also used for characterization of pharmaceutical solids. Despite constant surface area (0.5 cm^2 according to Ph. Eur. and USP), IDR is determined on samples with theoretically zero but practically minimal porosity which is a result of substance compaction before dissolution test. During the test the amount of dissolved drug per time unit and per exposed area is specified and usually expressed in milligrams per minutes per square centimeter ($mg \cdot min^{-1} \cdot cm^{-2}$). The factors influencing IDR and solubility are similar and intrinsic factors such as crystal habit, amorphism, polymorphism, as well as extrinsic factors e.g. temperature, pH, ionic strength, hydrodynamics can be distinguished. Additionally, some changes may occur on the sample surface, i.e. transformation to the stable but less soluble polymorphic form, recrystallization of amorphous form, conversion of salts into acid/base form or anhydrous form hydration. This phenomena can be monitored and quantitatively assessed in time points during IDR, before saturation is reached, what is the beneficial aspect of this method. It is worth mentioning that during sample compaction, recrystallization of some amorphous substances may occur. The die with compressed sample are the elements of rotating disk—the rotating-disk system known as "Wood's apparatus" (USP, Ph. Eur.) or are mounted at the bottom of vessel in dissolution apparatus 2—the stationary disk system (USP) (Fig. 3.4). Due to close relationship between intrinsic dissolution rate and solubility, this method can be an alternative for solubility estimation when equilibrium solubility cannot be experimentally obtained in a simple way.

Among the methods used to determine the drug dissolution rate, the Noyes-Whitney, Nernst and Higuchi models are well known, however the dissolution process of binary systems consisting of drug within a carrier is not simple to explain in this way [32]. It was emphasized that supersaturation of drug must be kept during dissolution of solid dispersion, however it can change in time. Due to the fact that SDs are more complex systems, different models describing dissolution process have been introduced [16]. The dissolution rate of drug molecules was described as drug or carrier mediated. In the case of drug-controlled dissolution the physical properties of drug are dominated. The drug diffuses through highly concentrated polymer diffusion layer to the dissolution medium at rate proportional to aqueous solubility of drug. On the other hand, in case of carrier-controlled dissolution, the drug molecularly dispersed in polymer diffusion layer and diffuse slowly to the dissolution medium at rate determined by the carrier. However, when this process runs too rapidly, and dissolution of drug is too slow, particles agglomerate, what results in crystallization and leads to decrease in dissolution rate.

Van Drooge et al. [32] introduced the model that considers phase transition during dissolution. Based on diazepam solid dispersion tablets with amorphous disaccharide carriers, three-phases dissolution process was described. The first phase was characterized by slow, but gradual drug release and nonlinear drug and carrier dissolution profile, which was followed by the faster release in second phase. In third one, slow dissolution of crystalline drug was identified. It should be underline, that during dissolution of ASDs, solution-mediated transformation may also occur. In contact with aqueous media, amorphous APIs have a tendency to crystallize by solid to solid transition. Alonzo et al. [19] summarized that dissolution, nucleation and crystallization are important processes during dissolution of amorphous systems. In dissolution path presented in that work, the dissolution rate (dc/dt) is proportional to the surface of the drug particles (A) and the difference between solution (C) and equilibrium concentration (C_{eq}) can be described by the modified Noyes-Whitney equation. In the nucleation and growth path, the nucleation rate is proportional to the degree of supersaturation (S), whereas crystal growth to the difference between solution an equilibrium concentration.

The increase of crystal growth can be expressed by the following equation:

$$\frac{d_r}{d_t} = \left[\frac{D \times v \times N_A}{(r + D)/k_+}\right] \times \left(C - C_{eq}\right) \tag{3.6}$$

where D is diffusion coefficient of molecule, v is the frequency of atomic or molecular transport at the nucleus-liquid interface, N_A is Avogadro's constant, k_+ is surface integration factor, $C - C_{eq}$ corresponds to the difference between bulk concentration and the concentration in the liquid layer surrounding the crystals. The surface integration factors change in the presence of drug molecules adsorbed onto crystal surface [33].

The mean dissolution behavior of drug particles can be expressed by the value of mean dissolution time (MDT). In the case of diffusion layer model determined by Noyes-Whitney equation, MDT is applied when the dose is equal to the amount needed to saturate dissolution medium whereas, for incompletely dissolved drug the mean dissolution time for saturation is used [34].

Particle size reduction and improvement of wettability ensure better contact of solvent with drug particles what results in enhanced dissolution. The viscosity of polymers used in ASDs increasing with molecular weight has also significant effect on the dissolution properties. Moreover, the drug molecular weight and viscosity of physiological fluids influence the value of diffusion coefficient, thus intra- and intersubject variabilities can be observed. The saturation solubility can be increased by changing the physical state of the drug. Baghel et al. [15] presented three scenarios for dissolution behavior of polymer ASDs:

- rapid dissolution of ASDs particles leading to generation of highly supersaturated solution;
- gradual release of drug and polymer from ASDs with retaining amorphous form in undissolved particles;
- gradual release of drug and polymer combined with drug particles crystallization due to plasticizing effect of water.

Alonzo et al. [19] presented two pathways through which drug can crystallize during dissolution process. The crystallization of amorphous solids may take place either after direct contact with dissolution medium or through rapid crystallization in saturated solution. This process can happen in both, in vitro and in vivo dissolution by integration of diffused molecules into crystal lattice. This kind of transformation confirms the loss of advantage of amorphous form and basically has impact on the drug product performance.

The improvement of physicochemical drug properties leads to the optimization of amorphous drug formulation. This pharmaceutical profile can help to identify the risk of suitable developability characteristic of formulation, but the lack of detailed information may lead, on the one hand to variables between batches, on the other, to decrease in in vivo efficacy. Thus, the monitoring of solid state feature for assessment of differences in amorphous/crystalline form solubility is the most important. Tables 3.1 and 3.2 show an example of the effect of amorphous forms of drug on their solubility and dissolution rate.

Due to the fact that each active pharmaceutical ingredient has own physicochemical properties and the polymers used as carriers have a limitations dependent on their features, the ASD cannot be universal formulation. The development of ASD with desired API requires proper choice of polymer, drug to polymer ratio and technology to obtain absorption efficiency. Dissolution of drug molecules in the gastrointestinal fluids has become important to be considered in optimization of drug formulation because the dissolution process can affect the rate and extent of oral drug absorption. On the other hand, both, solubility and drug dissolution rate in gastrointestinal environment are affected by several physiological parameters such as: physicochemical properties of fluids, available volume in different regions of gastrointestinal tract, agitation due to the gastrointestinal motility, fasting and fed conditions. The knowledge about the influence of endo- and exogenous components on degree of supersaturation in intestinal lumen is insufficient. Fluid volume, ionic

Table 3.1 Solubility improvement of amorphous form of drug

Drug	Forms	Solubility ratio	Reference
Pure drugs			
	State		
Bicalutamide	Amorphous/crystal	1.1	[35]
Glibenclamide	Amorphous/crystal	14	[5]
Griseofulvin	Amorphous/crystal	1.4	[5]
Ketoconazole	Amorphous/crystal	1.1	[35]
Loperamide	Amorphous/crystal	67.5	[35]
Indomethacin	Amorphous/crystal	4–17	[36]
Pranlukast	Amorphous/crystal	19.4	[37]
Tadalafil	Amorphous/crystal	11.5	[38]
Telmisartan	Amorphous/crystal	4.6	[39]
Terfenadine	Amorphous/crystal	1.4	[35]
Solid dispersion			
	Carrier		
Carbamazepine	PEG 6000	1.2	[40]
Hesperidin	porous silica	90	[41]
Ibuprofen	poloxamer P188	5.8	[42]
Itraconazole	HPMC	9.6	[43]
Nifedipine	PVP	11	[44]
Praziquantel	PVP	4.16	[45]
Tacrolimus	HPMC	25	[46]
Telmisartan	HPMC/PVP	6–38	[47]

strength, osmolarity, pH and surface tension are the most important variables for the fasting and fed states in stomach, whereas in upper small intestine particularly flow rate and amount of bile acids play an crucial role.

For prognosis the influence of drug dissolution on oral bioavailability all factors should be taken into account, however, it is difficult to fully predict in vivo dissolution. Most drugs are absorbed in small intestine therefore, it is important to understand the behavior of amorphous systems in conditions simulating physiological medium [56]. Thus, in early drug development stage the influence of drug dissolution parameters on oral absorption should be considered in details. Usually the dissolution affects absorption rate of drug rather than the solubility. If the dissolution rate is to slow to obtain complete drug dissolution, it can limit the absorption rate. If solubility of drug provides its appearance in absorption region in short time, the prediction of in vivo absorption can be made based on solubility results.

To avoid absorption problems, the solubility of drug above 10 mg/mL at pH 1–7 is suggested. However, the drugs with low solubility, below 100 μg/mL, can be completely absorbed when are administered in dissolution enhancing formulation such as ASDs or in appropriate dosage form. To estimate dissolution/solubility effect on drug absorption properly, permeability data should be involved because the

Table 3.2 List and description of dissolution enhancement of solid dispersions with poorly soluble drugs

Drug	Carrier	Dissolution improvement ASD/C	Comments	Reference
Carvedilol	Eudragit® R, E, PO	16	After 5 min, 900 mL 0.1 M HCl paddle 50 rpm	[48]
Cefuroxime	–	2.2	After 150 min, 900 mL 0.1 N HCl paddle 100 rpm	[49]
Celecoxib	PVP	7.3	After 30 min 500 mL biorelevant medium (FaSSIF) paddle 100 rpm	[50]
Felodipine	HPMCAS AS-LF	8.5	After 30 min 700 mL phosphate buffer (pH 6.8) paddle 50 rpm	[51]
Nifedipine	Zeolite beta	9	After 5 min, 900 mL 0.1 N HCl + 1% SLS paddle 50 rpm	[52]
Nimodipine	PVP/VA	17	After 1 h, 900 mL buffer pH 4.5 + 0.05% SDS, 75 rpm	[53, 54]
Oridonin	PVP K17	3.3	After 5 min, Chinese Pharmacopoeia Release Method III 200 mL distilled water 100 rpm	[55]

intestinal permeability determines the fraction of absorbed drug. In such consideration the dose of the drug as well as the extent of drug degradation in physiological conditions, interaction between drug with intestinal components and formation of non-absorbable components should be also taken into account. Solubility/dissolution and permeability are of key importance in the drug selection. Thus, especially in case of drug with poor aqueous solubility, the important decision in oral formulation development relying on the choice of suitable drug substance modification is of crucial significance. Drug amorphisation by ASD formulation is essential tool in design drug products with adequate drug absorption. Numerous studies have proved that ASD formulations are able to improve pharmacokinetic profiles, resulting in changes of AUC, C_{max} and T_{max} parameters [50, 57–59]. All ADS formulations listed in Table 3.3 show 2- to 22-fold and 2.1- to 14.9-fold enhancement in AUC and C_{max} values, respectively, compared with crystalline API or physical mixture.

Table 3.3 PK parameters comparison on examples of investigated solid dispersion and pure drug

Drug	Carrier	PK parameters after oral administration			References
		Cmax	AUC	Comments	
Albendazole	HPMC, HPMCP	2.8-fold↑	3.9-fold↑	vs. physical mixture (rabbits)	[60]
Carbamazepine	PEG 6000	3.5-fold↑	2.0-fold↑	vs. API (rabbits)	[40]
Cyclosporin A	HPC(SSL)	5.1-fold↑	5.2-fold↑	vs. amorphous API (rats)	[61]
Danazol	PVP K-15	2.1-fold↑	2.3-fold↑	vs. physical mixture (mice)	[62]
Ibuprofen	PEG 8000	10.0-fold↑	10.2-fold↑	vs. API (rats)	[63]
	Poloxamer P188	–	17.6-fold↑	(rats)	[42]
Itraconazole	TPGS/Aerosil® 200	11.7-fold↑	9.8-fold↑	vs. API (rats)	[61]
	HPMC	–	2.48-fold↑		[43]
Nifedipine	PVP	–	2.92-fold↑	dogs	[44]
Nimodipine	Eudragit® E-PO	2.7-fold↑	2.9-fold↑	vs. API (dogs)	[53, 54]
Ritonavir	Gelucire® 44/14	14.9-fold↑	6.1-fold↑	vs. API (rats)	[61]
	PEG 8000	13.7-fold↑	22-fold↑	vs. API (dogs)	[64]
Tacrolimus	HPMC	10.0-fold↑	9.9-fold↑	vs. API (dogs)	[61]

3.4 Case Studies

Taking into account increasing demand for systems providing efficient therapy with lipophilic drugs, it is not surprising to note an increasing number of research papers focused on dissolution improvement of poorly soluble drugs. To solve solubility challenges, the studies have been mostly directed towards amorphous solid dispersions, therefore, various preparation methods and carriers have been compared. Recently, the prevalence of solubility/dissolution rate enhancement methods is identified in both, scientific literature and in marketed product elaboration.

Zheng et al. [53, 54] reported solubility and dissolution rate improvement of nimodipine, a dihydropirydine calcium channel blocker, from ASDs prepared by hot-melt extrusion. The results demonstrated the usefulness of method preparation to obtain solid dispersion with three types of carriers PVP/VA, HPMC and Eudragit®

EPO by DSC, X-ray powder diffraction and SEM analysis confirmed better miscibility of PVP/VA and Eudragit® EPO with nimodipine than with HPMC. XPRD of extrudates indicated that nimodipine was present in amorphous form in all drug-polymer studied systems. In comparison to drug alone, the solubility increase was approximately 20-fold and 3-fold greater from PVP/VA and Eudragit® EPO solid dispersions, respectively. Formulation of ASDs with hydrophilic polymers PVP/VA and HPMC markedly increased drug dissolution. The best results, 88.48% of nimodipine dissolved after 1 h was obtained for 50% drug loading. This value was approximately 17 times greater in comparison to drug alone. The dissolution of nimodipine solid dispersions was predominantly diffusion-controlled. High viscosity of used carriers was the factor that controlled drug dissolution rate. It was also evident that the enhancement of dissolution rate depended on the drug to polymer ratio as higher polymer concentration inhibited the diffusion of the medium into the matrix.

Felodipine and indomethacin as poorly soluble drugs were selected to investigate the phase behavior of amorphous solids during dissolution at 25 °C and 37 °C, respectively [19]. Three polymers HPMC, HPMC-AS and PVP were used to inhibit drug crystallization from supersaturated solution. Amorphous felodipine crystallized rapidly, however at 25 °C the small extent of supersaturation was generated. The presence of polymers in solution reduced the tendency of amorphous drug to crystallization. It subsequently resulted in solution concentration peaking in range 8–12 μg/mL what was in agreement with estimated value for amorphous solubility around 9 μg/mL. The ratio of the solubility of amorphous to crystalline forms were determined according to the following equation:

$$\frac{\sigma^{amorph}}{\sigma^{crystal}} = e^{\frac{\Delta G}{RT}} \tag{3.7}$$

where ΔG as the free energy reference between both forms estimated from the Hoffman equation, R is the universal gas constant and T is the temperature in K.

The experimental results for felodipine clearly pointed that HPMC and HPMC-AS were particularly able to inhibit solution crystallization, whereas PVP was much less effective. In comparison to felodipine amorphous indomethacin crystallized much slower in dissolution medium and supersaturation was generated at both investigated temperatures. By adding small amount of HPMC and PVP to the solution, the inhibition of indomethacin crystallization was successfully achieved at 25 °C. At 37 °C indomethacin agglomerated extensively leading to delayed dissolution.

The same polymers were used to estimate the effectiveness in inhibition of celecoxib crystallization from supersaturated solution by performing the nucleation induction time measurements [25]. Celecoxib ASDs as binary and ternary systems containing 50% wt of the drug were formulated by evaporation method. The dissolution were performed at sink conditions with respect to crystalline solubility and non-sink conditions with respect to amorphous solubility. High drug loading of amorphous solids led to supersaturated solution, however the differences in dissolution profiles as well as solution crystallization behavior were noted. The greater effectiveness of cellulose derivatives as nucleation inhibitors in comparison to PVP was explained by its greater ability to interact with dense drug liquid cluster. The higher tendency to crystallization from PVP-based solid dispersions was partially inhibited by adding small amount of cellulose derivatives as crystallization inhibitors. Therefore, combining a cellulose derivatives and PVP enabled drug dissolution improvement and resistance to crystallization. By inclusion of crystallization inhibitor as a minor component in ternary ASDs, prolonged supersaturation following dissolution was achieved. Mura et al. [65], based on results of ternary solid dispersions with ketoprofen, PEG and surfactant also stated that the crystallinity degree was lower than in case of binary systems because surfactants, dependently on their type, can improve drug-polymer miscibility.

The polymer combinations can provide a synergistic advantage in the development of drug formulation with better stability and modified dissolution profiles. Thus, combination of Kollidon® VA64, Soluplus® and Eudragit® EPO were proposed as matrices to prepare ASDs by hot-melt extrusion with thermally unstable carbamazepine [66]. The solubility parameters were utilized to indicate drug-polymer interactions. The miscibility of drug with polymeric carriers was improved by combining Eudragit EPO with one of other polymers. In solid dispersion with Soluplus and Eudragit EPO (1:1wt) containing 30% of carbamazepine, the drug was mainly presented in an amorphous form with a small amount of microcrystalline drug, which significantly improved its dissolution. Approximately 90% of carbamazepine was dissolved after 5 min, whereas 85% of pure drug within 1 h. This polymer combination not only enhanced components miscibility using HME technology below the melting point of carbamazepine but also improved stability of formulation for at least 3 months.

ASD formulations with high concentration of drug are of great significance. In order to achieve successful formulation, rational selection of polymers is particularly important because the polymers play a dominant role in drug amorphization, solubility and stability enhancement. In case of high drug loading ASDs, miscibility between drug and polymer is a key factor, because the risk of drug crystallization is very high. For curcumin, poorly soluble antioxidant, anticancer drug, four polymers HPMC, PVP, K90, PEG8000 and Eudragit® EPO were chosen to solid dispersion (70:30 w/w) prepared by solvent evaporation [26]. Based on miscibility, molecular interactions between drug and polymers, dissolution improvement and the effectiveness of polymers was ranked as follow: Eudragit® EPO > HPMC > PVPK90 > PEG8000. Both, Eudragit® EPO and HPMC, formed strong molecular interactions affecting dissolution behavior of solid dispersions.

Chauhan et al. [67] also confirmed the significant role of polymers in crystallization inhibition of amorphous indomethacin in solid state, showing the importance of molecular interactions. The rank order of such effect was found to be PVP K90 > Eudragit®100 > HPMC.

ASDs have become a topic of scientific interest for its potential in improving oral bioavailability. The formulations should have adequate pharmacokinetic properties and acceptable safety profile.

On example of artesunate, antimalarial drug, the improvement of aqueous solubility, dissolution rate and subsequently bioavailability enhancement from solid dispersions obtained by hot melt extrusion with Soluplus® or Kollidon® VA64 and surfactant were demonstrated [68]. Significantly faster release of drug in comparison to drug alone was pointed, furthermore, it was shown that the dissolution rate of drug increased with increasing the ratio of polymer and surfactant to the artesunate. The formulation artesunate-Soluplus®-PEG400 and artesunate-Kollidon® VA64-PEG400 (1:2.8:0.2) were selected to in vivo studies. After administration of each formulation to the rats by oral gavage, the amorphous drug in solid solution quickly dissolved and reached blood circulation after absorption. In comparison to pure drug, solid dispersion with Soluplus® was characterized by 66.44 and 16.60 times higher values of plasma drug concentration-time curves (AUC) and maximum drug plasma concentration (C_{max}),respectively. The results of formulation with Kollidon® VA64 showed a 62.20 times increase in the $AUC_{(0–72)}$ and 13.40 times increase in C_{max} compared to that of pure drug. For both artesunate solid solutions the T_{max} values were six times shorter than in case of artesunate. Moreover, all these factors increased by several folds as compared to marketed formulation Gsunate®. Authors stated that both polymers and added surfactant also act as crystallization inhibitors during dissolution process.

Similar results were obtained for two another antimalarial drugs, artemether and lumefantrine, also prepared by hot melt extrusion with Soluplus® and PEG 400 [69]. In all solid dispersions the drug retained its amorphousness even after stability studies conducted over 6 months. The improvement of the solubility and in vitro dissolution rate in comparison to pure drugs were determined, differences between in vivo results were also noticed.

Freeze-drying method was used to obtain solid dispersions of docetaxel (BCS IV class) with poloxamer F68 and poloxamer F68/P85 in different drug polymer(s) ratio [70]. Solubility, dissolution rate as well as permeability across intestinal segment and oral pharmacokinetics parameters in rats were evaluated. It was shown that the extent and dissolution rate of docetaxel from solid dispersion were significantly higher in comparison to pure drug. After 45 min more than 85% of drug dissolved from solid dispersions containing one or both polymers, whereas 25% of pure drug dissolved at the same time. The dissolution rate from all formulations were not significantly different. Despite that dissolution improvement from solid dispersion prepared with poloxamer F68 alone, only limited intestinal permeation was determined, leading finally to 1.39-fold increase in oral bioavailability. By utilization both kinds of poloxamers as carriers of solid dispersion, enhanced dissolution rate and intestinal permeation of docetaxel improved bioavailability by the factor of 2.97 was noticed. These results can suggest the importance of intestine permeation on drug bioavailability and generally on the design an oral formulation.

To evaluate modified evaporation method for solid dispersions containing tacrolimus with three different polymers: PEG 6000, PVP and HPMC, the physicochemical, dissolution and in vivo absorption studies were performed [46]. Based on results of investigations, HPMC was selected as the most appropriate water soluble carrier. This HPMC-based ASD or crystalline powder of tacrolismus were administered orally to beagle dogs. The markedly increase in pharmacokinetic parameters for ASD were noticed. The AUC_{0-8h} and C_{max} values were approximately 10-fold greater than those of crystalline drug.

To improve the solubility and oral bioavailability of ezetimibe, poorly soluble hypolipidemic drug, the binary and ternary solid dispersions were prepared by spray-drying method with HPC or HPC/Tween 80 [71]. In case of all solid dispersions, significant enhancement of solubility and dissolution rate were identified. The binary 1:10 and ternary 1:10:0.1 ASDs were selected as the optimal formulations to pharmacokinetic studies in rats. The C_{max} values of ezetimibe after administration of binary and ternary solid dispersions were 1.5- and 2.4-fold higher, respectively, as compared to ezetimibe powder, whereas AUC values were 1.6- and 1.8-fold higher, respectively. No significant differences in T_{max} and k_{el} were estimated. Thus, the enhancement of drug oral bioavailability was assigned to its amorphous state in solid dispersion.

The biopharmaceutical advantages of the ASDs with celecoxib over crystalline and amorphous form of pure drug were evaluated in vivo in rats [50]. ASDs with hydrophilic polymers: HPMC and PVP containing four different doses were found to improve in vitro dissolution profiles as well as plasma drug concentration-time curves (AUC) for all doses. Despite drug crystallization upon the contact with dissolution medium, long supersaturation state assured significant enhancement of AUC_{0-24} values for all doses. The crystallization process intensified when doses increased and therefore the absorption of celecoxib was solubility-limited.

The ASDs as dosage forms are not new to the pharmaceutical industry. The first product in tablets form with Griseofulvin-PEG® solid dispersion was introduced into the market in 1975, 10 years later the next one with nabilone. However, after three decades of intensive studies not too much pharmaceutical products have been approved and commercialize. Nevertheless, since 2010 the amount of ASDs drug products in tablets or capsule forms arose. Table 3.4 gives overview of polymers used in marketed formulations. Most of them are manufactured based on cellulose derivatives and synthetic polymers.

In literature, cellulose derivatives are well established and developing mainly as the carriers of amorphous solid dispersions. Most of them, particularly ether- and ester-derivatives cannot be absorbed in gastrointestinal tract and the products of biochemical catalysis are usually supplied endogenously or dietary. They are considered to be even safer carriers than synthetic polymers. Three types of cellulose derivatives: HPC, HPMC, and HPMCAS were used in commercially available products, manufactured by hot melt extrusion or spray-drying method (Table 3.4). Synthetic polymers such as PEG and PVP are often used to prepare ASDs, however it was reported that solubility enhancement from solid dispersions with PEG is something less in comparison to PVP-based ASDs [72]. Among the marketed products, PEG was used to produce tablets with giseofulvin-PEG solid dispersion (Gris–PEG™), and together with poloxamer 188 was the matrix of fenofibrat solid dispersion (Fenogilde™). PVP was utilized in ASD drug formulation with nabilone (Cesamet™), whereas hydrophobically modified PVPVA of reduce hygroscopicity, was the carrier of solid dispersion forms with lopinavir/

Table 3.4 Example of different polymers and preparation methods used in commercial products

Carrier	Drug	Product	Preparation method
HPMCAS	Duloxetine	Cymbalta®	–
HPMC	Etravirine	Intelence®	Spray drying
HPMC	Everolimus	Certican®	Spray drying
PEG	Fenofibrate	Fenoglide®	Spray melting
PEG	Griseofulvin	Gris-PEG®	Melt extrusion
HPMC	Itraconazole	Sporanox®	Spray drying
HPMC	Itraconazole	Onmel™	Melt extrusion
HPMCAS/SLS	Ivacaftor	Kalydeco®	Spray drying
PVP	Nabilone	Cesamet®	–
HPMC	Nivaldipine	Nivadil®	–
HPMCAS	Posaconazole	Noxafil®	Melt extrusion
HPMC	Rosuvastatin	Crestor®	Spray drying
HPMC	Tacrolimus	Prograf®	Spray drying
HPMCAS	Telaprevir	Incivek®	Spray drying
–	Tolvaptan	Samsca®	Granulation
HPMC	Troglitazone	Rezulin®	–
PVP-VA	Lopinavir/ritonavir	Kaletra®	Melt extrusion
HPMCAS/SLS	Lumacaftor/ivacaftor	Orkambi®	Spray drying
PVP-VA/TPGS	Ombitasvir/paritaprevir/ritonavir	Viekirax®	Melt extrusion
HPMCAS	Vemuralenib	Zeboral®	Co-precipitation
HPC	Vildagliptin/metformin	Eucreas®	–

HPC hydroxypropyl cellulose, *HPMC* hydroxypropyl methylcellulose, *HPMCAS* hydroxypropyl methylcellulose acetate succinate, *PEG* polyethylene glycol, *PVP* polyvinylpyrrolidone, *PVP-VA* polyvinylpyrrolidone/vinylacetate copolymer, *SLS* sodium lauryl sulfate, *TPGS* d-α-tocopheryl polyethylene glycol succinate

ritonavir (Kaletra®) and ritonavir (Novir®). Ternary solid dispersion systems with PVAVA and TPGS were manufactured with ombitasvir/paritaprevir/ritonavir tablets (Orkambi®). All solid dispersions with synthetic polymers were produced by hot melt extrusion.

Due to the fact that all polymers used in ASD formulations have some limitations, the elaboration and production of new polymers or derivatives of existing ones having better properties for solid dispersion manufacturing is also an important goal in future oral dosage forms formulation.

3.5 Conclusion

During ASDs formulation unpredicted solid state changes of drugs may occur under the influence of different factors such temperature, pressure and moisture. It is the reason why monitoring of phase transformation during technological processing and proper knowledge of the factors determining eventual conversion is important.

For oral drugs containing ASDs the physicochemical properties and dissolution behavior is important to the dissolution rate performance of poorly soluble drugs, Therefore, properly designed dissolution test is a measure of critical quality attributes of ASDs composition and physicochemical changes during dissolution as well as upon aging. Significance of factors discussed above such as solubility, dissolution rate, permeability and dose arises from their influence on in vivo performance, drug product assessment as well as determination control quality attributes (CQA). It makes ASDs important in pharmaceutical development due to the need for improvements in manufacturing techniques and the growing number of poorly soluble drug candidates.

References

1. Hörter D, Dressman JB (2001) Influence of physicochemical properties on dissolution of drugs in the gastrointestinal tract. Adv Drug Deliv Rev 46(1–3):75–87
2. Kawabata Y, Wada K, Nakatani M et al (2011) Formulation design for poorly water-soluble drugs based on biopharmaceutics classification system: basic approaches and practical applications. Int J Pharm 420(1):1–10
3. Brough C, Williams III RO (2013) Amorphous solid dispersions and nano-crystal technologies for poorly water-soluble drug delivery. Int J Pharm 453(1):157–166
4. DiNunzio JC, Miller DA, Yang W et al (2008) Amorphous compositions using concentration enhancing polymers for improved bioavailability of itraconazole. Mol Pharm 5(6):968–980
5. Hancock BC, Parks M (2000) What is the true solubility advantage for amorphous pharmaceuticals? Pharm Res 17(4):397–404
6. Pastrano GL, Ghaly ES (2012) Physicochemical characterization of spray dried for mulation containing amorphous drug. Int J Pharm Pharm Sci 4(4):563–570
7. Vasconcelos T, Marques S, das Neves J et al (2016) Amorphous solid dispersions: rational selection of a manufacturing process. Adv Drug Deliv Rev 100:85–101
8. Sekiguchi K, Obi N (1961) Studies on absorption of eutectic mixture. I. A comparison of the behavior of eutectic mixture of sulfathiazole and that of ordinary sulfathiazole in man. Chem Pharm Bull 9(11):866–872
9. Chiou WL, Riegelman S (1971) Pharmaceutical applications of solid dispersion systems. J Pharm Sci 60(9):1281–1302
10. Goldberg AH, Gibaldi M, Kanig JL (1966) Increasing dissolution rates and gastrointestinal absorption of drugs via solid solutions and eutectic mixtures III: experimental evaluation of griseofulvin—succinic acid solid solution. J Pharm Sci 55(5):487–492
11. Vasconcelos T, Sarmento B, Costa P (2007) Solid dispersions as strategy to improve oral bioavailability of poor water soluble drugs. Drug Discov Today 12(23–24):1068–1075
12. Vo CL-N, Park C, Lee B-J (2013) Current trends and future perspectives of solid dispersions containing poorly water-soluble drugs. Eur J Pharm Biopharm 85(3 Part B):799–813
13. Laitinen R, Priemel PA, Surwase S et al (2014) Theoretical considerations in developing amorphous solid dispersions. In: Shah N, Sandhu H, Choi DS et al (eds) Amorphous solid dispersions—theory and practice. Springer, New York
14. Leuner C, Dressman J (2000) Improving drug solubility for oral delivery using solid dispersions. Eur J Pharm Biopharm 50(1):47–60
15. Baghel S, Cathcart H, O'Reilly NJ (2016) Polymeric amorphous solid dispersions: a review of amorphization, crystallization, stabilization, solid-state characterization, and aqueous solubilization of biopharmaceutical classification system class II drugs. J Pharm Sci 105(9):2527–2544

16. Craig DQM (2002) The mechanisms of drug release from solid dispersions in water-soluble polymers. Int J Pharm 231(2):131–144
17. Sharma A, Jain CP (2011) Solid dispersion: a promising technique to enhance solubility of poorly water soluble drug. Int J Drug Deliv 1(2):149–170
18. Jain S, Patel N, Lin S (2015) Solubility and dissolution enhancement strategies: current understanding and recent trends. Drug Dev Ind Pharm 41(6):875–887
19. Alonzo DE, Zhang GGZ, Zhou D et al (2010) Understanding the behavior of amorphous pharmaceutical systems during dissolution. Pharm Res 27(4):608–618
20. Elder D, Holm R (2013) Aqueous solubility: simple predictive methods (in silico, in vitro and bio-relevant approaches). Int J Pharm 453(1):3–11
21. Sun DD, Lee PI (2015) Haste makes waste: the interplay between dissolution and precipitation of supersaturating formulations. AAPS J 17(6):1317–1326
22. Brouwers J, Brewster ME, Augustijns P (2009) Supersaturating drug delivery systems: the answer to solubility-limited oral bioavailability? J Pharm Sci 98(8):2549–2572
23. Huang L-F, Tong W-Q (2004) Impact of solid state properties on developability assessment of drug candidates. Adv Drug Deliv Rev 56(3):321–334
24. Lipinski CA, Lombardo F, Dominy BW et al (2012) Experimental and computational approaches to estimate solubility and permeability in drug discovery and development settings. Adv Drug Deliv Rev 64(Suppl):4–17
25. Xie T, Taylor LS (2016) Dissolution performance of high drug loading celecoxib amorphous solid dispersions formulated with polymer combinations. Pharm Res 33(3):739–750
26. Meng F, Trivino A, Prasad D et al (2015) Investigation and correlation of drug polymer miscibility and molecular interactions by various approaches for the preparation of amorphous solid dispersions. Eur J Pharm Sci 71:12–24
27. Lipinski CA, Lombardo F, Dominy BW et al (1997) Experimental and computational approaches to estimate solubility and permeability in drug discovery and development settings. Adv Drug Deliv Rev 23(1–3):3–25
28. Butler JM, Dressman JB (2010) The developability classification system: application of biopharmaceutics concepts to formulation development. J Pharm Sci 99(12):4940–4954
29. Byrn SR, Henck JO (2012) Optimizing the physical form—opportunities and limitations. Drug Discov Today Technol 9(2):e73–e78
30. Van den Mooter G (2012) The use of amorphous solid dispersions: a formulation strategy to overcome poor solubility and dissolution rate. Drug Discov Today Technol 9(2):e79–e85
31. Dressman JB, Vertzoni M, Goumas K et al (2007) Estimating drug solubility in the gastrointestinal tract. Adv Drug Deliv Rev 59(7):591–602
32. Van Drooge DJ, Hinrichs WLJ, Frijlink HW (2004) Anomalous dissolution behaviour of tablets prepared from sugar glass-based solid dispersions. J Control Release 97(3):441–452
33. Janssens S, Van den Mooter G (2009) Review: physical chemistry of solid dispersions. J Pharm Pharmacol 61(12):1571–1586
34. Rinaki E, Dokoumetzidis A, Macheras P (2003) The mean dissolution time depends on the dose/solubility ratio. Pharm Res 20(3):406–408
35. Colclough N, Ruston L, Tam K (2008) Aqueous solubility in drug discovery chemistry, DMPK, biological assays. In: van de Waterbeemd H, Testa B (eds) Drug bioavailability: estimation of solubility, permeability, absorption and bioavailability. Methods and prickles in medicinal chemistry, vol 40, 2nd edn. Wiley, Hoboken, NJ, pp 10–28
36. Ojarinta R, Heikkinen AT, Sievänen E, Laitinen R (2017) Dissolution behavior of co-amorphous amino acid-indomethacin mixtures: the ability of amino acids to stabilize the supersaturated state of indomethacin. Eur J Pharm Biopharm 112:85–95
37. Xiong X, Xu K, Du Q et al (2017) Effects of temperature and solvent on the solid-state transformations of pranlukast during mechanical milling. J Pharm Sci 106(6):1680–1687. https://doi.org/10.1016/j.xphs.2017.02.020
38. Wlodarski K, Sawicki W, Paluch KJ et al (2014) The influence of amorphization methods on the apparent solubility and dissolution rate of tadalafil. Eur J Pharm Sci 62:132–140

39. Lepek P, Sawicki W, Wlodarski K et al (2013) Effect of amorphization method on telmisartan solubility and the tableting process. Eur J Pharm Biopharm 83:114–121
40. Zerrouk N, Chemtob C, Arnaud P et al (2001) In vitro and in vivo evaluation of carbamazepine-PEG 6000 solid dispersions. Int J Pharm 225:49–62
41. Wei Q, Keck CM, Müller RH (2017) Oral hesperidin—amorphization and improved dissolution properties by controlled loading onto porous silica. Int J Pharm 518:253–263
42. Newa M, Bhandari KH, Li DX et al (2007) Preparation, characterization and in vivo evaluation of ibuprofen binary solid dispersions with poloxamer 188. Int J Pharm 343:228–237
43. Miller DA, McConville JT, Yang W et al (2007) Hot-melt extrusion for enhanced delivery of drug particles. J Pharm Sci 96:361–376
44. Sugimoto I, Kuchiki A, Nakagawa H et al (1980) Dissolution and absorption of nifedipine from nifedipine–polyvinylpyrrolidone coprecipitate. Drug Dev Ind Pharm 6:137–160
45. Costa ED, Orlandi S, Leonardi D et al (2016) Unexpected solvent impact in the crystallinity of praziquantel/poly(vinylpyrrolidone) formulations. A solubility, DSC and solid-state NMR study. Int J Pharm 511:983–993
46. Yamashita K, Nakate T, Okimoto K et al (2003) Establishment of new preparation method for solid dispersion formulation of tacrolimus. Int J Pharm 267(1–2):79–91
47. Park J, Cho W, Cha K et al (2013) Solubilization of the poorly water soluble drug, telmisartan, using supercritical anti-solvent (SAS) process. Int J Pharm 441:50–55
48. Ferguson J, Pataki H (2013) Solvent-free melt electrospinning for preparation of fast dissolving drug delivery system and comparison with solvent-based electrospun and melt extruded systems. J Pharm Sci 102:508–517
49. Dhumal RS, Biradar SV, Yamamura S et al (2008) Preparation of amorphous cefuroxime axetil nanoparticles by sonoprecipitation for enhancement of bioavailability. Eur J Pharm Biopharm 70:109–115
50. Knopp MM, Chourak N, Khan F et al (2016) Effect of polymer type and drug dose on the in vitro and in vivo behavior of amorphous solid dispersions. Eur J Pharm Biopharm 105: 106–114
51. Sarode AL, Wang P, Obara S et al (2014) Supersaturation, nucleation, and crystal growth during single- and biphasic dissolution of amorphous solid dispersions: polymer effects and implications for oral bioavailability enhancement of poorly water soluble drugs. Eur J Pharm Biopharm 86:351–360
52. Karavasili C, Kokove L, Kontopoulou I et al (2016) Dissolution enhancement of the poorly soluble drug nifedipine by co-spray drying with microporous zeolite beta. J Drug Deliv Sci Technol 35:91–97
53. Zheng X, Yang R, Tang X et al (2007) Part I: Characterization of solid dispersions of nimodipine prepared by hot-melt extrusion. Drug Dev Ind Pharm 33(7):791–802
54. Zheng X, Yang R, Zhang Y et al (2007) Part II: Bioavailability in beagle dogs of nimodipine solid dispersions prepared by hot-melt extrusion. Drug Dev Ind Pharm 33:783–789
55. Li S, Liu Y, Liu T et al (2011) Development and in-vivo assessment of the bioavailability of oridonin solid dispersions by the gas anti-solvent technique. Int J Pharm 411:172–177
56. Jensen LG, Skautrup FB, Müllertz A, Abrahamsson B, Rades T, Priemel PA (2017) Amorphous is not always better—A dissolution study on solid state forms of carbamazepine. Int J Pharm 522(1–2):74–79
57. LaFountaine JS, Prasad LK, Miller DA et al (2017) Mucoadhesive amorphous solid dispersions for sustained release of poorly water soluble drugs. Eur J Pharm Biopharm 113:157–167
58. Moes J, Koolen S, Huitema A et al (2013) Development of an oral solid dispersion formulation for use in low-dose metronomic chemotherapy of paclitaxel. Eur J Pharm Biopharm 83(1): 87–94
59. Otsuka M, Maeno Y, Fukami T et al (2016) Solid dispersions of efonidipine hydrochloride ethanolate with improved physicochemical and pharmacokinetic properties prepared with microwave treatment. Eur J Pharm Biopharm 108:25–31

60. Galia E, Horton J, Dressman JB (1999) Albendazole generics—a comparative in vitro study. Pharm Res 16:1871–1875
61. Onoue S, Sato H, Ogawa K et al (2010) Improved dissolution and pharmacokinetic behavior of cyclosporine A using high-energy amorphous solid dispersion approach. Int J Pharm 399: 94–101
62. Vaughn JM, McConville JT, Crisp MT et al (2006) Supersaturation produces high bioavailability of amorphous danazol particles formed by evaporative precipitation into aqueous solution and spray freezing into liquid technologies. Drug Dev Ind Pharm 32:559–567
63. Newa M, Bhandari KH, Kim JO et al (2008) Enhancement of solubility, dissolution and bioavailability of ibuprofen in solid dispersion systems. Chem Pharm Bull 56:569–574
64. Law D, Schmitt EA, Marsh KC et al (2004) Ritonavir—PEG 8000 amorphous solid dispersions: in vitro and in vivo evaluations. J Pharm Sci 93:563–570
65. Mura P, Moyano JR, González-Rodríguez ML et al (2005) Characterization and dissolution properties of ketoprofen in binary and ternary solid dispersions with polyethylene glycol and surfactants. Drug Dev Ind Pharm 31(4-5):425–434
66. Liu J, Cao F, Zhang C et al (2013) Use of polymer combinations in the preparation of solid dispersions of a thermally unstable drug by hot-melt extrusion. Acta Pharm Sin B 3(4): 263–272
67. Chauhan H, Kuldipkumar A, Barder T et al (2014) Correlation of inhibitory effects of polymers on indomethacin precipitation in solution and amorphous solid crystallization based on molecular interaction. Pharm Res 31(2):500–515
68. Fule R, Paithankar V, Amin P (2016) Hot melt extrusion based solid solution approach: exploring polymer comparison, physicochemical characterization and in-vivo evaluation. Int J Pharm 499(1-2):280–294
69. Fule R, Dhamecha D, Maniruzzaman M et al (2015) Development of hot melt co-formulated antimalarial solid dispersion system in fixed dose form (ARLUMELT): evaluating amorphous state and in vivo performance. Int J Pharm 496(1):137–156
70. Song CK, Yoon I-S, Kim D-D (2016) Poloxamer-based solid dispersions for oral delivery of docetaxel: differential effects of F68 and P85 on oral docetaxel bioavailability. Int J Pharm 507(1-2):102–108
71. Rashid R, Kim DW, Din FU et al (2015) Effect of hydroxypropylcellulose and Tween 80 on physicochemical properties and bioavailability of ezetimibe-loaded solid dispersion. Carbohydr Polym 130:26–31
72. Liu H, Taylor LS, Edgar KJ (2015) The role of polymers in oral bioavailability enhancement; A review. Polymer 77:399–415

Chapter 4
Amorphous Drug Preparation Methods

4.1 Hot-Melt Extrusion

Hot-melt extrusion (HME) as continuous melt manufacturing process is preferable and industrially applicable. Single- or twin-screw extrusion and hot-melt co-extrusion are widely adopted techniques in pharmaceutical technology. Conveying of solids, melting, mixing, devolatilization, pumping and pressurization are main stages of HME. In principle, extrusion equipment usually consists of motor as a drive unit, an extrusion barrels enclosing rotating screw(s), an extrusion die and electronic control unit [1]. For better dispersive mixing specialized elements are also used. The barrel can be independently heated and cooled by control system. The design variables concerns extruder, screw and die. The twin-screw extruder has two agitator assemblies mounted on parallel shafts which can rotate together in the same (co-rotating) or opposite directions and can be fully intermeshing. The diameter of screws which determine the size of equipment, and length of screws to diameter ratio (L/D), usually ranging between 20–40:1, are primarily defined. Modification of screw configuration affects the modification of manufacturing method leading to the process optimization for the planned application. Some examples of extruders modification are depicted schematically in Fig. 4.1. Commercial extruders have modular design that makes possible modification of the process under particular requirements [2].

In the most common setup the final product is formed by conveying drug-polymer materials from the feeder through a heated barrel with rotating one or two screws under controlled conditions i.e. temperature, feed-rate, screw speed and pressure. The temperature is important parameter in extrusion because the process should be conducted above the polymer glass transition temperature (T_g) and below its degradation temperature (T_{deg}) and above or below API melting temperature (T_m). The screw speed and feeding rates have to be correlated in order to proper filling of the barrels. The relationship between both parameters are particularly important during scaling up. Other processing parameters such as degassing may also have impact on quality of prepared formulations [3]. During extrusion, molten

M. Rams-Baron et al., *Amorphous Drugs*,
https://doi.org/10.1007/978-3-319-72002-9_4

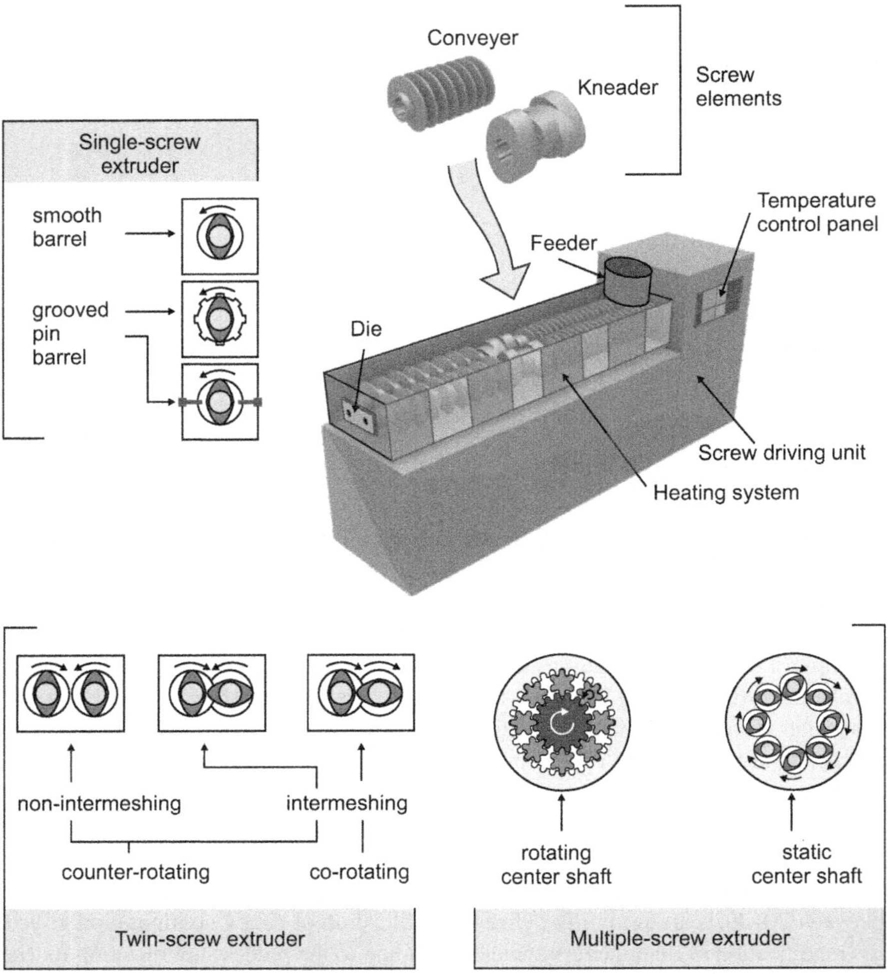

Fig. 4.1 The examples of extruders modification

blend is forced through the die with desired shape. Depending on the miscibility of drug-carrier composition, API can gradually dissolve or be homogenously dispersed in the process stream. The collected extrudates are milled or subjected to pelletization or compaction. The quality of the product relates to the materials properties, process parameters and extruders design variables. Liu at al. [4] reported the influence of extrusion temperature, screw speed and residence time on the drug dissolution.

The main advantage of HME technology is continuing performed operations, simplicity of transferring to the larger scale and low cost. Solvent-free process makes the technology environmentally friendly. Moreover, the HME is considered as valuable process for drugs sensitive on temperature, moisture and oxidation.

Currently, the twin-screw melt extrusion is of great interest in drug formulation development and product design, however the hot-melt co-extrusion is also developed and evaluated to order to obtain multi-layered extrudates with drugs incorporated in different polymeric matrices, providing a dual release of single drug [5]. Vynckier et al. [6] used two co-rotating twin-screw extruders connected to the co-extrusion die to elaborate a fixed dose combination product containing naproxen and esomeprazole magnesium.

HME gives an opportunity for the dissemination of different concepts of solid dispersion formation. The drug properties affect the way of ASDs formulation, while the polymers as a functional components are used as a drug carriers, solubilizers and inhibitors of the crystallization. For HME technology different polymer attributes should be considered: melting point (T_m), glass transition temperature (T_g), melt viscosity, molecular weight, miscibility, solubilization capacity and hygroscopicity as well. To select the suitable polymer for defined drug, the following criteria should be taken into account: thermoplastic properties, solubility in water and at low pH medium, if the immediate release is important, interaction with chosen drug. The miscibility of drug and polymer is important to obtain one-phase ASD but it depends also on melt polymer viscosity. The high polymer melt viscosity, strongly bonded with T_m and T_g parameters, may influence the miscibility of both components and diffusivity during manufacturing. In order to improve process efficiency, the process temperature should be lower than crystalline drug T_m and higher than polymer T_m and T_g. The application of supercritical fluids during hot-melt extrusion is also useful for temperature lowering. By reduction of processing temperature and materials residence time, the degradation of API can be avoided. Usually, polymers with low T_m and T_g values are chosen for HME i.e. various grades of PEG, PVP, polyvinylacetate, crosslinked polyvinylpyrrolidone and poly-methacrylate [7–9]. Despite mentioned, also several another polymers such as specific grades HPMC and Soluplus® are developed as specially dedicated to hot-melt extrusion process. For the processing and final product improvement, the addition of plasticizers to formulation is also often required as well as other processing aids such as carbon dioxide. Small molecules of plasticizers are included in the polymers, resulting in increasing molecular volume of polymeric chains. Due to plasticization effect, the flow of molten materials through extruder is easier because the processing temperature and shear forces is lower and thereby better product performance can be obtained. The usefulness of low molecular weight PEGs, poloxamer, Tween®, Myrj® as the plasticizers was confirmed in several investigations and have been published in numerous papers.

The HME technology is arousing the interest amongst researchers of academia as well as in industry. Numerous papers have been published on ASDs, successfully prepared by hot-melt extrusion (Table 4.1). The most of them are characterized by viscoelastic behavior. The mechanical properties of ASD extrudates are important for next processes i.e. milling, spheronization or compaction. The differentiation of particle size, their surface area, mechanical properties as well as swelling and gelling abilities has significant impact on drug dissolution, powder flow during tableting and mechanical properties of prepared final dosage forms i.e. tablets or pellets.

Table 4.1 Examples of solid dispersions prepared by hot melt extrusion

Drug	Carrier	Ratio	Type of extruder	Comment	Ref.
Carbamazepine	Kollidon® VA64—nicotinamide	40:20:40	Co-rotating twin screw extruder (screw speed 30 rpm at 160 °C)	Complete drug dissolution within 20 min	Liu et al. [10]
	Kollidon® 17PF	25:75	Co-rotating twin screw extruder (screw speed 100 rpm)	92.16% drug dissolved within 90 min	Alshahrani et al. [3]
	Kollidon® VA64—Eudragit EPO (1:1)	30:70	Co-rotating twin screw extruder (5–14 mm), 2 mm round opening die	95% drug dissolved within 5 min	Liu et al. [10]
	Soluplus®—Eudragit EPO (1:2)			90% drug dissolved within 5 min	
Felodipine	PVP	1:1	Co-rotating twin screw extruder (screw speed 100 rpm)	Slower drug release, complete drug dissolution within 75 min	Mahmah et al. [11]
	HPMCAS	1:1		Complete drug dissolution within 45 min	
17-Estradiol hemihydrate	PVP—Gelucire® 40/14	10:50:40	Single screw extruder with single rod die	57.2% drug dissolved within 60 min, 30-fold increase in dissolution rate	Hülsmann et al. [12]

Nisoldipine is an interesting example of poorly soluble drug with solubility in phosphate buffer below 0.5 μg/mL [13]. The Kollidon® VA64-based ASD with nisoldipine was successfully formulated by HME using twin screw extruder. The drug dissolution from ASD increased with the Kollidon® VA64 content up to 1:10 ratio. Further increase in the amount of the carrier content did not significantly improved this process. The complete drug dissolution was achieved after 30 min, whereas less than 1% of bulk drug was dissolved after 1 h. Moreover, optimized ASD formulation exhibited higher dissolution profile than those of commercial tablets and physical mixtures.

The extrusion process performed at 153 °C and at screw rate 2.5 Hz ensured formation of intermolecular hydrogen bond between carbonyl group in the nisoldipine and hydroxyl group in the Kollidon®. Based on PXRD and DSC results the amorphous state was confirmed. There neither characteristic peaks of crystalline nisoldipine in diffractogram of final ASD, nor the endothermic peak in thermograms. The pharamcokinetic studies in beagle dogs confirmed nisoldipine ASD bioequivalence to commercial tablets, making the new formulation promising alternative to marketed product. It was surprised that the Soluplus®, well known as suitable polymer candidate for HME formulation, did not markedly increased the dissolution rate of nisoldipine. After 1 h only 32.7% of drug was dissolved probably due to limited drug-carrier interaction.

The same carriers were used to develop itraconazole ASD by HME method [14]. Apart from them, Kollidon® 12 PF and Affinisol® HPMC were also taken into account. To obtain optimized itraconazole ASD, a complete study of extrudate formulation design was successfully conducted. Four ASD extrudated formulations with all mentioned carriers, containing 25 wt% of drug presented itraconazol in amorphous state. The factor having the greatest impact on dissolution profile in biphasic, biorelevant medium was the extrusion temperature. Only temperature of 155 °C assured amorphous solid dispersion, independently on the type of polymer. Differential scanning calorimetry confirmed ability of all polymers to stabilize amorphous form of drug. After screening experiments and stability studies at 25 °C and 60% relative humidity, Soluplus® was chosen as the main polymer due to its ability to enhance dissolution rate of itraconazole, and its capacity to stabilize effectively amorphous form. Moreover, it was suitable to HME processability, however the Kollidon® VA64 was easiest for formulation. Affinisol® HPMC as non-thermoplastic polymer was the hardest to coprocess. In order to optimize formulation and to improve dissolution, the excipients such as disintegrants, poloxamer or sodium bicarbonate were additional components of the extrudates. The best results, 80% of dissolved drug after 8 h was achieved by using AciDiSol®. Due to superdisintegrant properties, its great power to break down the Soluplus® gelling network could be an explanation of this phenomena, particularly in light of no impact of other excipients on dissolution profile. Finally, the optimal ASD formulation itraconazole-Soluplus® containing 2.5% of AciDiSol® was produced by HME at 155 °C and 100 rpm. Due to design space determination it was stated that the robustness of the process could be ensured because the extrusion temperature in the range 155–170 allows flexibility on screw speed from 75 to 130 rpm.

Eudragit® RS PO and RL PO, insoluble but water dispersible belong to the group of extrudable polymers [15]. They have similar characteristics, but differ in the content of functional ammonium groups, which control polymers permeability. They were selected as polymeric matrices to prepare melt-extruded formulations with water insoluble carbamazepine and water soluble theophylline in different w/w combination. Each drug-polymer blend was processed under employed conditions of HME. The DSC studies of the milled extrudates confirmed presence of amorphous form of both model drugs in extrudates. The PXRD analysis showed the changing crystalline form of both drugs to amorphous one upon melt extrusion. The results of drug release studies showed the enhancement of dissolution process for both APIs with increasing in the amount of Eudragit® RL PO, regardless of the drug loading amount. An increase in drug-Eudragit® RS PO ratio resulted in sustained drug release for up to 12 h. The intense mixing during melt-extrusion resulted in proper content uniformity and provided very high homogeneity independently on drug loading and their properties. Park et al. [15] for the first time used FT-IR chemical imaging technique to determine drug content uniformity and distribution. This technique allows to demonstrate visual distribution, as well as drug concentration according to their unique spectroscopic signature. The obtained results for studied Eudragit® extrudates corresponded well with data obtained by HPLC analysis. It is interesting proposition for Eudragit® ASDs optimization for in-line processing.

For a number of thermally labile drugs, the hot-melt extrusion requires the addition of the plasticizers in order to lower process temperature and viscosity of melt. Based on literature data, the concentration of plasticizers in extruded mass depends on drug-polymer composition and varies from 5% to 30%. Moreover, an additional component can increase the total mass of formulation, making the dosage form of unacceptable mass. Apart from different types of polymeric plasticizers, carbon dioxide has similar activity, reducing T_g value of amorphous or semi-crystalline polymers. The plasticization effect results from reduction of polymeric chain entanglement and melt viscosity due to carbon dioxide absorption between polymer chains and lubrication.

The concept of HME association with the use of carbon dioxide as a plasticizer deserves special acknowledgement. It is beneficial to have the plasticizers that lower the processing temperature without being the component of final extrudates [16]. The application of this procedure to *para*-amino salicylic acid/ethylcellulose solid dispersion resulted from the thermolabile properties of active substance which melts with decomposition. Below 110 °C it is thermally stable for at least 10 min. The results of experiments showed the extent of drug degradation less than 5% and reduction of processing temperature of approximately 30 °C.

The influence of injecting carbon dioxide on physicochemical properties of solid dispersion were investigated also with itraconazole which is practically insoluble in water. Results have been shown that dissolution of itraconazole can be controlled by presence of carbon dioxide injected under supercritical pressure during HME.

4.2 Technology of Three Dimensional Printing

The idea of three dimensional drug printing (3D printing) has completely changed the common opinion about drug manufacturing showing new possibilities of creating personalized products fabricated at the patient request [17]. This concept has caught the imagination of the academic and industrial community and has focused their attention on adapting 3D printing technology to the specific needs of the pharmaceutical market. In 2015 the U.S. Food and Drug Administration (FDA) approved the first fast-dissolving product manufactured using 3D printing technology. This anti-epileptic drug Spritam® (levetiracetam) was formulated using ZipDose® technology based on powder bed-liquid concept. During the printing procedure the liquid binder is applied in order to combine the powder layers into the desired form [18]. Among various 3D printings techniques, schematically presented in Fig. 4.2, the fused deposition modelling (FDM) deserves particular attention. In

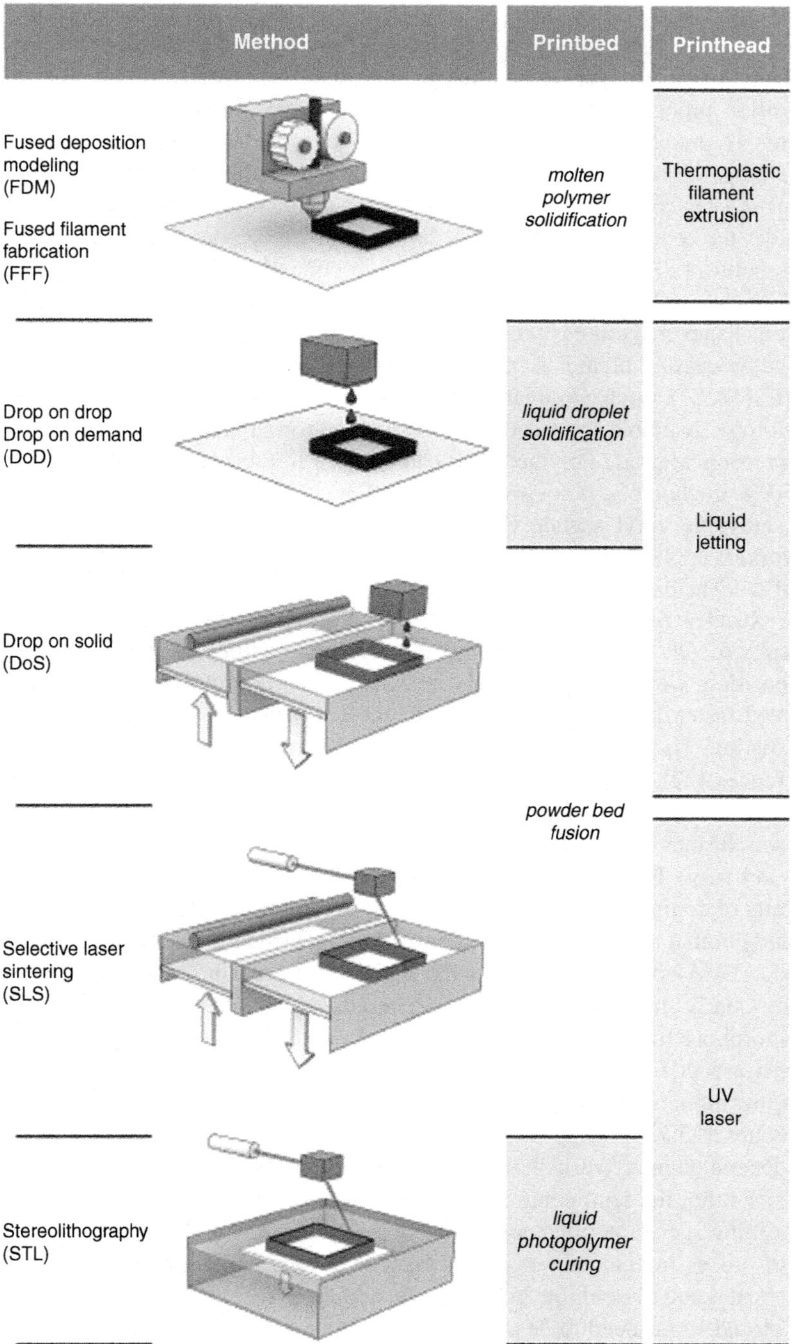

Fig. 4.2 Examples of 3D printing methods

this method, the printed material in a filament form (with a typical diameter of 1.75–3 mm) is passed through a heated nozzle [19]. Then, the molten filament is deposited layer-by-layer on the printing table until the desired shape is reached. The deposition procedure is digitally controlled and performed in accordance to the previously prepared pattern. At the design stage the 2D template is constructed containing information about the shape and dimensions of printed object. This model is later converted into STL file format, typical for stereolitography, which provides the coordinates of triangles that together create the surface of the designed 3D structure [20]. The printing program subsequently 'slices' this 3D model into machine assessable printable layers [17]. The unlimited possibilities of designing printed shapes make it easy to produce material with customized release profiles.

The processing filaments can be successfully produced using hot melt extrusion (HME) [18, 21]. Such conjugation of FDM technique with HME makes 3D printing technology available for various drug-polymer formulations. One can find many pharmaceutical grade polymers that can be used to fabricate drug-loaded filament *via* HME method e.g. polyvinyl alcohol (PVA) [22, 23] and polylactic acid (PLA) [21], ethylene vinyl acetate (EVA) [19], cellulose derivatives [24]. Jamróż and co-workers applied FDM 3D printing to fabricate thin films containing aripiprazole and PVA. The drug loaded-filaments were previously fabricated using a laboratory-scale extruder. According to DSC and XRD assessment the printed films were fully amorphous. On the contrary, samples prepared by film casting, with the same composition, were amorphous only in a part. What is important, 3D printed films revealed faster dissolution rate of aripiprazole in comparison to those obtained by film casting. The amorphous nature of the former may clearly explain the observed differences [22]. Chai and co-workers [24] used the HME technique to fabricate domperidone-loaded hydroxypropyl cellulose (HPC) filaments. The FDM technology was further applied to print hollow structured tablets suitable for intragastric drug delivery. Their XRD characterization indicated that during fabrication the majority of domperidone was converted into the amorphous form. Genina et al. [19] pointed out that the printed prototypes containing different grades of ethylene vinyl acetate (EVA) copolymers and γ-indomethacin were amorphous as well. These results clearly show that 3D printing technology constitutes an attractive approach for amorphous formulations containing poorly water soluble APIs and polymers. So far this aspect of 3D printed formulations is poorly explored and the degree of amorphization of the printed forms seems to be insufficiently controlled. The procedure of 3D printing is complex. The ratio of amorphous to crystalline content will depend on many variables (thermoplastic properties of filaments, applied drug/polymer ratio, the temperature of nozzle and the temperature of deposition plate, 3D printing speed, the thickness of the deposited layer etc. [20]). Proper optimization of processing parameters is necessary to produce fully amorphous drug forms with beneficial dissolution behavior. In addition, the long-term stability issues unexplored so far need to be addressed to further advance this attractive field.

4.3 Manufacturing Methods Based on Solvent Evaporation

Solvent-based methods assume the dissolution of a drug or drug and polymer in an appropriate solvent which is subsequently removed *via* evaporation leading to material amorphization. The undoubted advantage of such processing is the lack of API exposure to extremely high temperatures (such as during melting) which makes solvent evaporation methods useful for materials susceptible to thermal degradation. On the other hand the increase of solvent viscosity during the drying procedure may hinder the residual solvent evaporation which makes its complete removal the rate-limiting step. On the laboratory scale the vacuum drying or the use of rotary evaporator is sufficient to remove the solvent. But on industrial scale more advanced methods have to be applied (e.g. spray-drying) to make the drying procedure more effective and efficient. Here, the most common solvent-based methods of amorphous form preparation will be described in details along with novel and emerging approaches that have the potential to be implemented into industrial practice in the future.

4.3.1 Freeze-Drying

Freeze drying method, also known as lyophilization, over the years has been well recognized as an effective way to extend the shelf-time of various medical and biopharmaceutical products (e.g. vaccine formulations, peptides and proteins, labile pharmaceutical products) [25, 26]. In recent years progress in the field of pharmaceutical approaches based on lyophilization has been made [27–29]. There has been remarkable interest in the application of freeze-dried products as rapidly dissolving oral dosage forms, orally disintegrating tablets (ODTs), fast dissolving tablets (FDTs), wafers to drug delivery *via* the buccal mucosa, 3D scaffolds for tissue engineering, dry powder vaccine formulation for nasal delivery etc. [25]. Besides, the freeze-drying method has gained recognition as a possible manufacturing method for amorphous formulations and amorphous solid dispersions, in particular for materials prone to thermal degradation [28].

In general the typical freeze-drying cycle involves three processing stages: (1) freezing, (2) primary drying and (3) secondary drying [30]. During the first step the processing solution is cooled down until the ice nucleation will take place. A cold surface application, immersion in a cold bath, direct spraying into the liquid nitrogen, and utilization of liquefied gases are usually used during the freeze-drying process [31]. Then, the applied solvent, ideal when it can be water, is separated from the solute due to the subsequent growth of ice crystals. The value of nucleation temperature depends on processing variables (e.g. cooling rate [28], surface properties of a container [30]). The microstructure of ice crystals and the solute obtained during freezing determines the subsequent proceeding steps and is crucial for the quality of the final product [30]. For instance, faster cooling rates correspond to smaller crystals which are easier to sublimate but on the other hand their presence may negatively impact the timeframe of secondary drying. The crystal formation imposes the growth of solute concentration and increases the density of remaining solution [28]. If solute crystallization does not occur within the timescale of cooling the system will undergo a glass transition and will convert into amorphous phase. The corresponding value of the glass transition temperature will be lower than the value of the glass transition of the final freeze-dried product due to the presence of water, a well-known plasticizer which decreases the T_g of a system [30].

During the second stage the ice crystals are removed within the sublimation process. Then, the pressure is reduced and kept below the vapor pressure of ice. The ice crystals are transferred into the cold coils/plates of the condenser [27] leading to a so-called porous-cake structure. This stage is the most time-consuming and significantly impacts the economics of the whole process. As reported by Pikal an increase in drying temperature of 1 °C may reduce the time of primary drying by 13% [27]. On the other hand too high temperature will lead to a viscous flow of amorphous phase and may result in the collapse of the porous-like structure which will reduce the performance of the drying procedure. Thus, for amorphous systems it is recommended to keep the drying temperature below the glass transition temperature or slightly higher collapse temperature [28].

During the last stage, i.e. secondary drying, the unfrozen residual water is removed. The remaining water is then trapped in the glassy phase. Then, the diffusion of water molecules is hindered and depends predominantly on the porosity of the dried material [28]. The presence of small pores will hinder the diffusion of water, thus, rational selection of processing condition is recommended (it must be taken into account that water removal will increase the T_g of processing material).

Water is preferentially used as a solvent in the freeze drying method but its application is excluded in the case of poorly-water soluble APIs. Then some organic solvents or organic/water co-solvents have to be applied. However, due to the low freezing temperatures and low vapor pressures their removal from processing solutions may be difficult. Other issues concerning residual solvent level, their toxicity, special technical requirements, regulatory issues have to be considered as well [25, 32]. In the case of APIs with limited water solubility tert-butanol (TBA) and its water mixtures are the most commonly applied solvents in the freeze-drying procedure. Elgindy [33] and co-workers studied amorphous solid dispersions prepared *via* lyophilization using TBA as a co-solvent. They showed that complexes of flutamide with β-cyclodextrin and hydroxypropyl-β-cyclodextrin revealed enhanced solubility and dissolution rate. In this study, TBA was used to dissolve the highly hydrophobic flutamide while the hydrophilic carrier was dissolved in water. Subsequently, both solutions were lyophilized. It should be mentioned that TBA application has many advantages. First of all, TBA can be easily dissolved in water. Besides, it has a high vapor pressure (41.25 mm Hg at 25 °C), a high melting point (24 °C) and a low toxicity [33]. Moreover, the dissolution of TBA in water allows one to obtain crystals which are needle-shaped. So the resultant porous structure positively impacts the sublimation performance.

Laboratory and industrial-scale freeze-dryers may differ from each other but all of them contain a vacuum-tight chamber equipped with shelves, a separate condenser and a vacuum pump to evacuate non-condensable gases. Appropriate sensors necessary to control processing conditions are provided [34]. It has to be highlighted that the outcome of freeze-drying is very sensitive to any variations in processing conditions which may be related, for instance, to changes in dryer construction (e.g. variations in temperature distribution and heat radiation, variation in heat and mass transfer due to different shelf construction etc.) [25]. Thus, the transition between laboratory to industrial scale is problematic. Besides, the procedure of freeze-drying is relatively time and energy-consuming in comparison to other solvent-based method. Thus, the condition optimization and proper process control is necessary to make the freeze-drying method more profitable.

4.3.2 Spray-Drying

Spray-drying technology involves the transformation of a liquid material into the solid form within the drying of solution being previously separated into small droplets. The material processing involve several stages. First, the working solution is pumped into a drying chamber through the appropriately designed nozzle. It allows for liquid atomization before the liquid molecules are released in the chamber containing drying medium. Liquid atomization allows a high surface to volume ratio to be achieved over which heat and mass transfer take place during the drying stage. There are several types of nozzles that might be potentially used in the pharmaceutical industry, namely, two-fluid nozzles, pressure nozzles, rotary or ultrasonic atomizers [35]. In rotary nozzles the splitting of the liquid stream into fine droplets takes place due to the action of centrifugal force. In pneumatic (kinetic) atomization the drops of the liquid are formed due to the disruptive action of a high velocity gas upon a liquid stream. As two fluid streams are involved two-fluid nozzles need to be applied. In pressure nozzles the liquid is discharged through the circular orifice under high pressure, while in ultrasonic atomization the liquids are spread over a rapidly vibrating surface [36]. Recently, there has been marked interest in spray dryers which take advantage of vibrating mesh technology [37, 38]. The vibrations are generated by the piezoelectric crystal placed in a spray dryer head with a thin perforated membrane. The liquid penetrating an array of micron-size holes of mesh oscillating with ultrasonic frequency is transformed into a uniform mist in a highly controllable manner. This approach called nanospray drying (spray meshes with 4–7 μm apertures) allows the production of droplets with very a narrow size distribution [38].

In the next step the obtained droplets are mixed with a drying gas (air or nitrogen) at appropriate temperature what initiates the solvent evaporation. As a droplet is exposed to the drying medium the solvent content decreases while the viscosity of droplet increases leading subsequently to droplet solidification. This stage is crucial from the amorphization point of view. The properties of the formed amorphous state are governed by the kinetics of solvent evaporation and subsequent particle formation. Thus, processing parameters conditioning the drying rate, in particular the drying temperature, humidity and velocity of drying medium have to be suitably adjusted [39]. More details about the drying process, its thermodynamics and kinetics of droplet-particle transformation can be found in Chap. 6.

The drying chamber can be divided into tall (height to diameter exceeding 5:1) or small (2:1) [40]. The choice of suitable construction is related to the applied atomization device and determines the drying outcome. The contact between the spray droplets and a drying medium may vary depending on the mutual arrangement of the gas drying stream and the atomizer output. Depending on the design one can distinguish spray dryers with co-current, counter-current or mixed flow type [39]. Their detailed description can be found elsewhere [40]. The most popular is the co-current construction where both elements are placed at the top of the chamber providing gas and spray flow in the same direction (see Fig. 4.3) [41].

The last stage involves the collection of dried solid particles by appropriate separating and collecting devices. After drying, the particles are collected in the bottom part of the chamber. Sometimes to remove the product scrapping devices (e.g. vibratory devices, mechanical brushes, a stream of compressed air) have to be applied. Obviously, any supplementary mechanical treatment exerts an additional stress on amorphous particles and may improve the crystallization ability of a

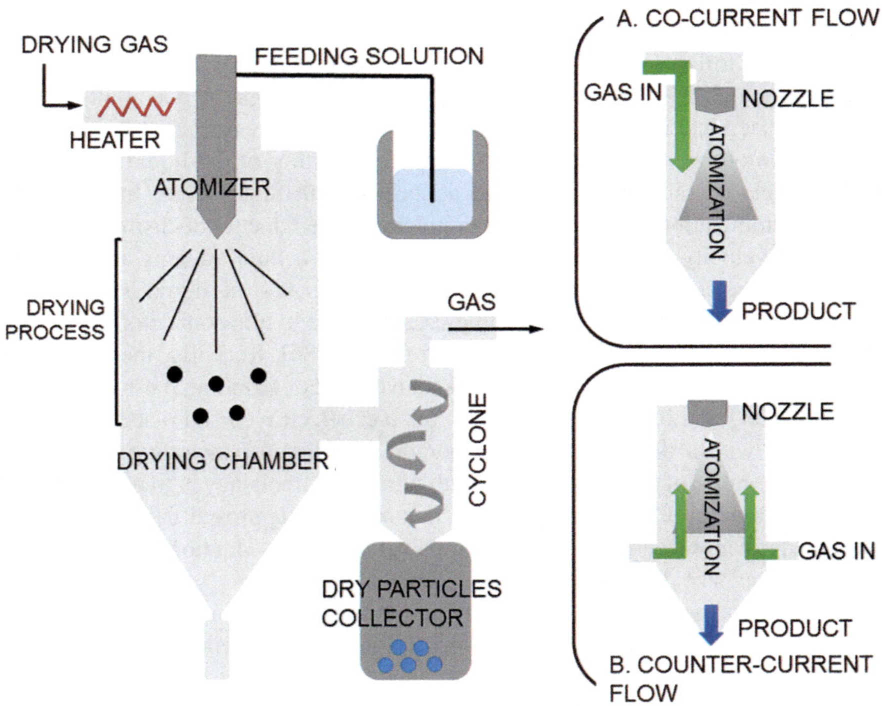

Fig. 4.3 The general concept of spray-drying method

processing material. To avoid this the bottom of the chamber should have a conical shape that facilitates the flow and collection of the dried product [39]. Another possibility assumes that the dried particles are removed from the chamber with air and then will be separated by means of cyclones or bag filters. In cyclone separators the solid and gas particles are subjected to an accelerated flow field generated by a rotating vortex, where inertia and gravitational forces affect the separation process. A textbook explanation assumes that solid particles with higher density are subject to a greater centrifugal force. Thus, they are moving towards the outer wall of a cyclone what allows for their separation (see Fig. 4.4b). Then, being outside the core flow they start falling from the upper cylindrical part to the conical bottom where their collection takes place [42]. At the end of the conical part the flow direction is reversed (reverse-flow cyclones). This allows for the separation of larger (denser) particles while smaller particles remain incorporated in the gas stream. The smaller the particles we want to separate, the higher the acceleration we need to apply. Alternatively, electrostatic collection can be implemented as was proposed by Buchi in a new generation of laboratory nano spray dryers [37, 38]. An electrostatic particle collector contains a stainless steel cylindrical anode and a star-shaped cathode placed inside the cylinder (see Fig. 4.4a). During the drying process

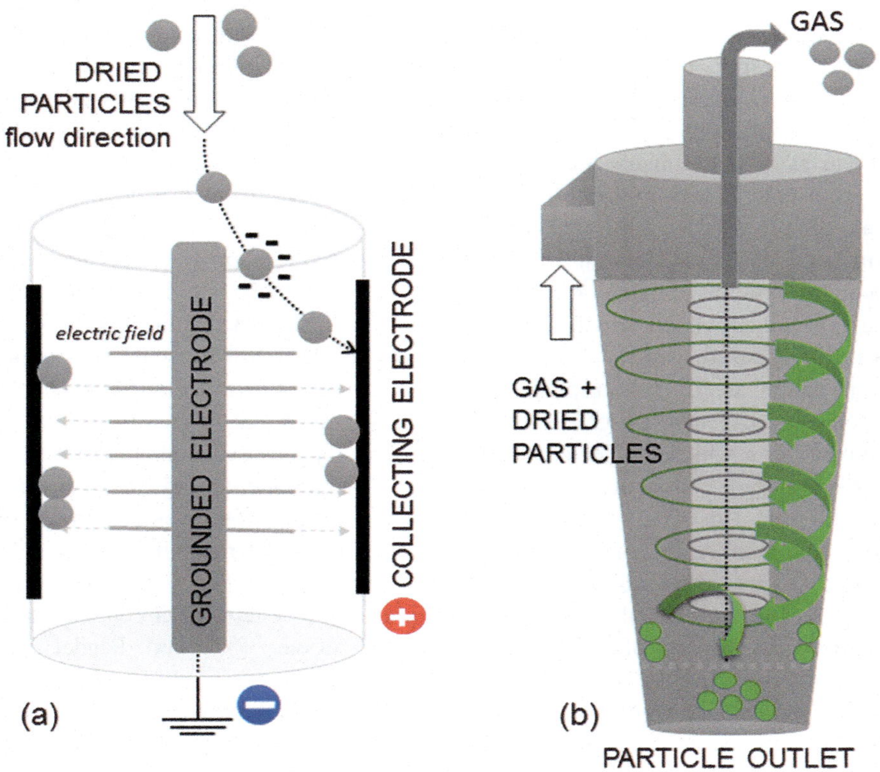

Fig. 4.4 Various methods of dry particle selection and separation during spay-drying procedures

Table 4.2 The main adjustable parameters determining the outcome of freeze-drying procedure [44]

Processing parameters		
Related with process	Related with processing solution	Related with equipment
Processing solution flow rate	Concentration (drug/polymer ratio)	Co-current flow
Drying gas flow rate	Viscosity	Counter-current flow
Gas inlet temperature	Density	Mixed-flow
Drying rate	Surface tension	Atomizer geometry
Pressure		
Type of gas		

a high voltage is applied and particles becomes electrically charged and deposited in the inner wall of the cylinder electrode. Finally they are collected by a rubber spatula [43].

Today spray drying technology is a powerful manufacturing platform for amorphous solid dispersions. The spray drying procedure involves several operating units. The characteristics of the final products are determined by a combination of several parameters summarized in Table 4.2.

Its scalable and continuous character allows for successful laboratory and industrial scale implementation [45]. These features ensure spray drying is in leading position among other solvent-based methods of amorphous form preparation. To get some fundamental insight into the physical stability and enhanced solubility of spray-dried amorphous formulations various systems containing APIs alone as well as binary drug-polymer systems have been deeply investigated. Recently, ternary or multicomponent systems which besides API and carrier contain some functional adjuvants (like surfactants, pH modifiers etc.) are of growing interest [41]. These additional excipients are intended to improve the performance and processability of spray-dried products. For instance, the addition of adsorbent (e.g. colloidal silicon dioxide Aerosil 200) to the spray-dried solid dispersion of simvastatin allows improved drug flow ability [46]. The low T_g components, like simvastatin, are difficult to spray-dry. The same applies to electrostatic powders. In such cases the presence of adsorbent allows to prevent sticking and blocking of material flow which is reflected in the higher drying yields. As another example the ternary system containing celecoxib, PVP and meglumine, an organic base commonly used as a pH-adjusting and solubilizing agent, can be given. The polymer acts as stabilizer and increases the drug solubility, while meqlumine forming a complex with a drug leads to further solubility enhancement [47]. The current state of the art in the field of multicomponent spray-dried systems is thoroughly summarized in papers of Singh and Van der Mooter [39] and Paudel and co-workers [41].

To prepare feed solution of poorly water soluble API or multicomponent system various organic solvents or solvent mixtures have to be used [39]. Then, the spray dryer works in a closed cycle employing a drying gas being continuously recirculated [48]. During solvent selection its volatility, toxicity, viscosity, solubility limits for drug and carrier have to be taken into account [35, 41]. We have to keep in mind that our choice will affect the structure and physicochemical features of the spray-dried product. Paudel and Van der Mooter investigated the impact of solvent composition on the properties of solid dispersion containing naproxen and PVP K25 prepared by co-solvent spray-drying [49]. They found that the proper solvent chemistry lead to a material with better miscibility, smaller crystallinity and higher physical stability. The solvent impact was also discussed for solid dispersion containing griseofulvin, poly [N-(2-hydro-xypropyl) methacrylate] (PHPMA) and PVP [50]. The application of an acetone-methanol mixture instead of an acetone-water mixture had significant impact on morphology, viscosity, stability and dissolution properties of the final spray-dried product. The solvent used to dissolve drug and excipients affected the materials hydrogen bonding pattern and conformational preferences and decided whether the globular or extended forms of polymer were preferred.

4.3.3 Supercritical Fluid Technology (SCF)

Supercritical fluid technology has been well recognized by the food industry and purification science, and has become a subject of growing interest in a pharmaceutical field. This attractive approach has been also examined in terms of possible application in the amorphous solid dispersion field. So far, supercritical technology has been successfully applied to obtain solid dispersions containing bicalutamide [51], indomethacin [52, 53] carbamazepine [54], piroxicam [55], felodipine [56]. Despite its commercial potential at this moment this concept is at the early stage of development and further steps regarding its scale-up have to be considered.

Supercritical fluid above its critical temperature and critical pressure reveals unique properties resulting from the disappearance of phase boundaries between the liquid and gaseous state (see Fig. 4.5). These extraordinary liquid-like and gas-like properties are beneficial for various pharmaceutical applications. For instance, in the supercritical state the material has liquid-like density, the viscosity typical for gaseous phase, and diffusivity lying between that of gas and that of a liquid [57]. It has been reported that liquid-like properties are important during extraction processes, drug solubilization, and polymer plasticization, and that gas-like features significantly facilitate mass transfer processes [58]. The gaseous properties of supercritical CO_2 allows for its efficient penetration into porous solids and makes its application attractive from the perspective of confined systems containing an API embedded into a porous matrix (e.g. porous silica). Recently, such possibility has been reviewed by Gurikov and Smirnova [59]. The properties of supercritical fluids can be further adjusted by changing the thermodynamic conditions. In particular, due to their large compressibility the small pressure changes may remarkably affect the density and dissolution ability of the supercritical fluid [60]. Consequently, the application of supercritical fluids in pharmaceutical processing gives tremendous

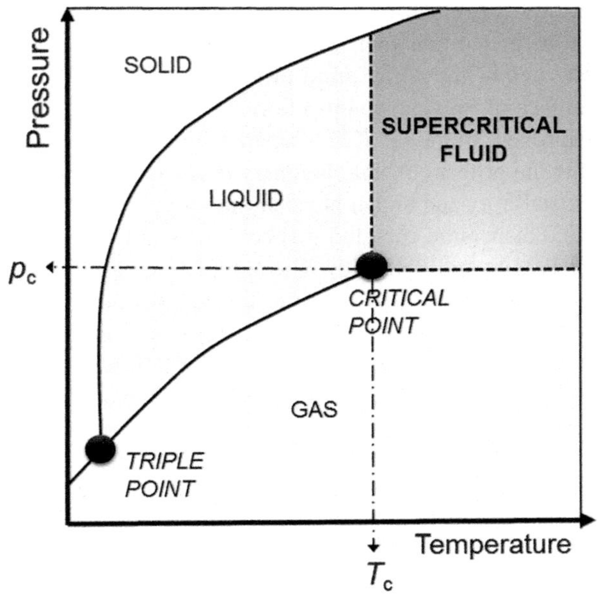

Fig. 4.5 A typical phase diagram of carbon dioxide. At certain pressure and temperature condition i.e. $T_c = 31.2$ °C and $p_c = 7.4$ MPa the liquid and gaseous phases become undistinguishable

possibilities. Depending on their role (solvent or anti-solvent) and processing mechanism various techniques involving supercritical fluids can be deduced. Readers interested in this topic should refer to excellent papers where these methods are described in details [58, 61–63].

In general any solvent can be transformed to the supercritical state only if particular thermodynamic conditions are fulfilled, however, only some of them are suitable for application in the pharmaceutical sector. In practice this area is almost completely dominated by the usage of carbon dioxide (CO_2). The accessible critical conditions ($T_c = 31.2$ °C and $p_c = 7.4$ MPa), low costs and environmentally safe handling make supercritical CO_2 ideal for this purpose. The genuine benefits associated with carbon dioxide application make other solvent usage too expensive and unprofitable.

The particular utility of supercritical CO_2 has been reported for drug-polymer formulations where supercritical CO_2 acts as polymer plasticizer and substantially impacts drug-polymer processing. On laboratory scale during the supercritical CO_2 treatment the physical mixture of API and polymer is placed inside a pressure vessel and pre-purged. Later, CO_2 is pumped into system as long as the proper thermodynamic conditions are obtained (the temperature is then internally regulated). The system is subsequently kept at the desired conditions for the time necessary for full blending. Afterwards, prompt decompression is performed [64]. The advantages associated with supercritical fluids application in the HME are discussed at the beginning of this chapter in the section concerning HME technique. The sorption of CO_2 by polymer leads to its swelling and influences its mechanical and physical properties, in particular decreases the value of T_g [65]. One can expect that the increased molecular mobility and higher free volume will facilitate drug diffusion in the polymer matrix leading to improved drug-polymer mixing. Ugaonkar and co-workers investigated how near-supercritical CO_2 impacts the crystallinity of drug-polymer mixtures containing carbamazepine and polymers of varying molecular weights [66]. They observed that only PVP with the lowest molecular weight allows for successful conversion of carbamazepine to the amorphous form. Such effect was attributed to the degree of plasticization of polymer molecules by supercritical CO_2. It was postulated that with increasing molecular weight of the polymer the chain flexibility decreases reducing the number of potential interaction sites between CO_2 and the polymer leading consequently to insufficient plasticization. The importance of polymer molecular weight on solubility of solid dispersions prepared using supercritical CO_2 has been also indicated. The enhanced aqueous dissolution of carbamazepine dispersed in PVP K10 in comparison to PVP K29 was observed [66]. Such effect was attributed to efficient plasticization of PVP K10 by supercritical CO_2 facilitating water penetration and drug diffusion [67]. A similar effect was observed for piroxicam-PVP systems mixed with the assistance of supercritical CO_2. The accelerated drug release was observed for systems containing PVP K15 having the smallest molecular weight among other polymers investigated in this study [55]. In the case of PVP K90 the drop of release was significant and the measured dissolution rate was even slower than that obtained for physical mixtures.

4.3.4 Electrospinning and Electrospraying

Electrospinning and electrospraying are manufacturing methods based on similar concepts of electric filed application in order to convert the liquid solution into solid fibers or particles. During processing the working solution is pumped into the syringe needle attached to a high voltage supplier. The interplay between electrostatic repulsion forces and surface tension leads to the deformation of a liquid drop into a so-called Taylor cone. At a particular critical voltage condition the jet of processing sample separates from the rest of the solution and starts moving towards the grounded collector. The collecting device is located a short distance from the needle tip, typically 10–20 cm. During crossing this gap solvent evaporation takes place leading to material solidification [68–70]. In electrospinning the resultant solid fibers are deposited on the surface of a collector, while in electrospraying the jet of processing solution is divided into small droplets.

The ability to fabricate objects, fibers and particles, with diameters in the nano- or micro-range makes electrospinning and electrospraying techniques attractive for various pharmaceutical applications. Their utilization in solid dispersion technology has also been investigated. The large surface area characterizing electrospinning and electrospraying materials is important for efficient drug release and at the same time allows a very fast evaporation rate to be achieved [71]. Since the solvent evaporation is so rapid, the amount of time when the drug-polymer system can crystalize is limited, which favors the amorphous state formation [72]. Verreck and co-workers pointed out that *via* electrospinning from dimethylformide or dimethylacetamide solutions the amorphous nanofibers containing polyurethane and itraconazole or ketanserin can be obtained [71].

The amount of drug and polymer in the working solution determines its viscosity and is crucial for the final morphology of the fabricated fibers. The higher the polymer content in solution is the larger and more regular structure of fibers is obtained [71, 73]. Besides, the resultant microstructure of fibers is determined by other processing variables like applied voltage or feed rate [74]. For example, the comparison of release profiles for amorphous itraconazole-HPMC formulations obtained by electrospinning revealed differences in in vitro drug release depending on the applied voltage during the manufacturing procedure. It was speculated that the observed variations were related to the differences in packing and tangling of fibers produced under different conditions [73].

Recently, so-called high speed electrospinning has been proposed to prepare amorphous solid dispersion containing itraconazole and PVPVA64 [75]. This approach was reported as much more efficient than the conventional route and thus much more appropriate from the perspective of potential commercialization. The application of spinneret connected with high speed rotary motor allows an increase in the process performance by up to 75 times due to combined electrostatic and high speed rotational jet generation. Nagy and co-workers reported that amorphous formulations with itraconazole obtained by conventional and high speed electrospinning had improved dissolution rates (>90% in 10 min). The dissolution enhancement was much better than those observed for samples with the same composition but produced *via* spray-drying method [75].

Despite the undoubted potential, the implementation of electrospinning to amorphous solid dispersion technology still requires a lot of work. It is necessary to look for clever solutions to increase its productivity (e.g. multi nozzles setups) before its transfer to industry takes place. Besides, the fundamental knowledge about the final product quality and stability is still very limited.

When the processing solution contains lower polymer content instead of continuous fibers an aerosol of charged particles can be produced [68]. As previously discussed, several parameters related to the solution composition as well as operating conditions will influence the features of the manufactured particles. Zhang and co-workers investigated coaxial electrospray formulations containing griseofulvin and Eudragit L-100 [76]. Core-shell solid dispersion containing fenofibrate and two polymers, Eudragit L-100 and PVP K12-17, were investigated by Kawakami and co-workers [77]. The application of coaxial setup allows particles with different composition in the inner and outer part to be produced, i.e. core-shelled particles. Then, the solutions of drug and polymer are supplied by two syringe pumps to coaxial nozzle to which a high positive voltage is applied (see Fig. 4.6, panel c). Then, the particles are deposited on a grounded steel plate located perpendicular a short distance from the nozzle [76, 77]. Such co-axial electrospray technology has great potential for many reasons: (1) it allows the dissolution properties of API to be improved, (2) the possibility of surface modification may resolve the problem of particle aggregation,

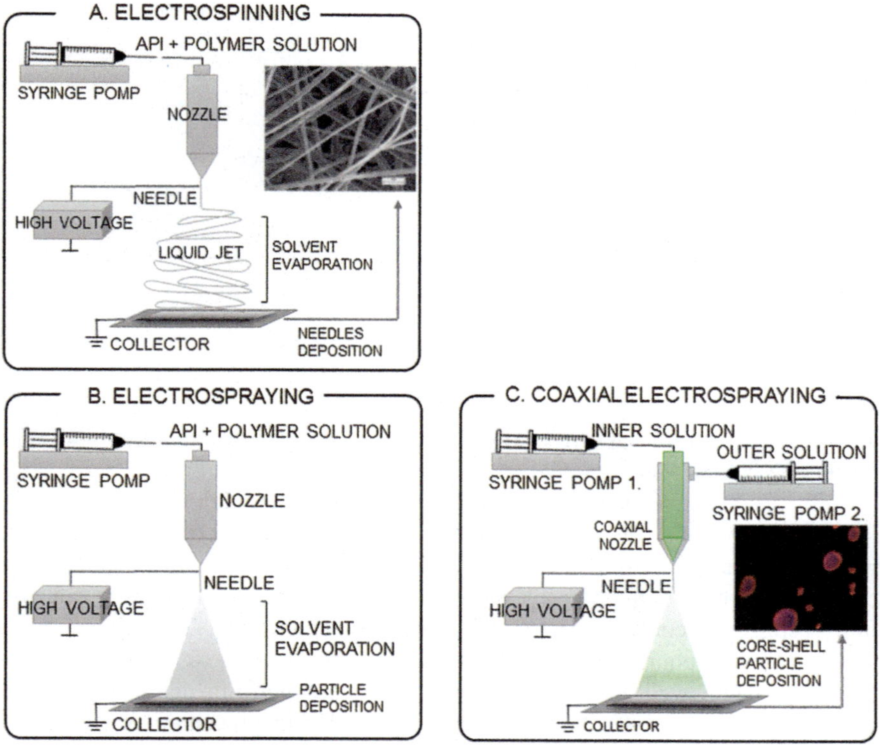

Fig. 4.6 The scheme of laboratory electrospinning (**a**), electrospraying (**b**) and coaxial electrosprying (**c**) setups. Adopted from [78]. In panel (**a**) the SEM picture of itraconazole/PU 40% w/w fibers is presented (reprinted from [73] with permission of Springer). In panel (**c**) the confocal images of griseofulvin and Eudragit L-100 are shown (reprinted with permission from Zhang S, Kawakami K, Yamamoto M, et al. Coaxial electrospray formulations for improving oral absorption of a poorly water-soluble drug. *Mol Pharm.* 2011; 8:807–813. Copyright 2011 American Chemical Society)

(3) it gives the possibility to explore novel approaches in particle design, and finally (4) it addresses the important issue of surface-enhanced crystallization.

4.3.5 Thin Films Manufacturing

The idea of thin film application as a fast dissolving drug delivery systems is well-recognized in the field of drug formulation. Such thin films containing APIs embedded in the polymer matrix are suitable for various administration routes (oral, buccal, sublingual, ocular, and transdermal). Among multiple advantages of thin film formulations their enhanced bioavailability and convenient administration

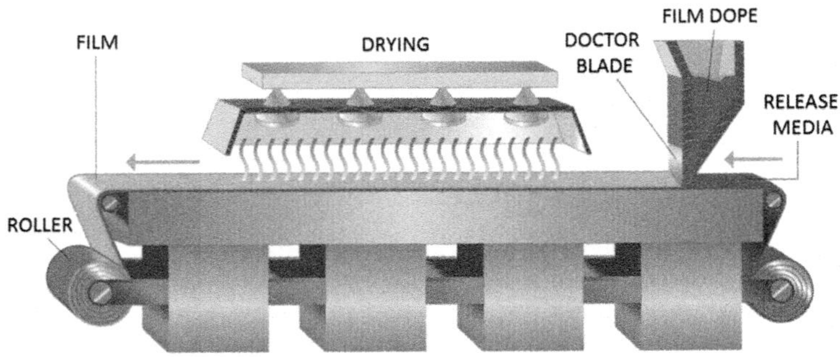

Fig. 4.7 Industrial scale film casting setup. Reproduced from [81]

are usually highlighted [79]. Thus, their implementation in the amorphous solid dispersion field is also attractive.

Film casting is one of the most common methods of thin film manufacturing. Its idea is very simple. In the first step the working solution containing drug and polymer is casted on a suitable substrate. Later, the casted solution is subjected to elevated temperature in order to evaporate the solvent. On the laboratory scale the film casting procedure is performed using film applicators. The drug-polymer solution is then poured onto the glass plate which in the next step is swept by film caster with well-defined gap which allows for uniform distribution of the processing solution [80]. In the second step a drying procedure is performed. During industrial manufacturing a continuous procedure is provided (see Fig. 4.7). The produced dried films are subsequently rolled and cut to the desired forms [79].

During film processing the rate of solvent removal is critical from the perspective of API amorphization. For sure, the evaporation rate obtained during film casting is much slower than in the case of spray-drying, thus, the risk of crystallization increases. Despite this, film casting has been chosen as a solid dispersion preparation method in several studies. Weuts used the film casting technique to prepare solid dispersions containing etravirine and HPMC [82]. Janssens and co-workers compared solid dispersions containing itraconazole and Eudragit E100 prepared *via* spray drying and film casting paying attention to the difference between the received formulations [83].

Another method allowing thin film manufacturing is called spin-coating. The solution containing the drug and polymer is dropped into the center of a spinning substrate. During spinning the solution is spread out on the substrate while the solvent becomes evaporated. Obviously, the operating parameters related to the processing solution (viscosity, polymer content, surface tension) as well as rotation parameters will determine the features of the final product [84]. Ng and co-workers investigated spin-coated films containing various APIs (felodipine, fenofibrate, carbamazepine, and celecoxib) and polymers (Eudragit E PO, PVP K30, Soluplus) [85]. Among investigated systems the films containing a drug layer with top polymer coatings have been successful. In comparison to other investigated films, e.g. those with drug dispersed in polymer matrix, the former were characterized by better physical stability. This was attributed to the presence of a polymer layer which protects the amorphous API against moisture uptake and suppresses the surface-induced drug recrystallization.

4.4 Preparing Amorphous Samples by Dehydration of Crystal Hydrates

One of the methods used to obtain amorphous solid samples is desolvation (dehydration) of solid crystal solvates (hydrates). This type of reactions has been under study since the beginning of the twentieth century in various aspects [86–94]. The possibility to control the state of the solid desolvation product is one of the main problems that remains unsolved, despite considerable efforts by many research groups. This pertains to the remaining inability to control whether a product is crystalline or amorphous, the particle size and shape or the product, and which of the polymorphs forms, if the product is crystalline. Early research focused mainly on crystal hydrates of inorganic salts and the possibility to use their dehydration products as catalysts and materials. Interest has recently shifted towards dehydration of pharmaceutical hydrates/desolvation of pharmaceutical solvates. This is of particular practical importance regarding polymorphism and polyamorphism of drugs [92–99].

In general, one can expect four possible types of dehydration product: [87]

1) Strictly pseudomorphous dehydrated samples. Uranyl oxalate, or zeolites are examples of this class. Covalent bonding and rigid network structures determine the properties.
2) Crystalline products in the thermodynamically stable state. The products of the high-temperature dehydration of ammonium alums are examples of this class.
3) Microcrystalline products that store energy partly in the form of internal disorder, but mainly as surface free energy. They appear amorphous when studied by X-ray diffraction because of the small particle size. However, while they are X-ray amorphous, they are not truly amorphous. In the presence of water vapour the size of the crystallites becomes larger (ripening occurs). As an example, one can mention cobaltous chloride monohydrate.
4) Truly amorphous dehydration products that store energy mainly in the form of internal disorder. The structure is sometimes conceived as a porous glass or highly disordered zeolite. In the presence of water vapour the product can crystallize. One of the first examples for which this type of dehydration products has been reported is manganous oxalate dihydrate.

Of these four classes the first one is rare, the second one is scientifically unrewarding, but the third and the fourth are of the greatest fundamental and practical interest. In contrast to amorphous samples that are obtained by mechanical treatment, the amorphous dehydration products can often be preserved for months, or even years, without recrystallization [87, 88]. Another obvious advantage is that the samples obtained by dehydration usually have a very narrow distribution of particle size and shape. The excess energy stored in the metastable amorphous dehydration products can be as high as 30–35 kJ/mol [100–102]. Amorphous dehydrated products are often less hygroscopic and less prone to rehydration on storage [88, 92, 99, 103].

A product of desolvation/dehydration is usually disordered since elimination of crystal solvent/water is accompanied by generation of mechanical stress. According to a wide-spread model, a stressed zeolite-like "lacunary phase" is formed as an intermediate dehydration product [104–107], which then either collapses into an amorphous sample, or recrystallizes into a crystalline product. A mathematical model relating stress, its relaxation and the size of crystallites formed on fragmentation of the product has been proposed and tested on a number of model systems [108–111]. However, modeling of true amorphization remains a challenge [98, 112]. In most cases, the control of amorphization is based on intuitive considerations, experimental experience and "common sense", which do not seem to have changed much since the early research of this problem started over a century ago.

The possibility that the dehydration product will be truly amorphous depends on several factors: temperature, the water vapour pressure, the starting particle size, the starting crystal structure, and chemical properties of atoms of the starting crystal hydrate (particularly the affinity of the anion to accept protons) [87, 88, 92, 93, 95, 98, 112–114].

The temperature and water vapour pressure determine the rate at which water is lost from the sample. Thus, these factors also determine the mechanical stress that arises over the course of this process, the possibility for the crystal structure to relax, and the mechanism of this structural relaxation. Fast dehydration favors amorphization. If water is removed slowly, the crystalline phase has time to nucleate and grow [88]. For many dehydration processes the non-monotonic dependence of the dehydration rate on the water vapour pressure (the Smith-Topley effect) has been observed [86–91, 114–119]. Despite considerable discussion on the origin of this non-monotonic dependence, no concensus has been reached. One explanation relates the region between the two extrema on the "dehydration rate—water vapour pressure" curve to the transition between an amorphous and the crystalline states of the intermediate dehydration product(s) [88]. However, some authors argued that the Smith-Topley effect is not related to the properties of a solid, since it was also observed for the evaporation of liquids [117, 119].

The starting particle size is a very important parameter for controlling the possibility of obtaining an amorphous dehydration product. As the size of the particles decreases, the heat of dehydration changes drastically, so that an endothermic process can become exothermic when the dehydrating particles become smaller than 10 nm [120]. This particle size is close to the estimated limit of the stability of crystalline particles [121]. Particles smaller than 5 nm tend to amorphize [122]. The dehydration of trehalose dihydrate was studied for different particle size fractions by a combination of hot-stage microscopy, X-ray powder diffraction, thermo-gravimetric analysis, and differential scanning calorimetry. An amorphous phase was formed on dehydration of small particles (< 45μm). At temperatures above the glass transition temperature of amorphous trehalose this phase transformed into a supercooled liquid, that then crystallized to give anhydrate. A solid-solid conversion from the dihydrate to the anhydrate has been observed at temperatures as low as 80 °C, if larger particles (>425 μm) were used. The liberated dihydrate water catalyzed the solid-solid rearrangement of the dehydrated phase to the anhydrate. The small particles have a large surface area-to-volume ratio. Because of this, the dehydration occurred prior to attaining the threshold temperature for a solid-solid rearrangement to a crystalline phase. Therefore, no solid–solid conversion was observed for these particles, and the product remained amorphous [123]. The starting particle size has been shown to be very important for the formation of amorphous product on dehydration of caffeine 4/5-hydrate at different relative humidities [114].

It is also documented that the amorphous state of the small particles can be stabilized by the presence of some residual water in amounts as low as 0.1%. Removal of this residual water results in spontaneous crystallization [124].

In most cases the dehydration is a multi-stage process, with the number of stages reaching 17 or more in some cases [125]. Amorphization is possible at some of the stages, depending on the structural relationships between the phases. The first stages of dehydration are accompanied by crystal fragmentation. One can therefore expect that, as the particle size gets smaller and smaller, the later dehydration stages are more likely to give amorphous products [88]. It is also important that no fluid phase—liquid water—is formed, as crystallization would therefore occur.

For inorganic crystal hydrates, the proton affinity of the anions has been shown to favour the formation of amorphous dehydration products. Thus the dehydration of the copper phosphate and selenite gives amorphous products at atmospheric water vapour pressures. These systems remain amorphous for several years on storage under ambient conditions, without special precautions on air [87]. In contrast, the dehydration products of copper sulphate hydrate remain crystalline at pressures higher than 2000 Pa [87]. If obtained as amorphous samples, they crystallize quickly. Dehydration of crystal hydrates the salts of strong acids with alkaline and earth-alkaline cations usually does not give amorphous products under ambient conditions [88].

The effect of the starting crystal structure on the possibility to form an amorphous dehydration product is rather straightforward. If chemical species in the starting structure are linked by robust directional bonds, which do not involve the water molecules that are to be removed on dehydration, then this framework can be preserved. If, however, the molecules are either loosely kept together by weak non-covalent interactions, or the hydrogen-bond network includes water molecules as an essential structure-forming element, then the structure is more likely to collapse on dehydration.

The method of obtaining amorphous solids by solid-state dehydration of crystal hydrates is recognized as advantageous as compared to mechanical treatment, anti-solvent precipitation, freeze-drying and spray-drying [95]. That said, the control of the dehydration process remains a challenge, especially for the case of crystal solvates of organic compounds. This is, in large, due to the fact that systematic research in this field is still missing, despite a century of experience with another class of hydrates of inorganic salts. An additional opportunity provided by organic solids is that, in addition to hydrates, they can form solvates with many different organic solvents. This adds an extra dimension to the problem and one more parameter (the solvent molecule) that can be used to control the amorphization. Further, the solvates of organic molecules are prone to polymorphism, and one can also study the role of the crystal structure of the starting solvates with identical chemical composition, in order to better rationalize factors that influence the amorphization on desolvation [126–129].

4.5 Preparing Amorphous Samples by Mechanical Treatment

Grinding, mixing and tableting are important operations in industry, particularly in the processing of drugs and the preparation of drug formulations. During these actions, the properties of drug substances, excipients and their mixtures can be modified. For example, new polymorphs, dehydrated forms, salts, cocrystals, solid dispersions, and amorphous forms can be produced [130–134]. These processes can be inadvertent and even not well-recognized, and can be a considerable issue. Crystalline phases, even if present as minor impurities immediately after treatment, can act as seeds to trigger subsequent physical and chemical transformations through the whole bulk, which may lead to a product inadvertently infringing on a patent of another company, or a product no longer conforming to the specifications and having unwanted pharmacological properties [135]. An amorphous phase can modify the properties of the product immediately after being formed, but also as a result of crystallization on storage. The latter process can be detrimental not only for mechanical properties of the tablets, or for their hygroscopicity, but also for the bioavailability of the active pharmaceutical ingredient, such as when a compound soluble and bioavailable as an amorphous form, but not in the crystalline state [132, 135–144]. It is because of amorphous samples having advantages over their crystalline analogues that they are often produced on purpose, in some cases *via* mechanical treatment.

There can therefore be two opposite targets when a solid is subjected to mechanical treatment: to *avoid* the formation of an amorphous state, or to *produce* an amorphous sample, which will be as free from any crystalline inclusions as possible, aiming to prevent its crystallization on storage for a prolonged time. Both tasks are challenging, and require the understanding of the mechanisms of amorphization under mechanical treatment and of the factors that help to preserve the metastable amorphous state for a prolonged time.

Amorphization is one of several different processes that can result from the mechanical treatment. Alternative (or parallel) phenomena are brittle fragmentation, local melting, chemical reactions and polymorphic transformations [145]. The main experimental parameters that can be varied in order to control amorphization of solids subjected to mechanical treatment, as well as the subsequent crystallization of the amorphous samples, include the temperature, the type and the protocol of mechanical treatment, the presence of additional solid phases in the system, and the presence of fluids in the system [145]. The latter is particularly noteworthy, being problematic when treating crystal solvates [146], when employing the use of Liquid Assisted Grinding [146–148], or when water is present in the atmosphere [149–151]. Water not only favor the crystallization of an amorphous phase, but in some cases can assist the amorphization of a crystalline phase on grinding [149–152]. Preliminary chemical treatment of the solid sample (etching or enzymatic treatment), or annealing defects by heating preceeding mechanical treatment can also be important [153].

In general, solids become more brittle on cooling [154]. Therefore, one could intuitively expect that grinding at low temperatures is more likely to result in brittle fragmentation, whereas at higher temperatures plastic deformation will account for disordering, accumulation of defects and amorphization. This is in fact true for some compounds, like inorganic salts, but the real situation is more complicated. The opposite phenomenon is observed more often than not: cryogrinding gives amorphous products, whereas grinding at room and more elevated temperatures gives various crystalline phases [155–157]. It has been documented that an amorphous product can be obtained by mechanical treatment only if the temperature of treatment is below the glass transition temperature of the sample, T_g. Decreasing the milling temperature leads to an increase of the amorphization tendency whereas milling above T_g can produce a crystal-to-crystal transformation between polymorphic varieties [133, 158–160]. These observations contradict the suggestion that milling transforms the physical state only by a heating effect, which induces local melting [132]. Equilibrium thermodynamics does not seem appropriate for describing the process. The driven alloys concept offers a more rational framework to interpret the effect of the milling temperature [133].

Quite often milling of a crystalline solid first results in the partial or complete amorphization of a material, but subsequent prolonged mechanical treatment of the already formed amorphous solid results in crystallization of the sample. As such, the final product becomes a stationary amorphous–crystalline mixture [132, 161, 162]. In contrast to such techniques as spray-drying, freeze-drying, or quench-cooling of a melt, mechanical treatment rarely gives pure amorphous samples, without any admixtures of crystalline or nanocrystalline inclusions. Higher temperatures and the presence of even traces of fluids often favor crystallization, and, even if an amorphous state is formed originally, it converts into a more stable crystalline form. Examples have been documented when milling of initially amorphous solids resulted in complete crystallization [132, 162–164].

The choice of a type of mechanical treatment, and, respectively, of a device can also be critically important to controllably induce either brittle fragmentation or plastic deformation and amophization. Devices that produce free impact with time of the impact significantly less than the time required for the relaxation of elastic strain ($t_{impact} \cong 3 \cdot 10^{-8}$s), such as jet mills, are likely to comminute the sample.

This process likely produces fine particles but does not distort much their inner structure. Devices in which the impact is restricted (ball mills, disc mills, impact devices), and its duration is sufficient for structural relaxation (0.02 s $< t_{impact} <$ 1 s) can be more suitable if a target product must be strongly deformed or completely amorphous [165]. Opposite examples have also been documented, when grinding in a jet mill caused larger distortions of the crystal structure and faster amorphization than treatment in a vibration mill [166]. This can be related to different intensity of treatment in the two devices. The frequency of impacts can be also important, when the structural rearrangement occurs to a large extent not during an impact, but between the impact pulses [145, 167–169].

Once formed, an amorphous phase produced by mechanical treatment is usually not preserved for a long time. Recrystallization is possible even below the glass transition temperature, T_g, of the same compound obtained as an amorphous sample by other techniques [170, 171]. It has been shown at least for some compounds, like griseofulvin, that the amorphous forms obtained by milling are more characteristic of strongly defective crystals rather than true amorphous forms. Therefore their recrystallization upon heating at temperatures well below the glass transition temperature (T_g) is a completely different process as compared to the crystallization of the amorphous compound, observed above T_g [172]. Amorphous states obtained by milling can differ much from long-lived amorphous states obtained by other techniques: quench-cooling of the melt, vapor deposition, freeze-drying or spray-drying [171–175]. For example, the cryo-milled amorphous samples of simvastatin recrystallized within 1 day upon storage at 20 °C, whereas the quench-cooled samples showed no signs of recrystallization even when stored at 40 °C for several days. An amorphous form for the cryo-milled simvastatin was spectrally distinguishable from the quench-cooled amorphous form. The thermodynamic parameters suggested that this form was less disordered compared to the quench-cooled form [174]. At the same time, for some compounds the difference in the properties of amorphous states obtained by different methods seems to be minimal [176]. This is similar to the fact that many crystalline compounds are not prone to polymorphism.

It has been already mentioned in Chap. 2 that an amorphous state for a compound is not unique, and that multiple amorphous states (polyamorphs) are possible, as well as the first-order phase transitions between these states [177]. The amorphous states obtained by mechanical treatment and by other techniques (vapor condensation, precipitation from solution, supercooling of melt) are usually different. They have different T_g and heat effects of crystallization, different densities, viscosities, vibrational spectra, and pair distribution functions. They can resist crystallization for various durations of time because of a large difference in the kinetics of nucleation and nuclei growth [171–174]. On crystallization, they often give different polymorphs [178–180]. These phenomena can be rationalized if one takes into account that the different amorphous states are obtained starting from different original structures. Mechanical treatment of crystalline phases is likely to destroy the perfect order and to introduce more and more disorder, but the main structure-forming units of a crystalline state can be preserved. Further, on mechanical treatment small crystalline particles may remain that can act as seeds for subsequent recrystallization, once the mechanical treatment is over and the sample is annealed or simply stored. Disordering can occur step-wise, via a series of mesophases described in Chap. 2 [132, 172, 181–183]. Cooling of a melt produces a solid that, to a large extent, preserves the original structure of this fluid phase (a "quenched liquid"). This fluid phase is not merely a "distorted crystal structure", but is radically different from the structure of a crystal (see Chap. 2). When an amorphous phase is formed on rapid crystallization from solution (spray-drying, freeze-drying), the structure of the product is again determined by clusters that nucleate fast and do not grow further [184, 185]. Since the amorphous states formed by different methods have different short-range structures, they give rise to different crystalline forms on crystallization, and the crystallization rate itself may be very different.

Amorphous states are metastable, but can be preserved for a very long time. One of the most common methods of extending the life-time of an amorphous solid produced by mechanical treatment of a crystalline precursor is to co-grind the target solid compound with another solid component: a polymer [132, 135, 145, 162, 186–198], an inorganic additive [198–202], an amino acid [203], α-lactose [198, 204], or another compound [149, 152, 198, 205]. The solid dispersions formed usually represent supramolecular complexes, also termed *mechanocomposites*, in which the molecules of the amorphous component are linked *via* multiple intermolecular interactions to the other component. As such, nucleation of the crystalline phase is hindered. The molecular structure in such a complex can be different from that in a crystalline phase—metastable isomers and coformers can be stabilized by linking them to the matrix of another component [206–208]. It is more efficient to co-grind the components together, than to grind one or all the components separately and then mix them [132, 135, 145, 162, 187, 202, 204–206].

References

1. Thiry J, Krier F, Evrard B (2015) A review of pharmaceutical extrusion: critical process parameters and scaling-up. Int J Pharm 479(1):227–240. https://doi.org/10.1016/j.ijpharm.2014.12.036
2. Breitenbach J (2002) Melt extrusion: from process to drug delivery technology. Eur J Pharm Biopharm 54:107
3. Alshahrani SM, Morott JT, Alshetaili AS, Tiwari RV, Majumdar S, Repka MA (2015) Influence of degassing on hot-melt extrusion process. Eur J Pharm Sci 80:43–52. https://doi.org/10.1016/j.ejps.2015.08.008
4. Liu H, Wang P, Zhang X, Shen F, Gogos CG (2010) Effects of extrusion process parameters on the dissolution behavior of indomethacin in Eudragit E PO solid dispersions. Int J Pharm 383(1–2):161–169. https://doi.org/10.1016/j.ijpharm.2009.09.003
5. Vynckier AK, Dierickx L, Saerens L et al (2014) Hot-melt co-extrusion for the production of fixed-dose combination products with a controlled release ethylcellulose matrix core. Int J Pharm 464(1–2):65–74. https://doi.org/10.1016/j.ijpharm.2014.01.028

6. Vynckier AK, De Beer M, Monteyne T et al (2015) Enteric protection of naproxen in a fixed-dose combination product produced by hot-melt co-extrusion. Int J Pharm 491(1–2):243–249. https://doi.org/10.1016/j.ijpharm.2015.06.010
7. Fule R, Dhamecha D, Maniruzzaman M, Khale A, Amin P (2015) Development of hot melt co-formulated antimalarial solid dispersion system in fixed dose form (ARLUMELT): evaluating amorphous state and in vivo performance. Int J Pharm 496(1):137–156. https://doi.org/10.1016/j.ijpharm.2015.09.069
8. Maniruzzaman M, Morgan DJ, Mendham AP, Pang J, Snowden MJ, Douroumis D (2013) Drug-polymer intermolecular interactions in hot-melt extruded solid dispersions. Int J Pharm 443(1–2):199–208. https://doi.org/10.1016/j.ijpharm.2012.11.048
9. Grymonpre W, Verstraete G, Van Bockstal PJ et al (2017) In-line monitoring of compaction properties on a rotary tablet press during tablet manufacturing of hot-melt extruded amorphous solid dispersions. Int J Pharm 517(1-2):348–358. https://doi.org/10.1016/j.ijpharm.2016.12.033
10. Liu J, Cao F, Zhang C, Ping Q (2013) Use of polymer combinations in the preparation of solid dispersions of a thermally unstable drug by hot-melt extrusion. Acta Pharm Sin B 3 (4):263–272. https://doi.org/10.1016/j.apsb.2013.06.007
11. Mahmah O, Tabbakh R, Kelly A, Paradkar A (2013) A comparative study of the effect of spray drying and hot-melt extrusion on the properties of amorphous solid dispersions containing felodipine. J Pharm Pharmacol 66(2):275–284. https://doi.org/10.1111/jphp.12099
12. Hülsmann S, Backensfeld T, Keitel S, Bodmeier R (2000) Melt extrusion–an alternative method for enhancing the dissolution rate of 17β-estradiol hemihydrate. Eur J Pharm Biopharm 49(3):237–242
13. Fu Q, Fang M, Hou Y et al (2016) A physically stabilized amorphous solid dispersion of nisoldipine obtained by hot melt extrusion. Powder Technol 301:342–348. https://doi.org/10.1016/j.powtec.2016.06.032
14. Thiry J, Lebrun P, Vinassa C et al (2016) Continuous production of itraconazole-based solid dispersions by hot melt extrusion: preformulation, optimization and design space determination. Int J Pharm 515(1–2):114–124. https://doi.org/10.1016/j.ijpharm.2016.10.003
15. Park J-B, Lee B-J, Kang C-Y, Tiwari RV, Repka MA (2017) Process analytical quality control of tailored drug release formulation prepared via hot-melt extrusion technology. J Drug Deliv Sci Technol 38:51–58. https://doi.org/10.1016/j.jddst.2017.01.007
16. Verreck G, Decorte A, Li H et al (2006) The effect of pressurized carbon dioxide as a plasticizer and foaming agent on the hot melt extrusion process and extrudate properties of pharmaceutical polymers. J Supercrit Fluids 38(3):383–391. https://doi.org/10.1016/j.supflu.2005.11.022
17. Norman J, Madurawe RD, Moore CMV, Khan MA, Khairuzzaman A (2017) A new chapter in pharmaceutical manufacturing: 3D-printed drug products. Adv Drug Deliv Rev 108:39–50. https://doi.org/10.1016/j.addr.2016.03.001
18. Goyanes A, Chang H, Sedough D et al (2015) Fabrication of controlled-release budesonide tablets via desktop (FDM) 3D printing. Int J Pharm 496(2):414–420. https://doi.org/10.1016/j.ijpharm.2015.10.039
19. Genina N, Holländer J, Jukarainen H, Mäkilä E, Salonen J, Sandler N (2016) Ethylene vinyl acetate (EVA) as a new drug carrier for 3D printed medical drug delivery devices. Eur J Pharm Sci 90:53–63. https://doi.org/10.1016/j.ejps.2015.11.005
20. Jonathan G, Karim A (2016) 3D printing in pharmaceutics: a new tool for designing customized drug delivery systems. Int J Pharm Sci 499:376–394. https://doi.org/10.1016/j.ijpharm.2015.12.071
21. Zhang J, Feng X, Patil H, Tiwari RV, Repka MA (2016) Coupling 3D printing with hot-melt extrusion to produce controlled-release tablets. Int J Pharm 519:186–197. https://doi.org/10.1016/j.ijpharm.2016.12.049
22. Jamróż W, Kurek M, Ewelina Ł et al (2017) 3D printed orodispersible films with aripiprazole. Int J Pharm 533:413–420. https://doi.org/10.1016/j.ijpharm.2017.05.052

23. Goyanes A, Kobayashi M, Martínez-Pacheco R, Gaisford S, Basit AW (2016) Fused-filament 3D printing of drug products: microstructure analysis and drug release characteristics of PVA-based caplets. Int J Pharm 514:290–295. https://doi.org/10.1016/j.ijpharm.2016.06.021
24. Chai X, Chai H, Wan X et al (2017) Fused deposition modeling (FDM) 3D printed tablets for intragastric floating delivery of domperidone. Sci Rep 7(2829):1–9. https://doi.org/10.1038/s41598-017-03097-x
25. Kasper JC, Winter G, Friess W (2013) Recent advances and further challenges in lyophilization. Eur J Pharm Biopharm 85(2):162–169. https://doi.org/10.1016/j.ejpb.2013.05.019
26. Franks F (1998) Freeze-drying of bioproducts: putting principles into practice. Eur J Pharm Biopharm 45(3):221–229. https://doi.org/10.1016/S0939-6411(98)00004-6
27. Tang X, Pikal MJ (2004) Design of freeze-drying processes for pharmaceuticals: practical advice. Pharm Res 21(2):191–200. https://doi.org/10.1023/B:PHAM.0000016234.73023.75
28. Craig DQM, Royall PG, Kett VL, Hopton ML (1999) The relevance of the amorphous state to pharmaceutical dosage forms: glassy drugs and freeze dried systems. Int J Pharm 179 (2):179–207. https://doi.org/10.1016/S0378-5173(98)00338-X
29. Siow CRS, Wan Sia Heng P, Chan LW (2016) Application of freeze-drying in the development of oral drug delivery systems. Expert Opin Drug Deliv 13(11):1595–1608. https://doi.org/10.1080/17425247.2016.1198767
30. Liu J (2006) Physical characterization of pharmaceutical formulations in frozen and freeze-dried solid states: techniques and applications in freeze-drying development. Pharm Dev Technol 11(1):3–28. https://doi.org/10.1080/10837450500463729
31. Sadikoglu H, Ozdemir M, Seker M (2006) Freeze-drying of pharmaceutical products: research and development needs. Dry Technol 24(7):849–861. https://doi.org/10.1080/07373930600734018
32. Teagarden DL, Baker DS (2002) Practical aspects of lyophilization using non-aqueous co-solvent systems. Eur J Pharm Sci 15(2):115–133. https://doi.org/10.1016/S0928-0987(01)00221-4
33. Elgindy N, Elkhodairy K, Molokhia A, Elzoghby A (2010) Lyophilization monophase solution technique for improvement of the physicochemical properties of an anticancer drug, flutamide. Eur J Pharm Biopharm 74(2):397–405. https://doi.org/10.1016/j.ejpb.2009.11.011
34. Pisano R, Fissore D, Barresi AA, Rastelli M (2013) Quality by design: scale-up of freeze-drying cycles in pharmaceutical industry. AAPS PharmSciTech 14(3):1137–1149. https://doi.org/10.1208/s12249-013-0003-9
35. Patel BB, Patel JK, Chakraborty S, Shukla D (2015) Revealing facts behind spray dried solid dispersion technology used for solubility enhancement. Saudi Pharm J 23(4):352–365. https://doi.org/10.1016/j.jsps.2013.12.013
36. Lefebvre AH, McDonell VG (2017) Atomization and sprays, 2nd edn. CRC, Boca Raton, FL
37. Lee SH, Heng D, Ng WK, Chan HK, Tan RBH (2011) Nano spray drying: a novel method for preparing protein nanoparticles for protein therapy. Int J Pharm 403(1–2):192–200. https://doi.org/10.1016/j.ijpharm.2010.10.012
38. Schmid K, Arpagaus C, Friess W (2011) Evaluation of the nano spray dryer B-90 for pharmaceutical applications. Pharm Dev Technol 16(4):287–294. https://doi.org/10.3109/10837450.2010.485320
39. Singh A, Van den Mooter G (2016) Spray drying formulation of amorphous solid dispersions. Adv Drug Deliv Rev 100:27–50. https://doi.org/10.1016/j.addr.2015.12.010
40. Cal K, Sollohub K (2010) Spray drying technique. I: hardware and process parameters. J Pharm Sci 99(2):575–586. https://doi.org/10.1002/jps.21886
41. Paudel A, Worku ZA, Meeus J, Guns S, Van Den Mooter G (2013) Manufacturing of solid dispersions of poorly water soluble drugs by spray drying: formulation and process considerations. Int J Pharm 453(1):253–284. https://doi.org/10.1016/j.ijpharm.2012.07.015
42. Cortés C, Gil A (2007) Modeling the gas and particle flow inside cyclone separators. Prog Energy Combust Sci 33(5):409–452. https://doi.org/10.1016/j.pecs.2007.02.001

43. Li X, Anton N, Arpagaus C, Belleteix F, Vandamme TF (2010) Nanoparticles by spray drying using innovative new technology: the Büchi Nano Spray Dryer B-90. J Control Release 147(2):304–310. https://doi.org/10.1016/j.jconrel.2010.07.113
44. Sosnik A, Seremeta KP (2015) Advantages and challenges of the spray-drying technology for the production of pure drug particles and drug-loaded polymeric carriers. Adv Colloid Interface Sci 223:40–54. https://doi.org/10.1016/j.cis.2015.05.003
45. Snyder HE (2012) Pharmaceutical spray drying: solid-dose process technology platform for the 21st century. Ther Deliv 3(7):901–912. https://doi.org/10.4155/tde.12.64
46. Ambike AA, Mahadik KR, Paradkar A (2005) Spray-dried amorphous solid dispersions of simvastatin, a low Tg drug: in vitro and in vivo evaluations. Pharm Res 22(6):990–998. https://doi.org/10.1007/s11095-005-4594-z
47. Gupta P, Bansal AK (2005) Spray drying for generation of a ternary amorphous system of celecoxib, PVP, and meglumine. Pharm Dev Technol 10(2):273–281. https://doi.org/10.1081/PDT-54460
48. Broadhead J, Edmond Rouan S, Rhodes C (1992) The spray drying of pharmaceuticals. Drug Dev Ind Pharm 18(11–12):1169–1206. https://doi.org/10.3109/03639049209046327
49. Paudel A, Van Den Mooter G (2012) Influence of solvent composition on the miscibility and physical stability of naproxen/PVP K 25 solid dispersions prepared by cosolvent spray-drying. Pharm Res 29(1):251–270. https://doi.org/10.1007/s11095-011-0539-x
50. Al-Obaidi H, Brocchini S, Buckton G (2009) Anomalous properties of spray dried solid dispersions. J Pharm Sci 98(12):4724–4737. https://doi.org/10.1002/jps.21782
51. Szafraniec J, Antosik A, Knapik-Kowalczuk J et al (2017) Planetary ball milling and supercritical fluid technology as a way to enhance dissolution of bicalutamide. Int J Pharm. 533:470–479. https://doi.org/10.1016/j.ijpharm.2017.03.078
52. Lim RTY, Kiong W, Tan RBH (2013) Dissolution enhancement of indomethacin via amorphization using co-milling and supercritical co-precipitation processing. Powder Technol 240:79–87. https://doi.org/10.1016/j.powtec.2012.07.004
53. Gong K, Viboonkiat R, Rehman IU, Buckton G, Darr JA (2005) Formation and characterization of porous indomethacin-PVP coprecipitates prepared using solvent-free supercritical fluid processing. J Pharm Sci 94(12):2583–2590. https://doi.org/10.1002/jps.20474
54. Sethia S, Squillante E (2004) Solid dispersion of carbamazepine in PVP K30 by conventional solvent evaporation and supercritical methods. Int J Pharm 272:1–10. https://doi.org/10.1016/j.ijpharm.2003.11.025
55. Banchero M, Manna L, Ronchetti S, Campanelli P, Ferri A (2009) Supercritical fluids supercritical solvent impregnation of piroxicam on PVP at various polymer molecular weights. J Supercrit Fluids 49:271–278. https://doi.org/10.1016/j.supflu.2009.01.008
56. Won DH, Kim MS, Lee S, Park JS, Hwang SJ (2005) Improved physicochemical characteristics of felodipine solid dispersion particles by supercritical anti-solvent precipitation process. Int J Pharm 301(1-2):199–208. https://doi.org/10.1016/j.ijpharm.2005.05.017
57. Brunner G (2005) Supercritical fluids: technology and application to food processing. J Food Eng 67(1–2):21–33. https://doi.org/10.1016/j.jfoodeng.2004.05.060
58. Deshpande PB, Kumar GA, Kumar AR et al (2011) Supercritical fluid technology: concepts and pharmaceutical applications. PDA J Pharm Sci Technol 65(3):333–344. https://doi.org/10.5731/pdajpst.2011.00717
59. Gurikov P, Smirnova I (2018) Amorphization of drugs by adsorptive precipitation from supercritical solutions: a review. J Supercrit Fluids 132:105–125
60. Goodship V, Ogar E-O (2004) Polymer processing with supercritical fluid. Rapra review reports 15(8). Rapra Technology, Shawbury. ISBN: 9781859574942
61. Girotra P, Singh SK, Nagpal K (2013) Supercritical fluid technology: a promising approach in pharmaceutical research. Pharm Dev Technol 18(1):22–38. https://doi.org/10.3109/10837450.2012.726998
62. Shah N, Sandhu H, Choi DS, Chokshi H, Malick AW (2014) Amorphous solid dispersions theory and practice. Springer, New York

63. Pathak P, Meziani MJ, Sun Y (2005) Supercritical fluid technology for enhanced drug delivery. Expert Opin Drug Deliv 2(4):747–761. https://doi.org/10.1517/17425247.2.4.747
64. Potter C, Tian Y, Walker G et al (2015) Novel supercritical carbon dioxide impregnation technique for the production of amorphous solid drug dispersions: a comparison to hot melt extrusion. Mol Pharm 12(5):1377–1390. https://doi.org/10.1021/mp500644h
65. Alessi P, Cortesi A, Kikic I, Vecchione F (2003) Plasticization of polymers with supercritical carbon dioxide: experimental determination of glass-transition temperatures. J Appl Polym Sci 88(9):2189–2193. https://doi.org/10.1002/app.11881
66. Ugaonkar S, Nunes AC, Needham TE (2007) Effect of n-scCO_2 on crystalline to amorphous conversion of carbamazepine. Int J Pharm 333(1–2):152–161. https://doi.org/10.1016/j.ijpharm.2006.12.010
67. Ugaonkar S, Needham TE, Bothun GD (2011) Solubility and partitioning of carbamazepine in a two-phase supercritical carbon dioxide/polyvinylpyrrolidone system. Int J Pharm 403 (1–2):96–100. https://doi.org/10.1016/j.ijpharm.2010.10.031
68. Williams GR, Chatterton NP, Nazir T, Yu DG, Zhu LM, Branford-White CJ (2012) Electrospun nanofibers in drug delivery: recent developments and perspectives. Ther Deliv 3:515–533. https://doi.org/10.4155/tde.12.17
69. Jahangiri A, Adibkia K (2016) Applications of electrospinning/electrospraying in drug delivery. BioImpacts 6(1):1–2. 10.15171/bi.2016.08
70. Chakraborty S, Liao I-C, Adler A, Leong KW (2009) Electrohydrodynamics: a facile technique to fabricate drug delivery systems. Adv Drug Deliv Rev 61(12):1043–1054. https://doi.org/10.1016/j.addr.2009.07.013
71. Verreck G, Chun I, Rosenblatt J et al (2003) Incorporation of drugs in an amorphous state into electrospun nanofibers composed of a water-insoluble, nonbiodegradable polymer. J Control Release 92:349–360. https://doi.org/10.1016/S0168-3659(03)00342-0
72. Brewster ME, Verreck G, Chun I et al (2004) The use of polymer-based electrospun nanofibers containing amorphous drug dispersions for the delivery of poorly water-soluble pharmaceuticals. Pharmazie 59(5):387–391
73. Verreck G, Chun I, Peeters J, Rosenblatt J, Brewster ME (2003) Preparation and characterization of nanofibers containing amorphous drug dispersions generated by electrostatic spinning. Pharm Res 20(5):810–817
74. Zong X, Kim K, Fang D, Ran S, Hsiao BS, Chu B (2002) Structure and process relationship of electrospun bioabsorbable nanofiber membranes. Polymer (Guildf) 43:4403–4412
75. Nagy ZK, Balogh A, Démuth B et al (2015) High speed electrospinning for scaled-up production of amorphous solid dispersion of itraconazole. Int J Pharm 480(1-2):137–142. https://doi.org/10.1016/j.ijpharm.2015.01.025
76. Zhang S, Kawakami K, Yamamoto M et al (2011) Coaxial electrospray formulations for improving oral absorption of a poorly water-soluble drug. Mol Pharm 8:807–813
77. Kawakami K, Zhang S, Singh R et al (2013) Preparation of fenofibrate solid dispersion using electrospray deposition and improvement in oral absorption by instantaneous post-heating of the formulation. Int J Pharm 450(1–2):123–128. https://doi.org/10.1016/j.ijpharm.2013.04.006
78. Bhushani JA, Anandharamakrishnan C (2014) Electrospinning and electrospraying techniques: potential food based applications. Trends Food Sci Technol 38(1):23–33. https://doi.org/10.1016/j.tifs.2014.03.004
79. Karki S, Kim H, Na S, Shin D, Jo K, Lee J (2016) Thin films as an emerging platform for drug delivery. Asian J Pharm Sci 11:559–574. https://doi.org/10.1016/j.ajps.2016.05.004
80. Parikh T, Gupta SS, Meena AK, Vitez I, Mahajan N, Serajuddin ATM (2015) Application of film-casting technique to investigate drug-polymer miscibility in solid dispersion and hot-melt extrudate. J Pharm Sci 104:2142–2152. https://doi.org/10.1002/jps.24446
81. Amin PM, Gangurde AB, Alai PV (2015) Oral film technology: challenges and future scope for pharmaceutical industry. Int J Pharm Pharm Res 3(3):184–203
82. Weuts I, Van Dycke F, Voorspoels J et al (2011) Physicochemical properties of the amorphous drug, cast films, and spray dried powders to predict formulation probability of success for solid dispersions: etravirine. J Pharm Sci 100(1):260–274. https://doi.org/10.1002/jps.22242

83. Janssens S, De Zeure A, Paudel A, Van Humbeeck J, Rombaut P, Van Den Mooter G (2010) Influence of preparation methods on solid state supersaturation of amorphous solid dispersions: a case study with itraconazole and Eudragit E100. Pharm Res 27(5):775–785. https://doi.org/10.1007/s11095-010-0069-y
84. Chatap VK, Wagh PN, Bari SB et al (2014) Novel spin coating technique for development of zolmitriptan mouth dissolving film. Int J Adv Chem Engg Biol Sci 1(1):110–113
85. Ng YC, Yang Z, McAuley WJ, Qi S (2013) Stabilisation of amorphous drugs under high humidity using pharmaceutical thin films. Eur J Pharm Biopharm 84(3):555–565. https://doi.org/10.1016/j.ejpb.2013.01.008
86. Garner WE (ed) (1955) Chemistry of the solid state. Academic Press, New York
87. Young DA (1966) Decomposition of solids; v.1: International encyclopedia of physical chemistry and chemical physics; Topic 21, Solid and surface kinetics. Pergamon, Oxford
88. Makatun VN (1985) Chemistry of inorganic hydrates. Nauka & Tehnika, Minsk
89. Dollimore D (1987) The thermal decomposition of oxalates. A review. Thermochim Acta 117:331–363. https://doi.org/10.1016/0040-6031(87)88127-3
90. Galwey AK, Brown ME (1999) Thermal decomposition of ionic solids: chemical properties and reactivities of ionic crystalline phases, vol 86. Elsevier, Amsterdam
91. Galwey AK (2000) Structure and order in thermal dehydrations of crystalline solids. Thermochim Acta 355(1–2):181–238. https://doi.org/10.1016/S0040-6031(00)00448-2
92. Petit S, Coquerel G (1996) Mechanism of several solid-solid transformations between dihydrated and anhydrous copper(II) 8-hydroxyquinolinates. Proposition for a unified model for the dehydration of molecular crystals. Chem Mater 8(9):2247–2258. https://doi.org/10.1021/cm9600438
93. Petit S, Coquerel G (2009) Contribution to the understanding of desolvation mechanisms: impact of crystal size, structural purity and process. JEEP 16. https://doi.org/10.1051/jeep/200900016
94. Willart JF, Descamps M (2008) Solid state amorphization of pharmaceuticals. Mol Pharm 5 (6):905–920. https://doi.org/10.1021/mp800092t
95. Li Y, Han J, Zhang GG, Grant DJ, Suryanarayanan R (2000) In situ dehydration of carbamazepine dihydrate: a novel technique to prepare amorphous anhydrous carbamazepine. Pharm Dev Technol 5(2):257–266. https://doi.org/10.1081/PDT-100100540
96. Morris KR, Griesser UJ, Eckhardt CJ, Stowell JG (2001) Theoretical approaches to physical transformations of active pharmaceutical ingredients during manufacturing processes. Adv Drug Deliv Rev 48(1):91–114. https://doi.org/10.1016/S0169-409X(01)00100-4
97. Pyne A, Chatterjee K, Suryanarayanan R (2003) Crystalline to amorphous transition of disodium hydrogen phosphate during primary drying. Pharm Res 20(5):802–803. https://doi.org/10.1023/A:1023445905372
98. Kachrimanis K, Griesser UJ (2012) Dehydration kinetics and crystal water dynamics of carbamazepine dihydrate. Pharm Res 29(4):1143–1157. https://doi.org/10.1007/s11095-012-0698-4
99. Petit S, Coquerel G (2006) The amorphous state. In: Hilfiker R (ed) Polymorphism in the pharmaceutical industry. Wiley-VCH, Weinheim, pp 259–285
100. Tsirenova S, Suponitsky, YuL Karapetyanz Mk (1974) A comparative study of thermal properties of oxygen-containing compounds of Sc and Y. Zh Fiz Khim 48(11):2705–2707
101. Angell CA, Tucker JC (1974) Heat capacities and fusion entropies of the tetrahydrates of calcium nitrate, cadmium nitrate, and magnesium acetate. Concordance of calorimetric and relaxational ideal glass transition temperatures. J Phys Chem 78(3):278–281. http://pubs.acs.org/doi/abs/10.1021/j100596a018?journalCode=jpchax
102. Guion J, Sauzade JD, Laügt M (1983) Critical examination and experimental determination of melting enthalpies and entropies of salt hydrates. Thermochim Acta 67(2–3):167–179. https://doi.org/10.1016/0040-6031(83)80096-3
103. Rani M, Govindarajan R, Surana R, Suryanarayanan R (2006) Structure in dehydrated trehalose dihydrate – evaluation of the concept of partial crystallinity. Pharm Res 23 (10):2356–2367. https://doi.org/10.1007/s11095-006-9058-6

104. Niepce JC, Watelle G, Brett NH (1977) Product crystallite size–reaction rate relationship in $M(OH)_2$–MO decomposition. Structural transformation mechanism. J Chem Soc, Faraday Trans 1 Phys Chem Condens Phases 74:1530–1537. http://pubs.rsc.org | doi:https://doi.org/10.1039/F19787401530
105. Mutin JC, Dusausoy Y (1981) Recherche d'une description structurale des decompositions endothermiques solide 1→ solide 2+ gaz. II. Caractéristiques structurales de la reaction 2 [$H_2C_2O_4$, BaC_2O_4, $2H_2O$] → $Ba(HC_2O_4)_2$, BaC_2O_4, $2H_2O$+ $H_2C_2O_4$+ $2H_2O$. J Solid State Chem 38(3):394–405. https://doi.org/10.1016/0022-4596(81)90070-0
106. Mutin JC, Watelle G, Dusausoy Y (1979) Study of a lacunary solid phase I—thermodynamic and crystallographic characteristics of its formation. J Solid State Chem 27:407–421. https://doi.org/10.1016/0022-4596(79)90183-X
107. Schoonover JR, Lin SH, Eyring L (1987) Time-resolved the thermal X-ray diffraction by synchrotron radiation: decomposition of $Cd(OH)_2$ powders. J Solid State Chem 218:214–218
108. Matvienko AA, Chizhik SA, Sidelnikov AA (2005) Factors controlling the morphology of the surface of BaC_2O_4 • $H_2C_2O_4$ • $2H_2O$ during its dehydration. Russ J Phys Chem A 79 (9):1478–1482
109. Chizhik SA, Sidelnikov AA (2007) The kinetics of solid state reactions accompanied by fracture: I. Reaction of ion exchange in lime-soda glass. Solid State Ionics 178 (23–24):1344–1352. https://doi.org/10.1016/j.ssi.2007.07.011
110. Chizhik SA, Sidelnikov AA (2007) The kinetics of solid state reactions accompanied by fracture: II. Model of stationary front with disordered fracture morphology. Solid State Ionics 178(27–28):1487–1492. https://doi.org/10.1016/j.ssi.2007.09.010
111. Chizhik SA, Sidelnikov AA (2008) The kinetics of solid state reactions accompanied by fracture: III. Model of stationary front with spatially ordered fracture morphology. Solid State Ionics 179(33–34):1823–1834. https://doi.org/10.1016/j.ssi.2008.05.002
112. Larsen AS, Rantanen J, Johansson KE (2017) Computational dehydration of crystalline hydrates using molecular dynamics simulations. J Pharm Sci 106(1):348–355. https://doi.org/10.1016/j.xphs.2016.10.005
113. Han J, Suryanarayanan R (1998) Influence of environmental conditions on the kinetics and mechanism of dehydration of carbamazepine dihydrate. Pharm Dev Technol 3(4):587–596. https://doi.org/10.3109/10837459809028643
114. Griesser UJ, Burger A (1995) The effect of water vapor pressure on desolvation kinetics of caffeine 4/5-hydrate. Int J Pharm 120(1):83–93. https://doi.org/10.1016/0378-5173(94)00416-3
115. Lallemant M, Watelle-Marion G (1967) Dégradation thermique du sulfate de magnésium heptahydraté sous pression de vapeur d'eau contrôlée. Mécanisme observé au-dessus de 50 torr. C R Acad Sci Paris C 264:2030–2033
116. Bertrand G, Lallemant M, Watelle-Marion G (1974) Variation anormale de la vitesse de decomposition d'un solide—I: Cas des deshydratations d'hydrates salins. J Inorg Nucl Chem 36:1303–1309. https://doi.org/10.1016/0022-1902(74)80068-0
117. Bertrand G, Lallemant M, Mokhlisse A, Watelle-Marion G (1978) Abnormal variation of the rate of decomposition of a solid—II: A property common to interfacial endothermic reactions. J Inorg Nucl Chem 40(5):819–824. https://doi.org/10.1016/0022-1902(78)80158-4
118. Lallemant M, Watelle-Marion G (1968) Anomalies presentees par la dissociation thermique, sous faible pression de vapeur d'eau, du sulfate du cuivre pentahydrate. C R Acad Sci Paris C 267(26):1775–1778
119. Chupakhin AP, Lyakhov NZ (1979) Dependence of the rate of water evaporation on the pressure of its vapor. Thermochim Acta 29:192–195. https://doi.org/10.1016/0040-6031(79)85033-9
120. Ferrier A (1966) Influence de l'état de division de la goethite et de l'oxyde ferrique sur leurs chaleurs de réaction. Revue de Chimie minérale 3:587–615
121. Yatsimirskii VK (1970) Minimal size of crystalline particles. Theor Exp Chem 6(5):587–615

122. Kimoto K, Nishida I (1973) Crystal structures of very small particles of chromium and iron. Thin Solid Films 17(1):49–58. https://doi.org/10.1016/0040-6090(74)90238-7
123. Taylor LS, York P (1998) Characterization of the phase transitions of trehalose dihydrate on heating and subsequent dehydration. J Pharm Sci 87(3):347–355. https://doi.org/10.1021/js970239m
124. Bregeault J-M, Pannetier G (1969) Etude de la dissociation thermique des sulfates et des sulfates basiques. Sur les polymorphisme du sulfate de zinc. Bull Soc Chim Fr 4:1061–1065
125. Walter LL, Quemneur E (1968) Sur la thermolyse des sulfates ferrique. Bull Soc Chim Fr 4:1061–1065
126. Bernstein J (2002) Polymorphism in molecular crystals. Clarendon Press/International Union of Crystallography Monographs on Crystallography, Oxford
127. Llinàs A, Burley JC, Prior TJ, Glen RC, Goodman JM (2008) Concomitant hydrate polymorphism in the precipitation of sparfloxacin from aqueous solution. Cryst Growth Des 8 (1):114–118. https://doi.org/10.1021/cg700908m
128. Minkov VS, Beloborodova AA, Drebushchak VA, Boldyreva EV (2014) Furosemide solvates: can they serve as precursors to different polymorphs of furosemide? Cryst Growth Des 14 (2):513–522. https://doi.org/10.1021/cg401257w
129. Beloborodova AA, Minkov VS, Rychkov DA, Rybalova TV, Boldyreva EV (2017) First evidence of polymorphism in furosemide solvates. Cryst Growth Des 17(5):2333–2341. https://doi.org/10.1021/acs.cgd.6b01191
130. Otsuka M, Kaneniwa N (1990) Effect of grinding on the crystallinity and chemical stability in the solid state of cephalothin sodium. Int J Pharm 62(1):65–73. https://doi.org/10.1016/0378-5173(90)90031-X
131. Ward GH, Schultz RK (1995) Process-induced crystallinity changes in albuterol sulfate and its effect on powder physical stability. Pharm Res 12(5):773–779. https://link.springer.com/article/10.1023/A:1016232230638
132. Shakhtshneider T, Boldyrev V (1999) Mechanochemical synthesis and mechanical activation of drugs. In: Boldyreva E, Boldyrev V (eds) Reactivity of solids. Chichester, Wiley, pp 271–311
133. Descamps M, Willart JF, Dudognon E, Caron V (2007) Transformation of pharmaceutical compounds upon milling and comilling: the role of Tg. J Pharm Sci 96(5):1398–1407. https://doi.org/10.1002/jps.20939
134. Descamps M (2016) Disordered pharmaceutical materials. Wiley-VCH, Weinheim
135. Boldyreva E (2016) Non-ambient conditions in the investigation and manufacturing of drug forms. Curr Pharm Des 22(32):4981–5000. http://www.ingentaconnect.com/contentone/ben/cpd/2016/00000022/00000032/art00009
136. Gubskaya AV, Lisnyak YV (1995) Effect of cryogrinding on physico-chemical properties of drugs. I. Theophylline: evaluation of particles sizes and the degree of crystallinity, relation to dissolution parameters. Drug Dev Ind Pharm 21(17):1953–1964. https://doi.org/10.3109/03639049509065880.
137. Hancock BC, Zografi G (1997) Characteristics and significance of the amorphous state in pharmaceutical systems. J Pharm Sci 86(1):1–12. https://doi.org/10.1021/js9601896
138. Mosharraf M, Sebhatu T, Nyström C (1999) The effects of disordered structure on the solubility and dissolution rates of some hydrophilic, sparingly soluble drugs. Int J Pharm 177(1):29–51. https://doi.org/10.1016/S0378-5173(98)00317-2
139. Hancock BC, Parks M (2000) What is the true solubility advantage for amorphous pharmaceuticals? Pharm Res 17(4):397–404. https://www.ncbi.nlm.nih.gov/labs/articles/10870982/
140. Mosharraf M, Nyström C (2003) Apparent solubility of drugs in partially crystalline systems. Drug Dev Ind Pharm 29(6):603–622. https://doi.org/10.1081/DDC-120021310
141. Yu L (2001) Amorphous pharmaceutical solids: preparation, characterization and stabilization. Adv Drug Deliv Rev 48(1):27–42. https://doi.org/10.1016/S0169-409X(01)00098-9

142. Brough C, Williams RO (2013) Amorphous solid dispersions and nano-crystal technologies for poorly water-soluble drug delivery. Int J Pharm 453(1):157–166. https://doi.org/10.1016/j.ijpharm.2013.05.061
143. Allesø M, Chieng N, Rehder S, Rantanen J, Rades T, Aaltonen J (2009) Enhanced dissolution rate and synchronized release of drugs in binary systems through formulation: amorphous naproxen-cimetidine mixtures prepared by mechanical activation. J Control Release 136 (1):45–53. https://doi.org/10.1016/j.jconrel.2009.01.027
144. Sharafutdinova D, Efremov YY, Rizvanov IH, Konygin GN, Rybin DS, Strelkov NS (2010) Composition and structure of calcium gluconate and its mechanoactivated (nanodispersed) form. J Struct Chem 51:S142–S144. doi:https://link.springer.com/article/10.1007%2Fs10947-010-0203-z?LI=true
145. Boldyreva E (2013) Mechanochemistry of inorganic and organic systems: what is similar, what is different? Chem Soc Rev 42:7719–7738. https://doi.org/10.1039/c3cs60052a
146. Losev EA, Boldyreva EV (2014) The role of a liquid in "dry" co-grinding: a case study of the effect of water on mechanochemical synthesis in a "l-serine–oxalic acid" system. CrystEngComm 16(19):3857. https://doi.org/10.1039/c3ce42321b
147. Karki S, Friščić T, Jones W, Motherwell WDS (2007) Screening for pharmaceutical cocrystal hydrates via neat and liquid-assisted grinding. Mol Pharm 4(3):347–354. https://doi.org/10.1021/mp0700054
148. Belenguer AM, Lampronti GI, Cruz-Cabeza AJ, Hunter CA, Sanders JKM (2016) Solvation and surface effects on polymorph stabilities at the nanoscale. Chem Sci 72:171–179. https://doi.org/10.1039/C6SC03457H
149. Bahl D, Bogner RH (2006) Amorphization of indomethacin by co-grinding with Neusilin US2: Amorphization kinetics, physical stability and mechanism. Pharm Res 23 (10):2317–2325. https://doi.org/10.1007/s11095-006-9062-x
150. Linol J, Morelli T, Petit M-N, Coquerel G (2007) Inversion of the relative stability between two polymorphic forms of (±) modafinil under dry high-energy milling: comparisons with results obtained under wet high-energy milling. Cryst Growth Des 7(9):1608–1611. https://doi.org/10.1021/cg0700723
151. Tumanov IA, Michalchuk AAL, Politov AA, Boldyreva EV, Boldyrev VV (2017) Inadvertent liquid assisted grinding: a key to "dry" organic mechano-co-crystallisation? CrystEngComm 19:2830–2835. https://doi.org/10.1039/C7CE00517B
152. Gupta MK, Vanwert A, Bogner RH (2003) Formation of physically stable amorphous drugs by milling with neusilin. J Pharm Sci 92(3):536–551. https://doi.org/10.1002/jps.10308
153. Politov A, Golyazimova O (2014) Increasing the energy yield of mechanochemical transformations: selected case studies. Faraday Discuss 170:345–356. https://doi.org/10.1039/c3fd00143a
154. Orowan E (1949) Fracture and strength of solids. Reports Prog Phys 12(1):185. https://doi.org/10.1088/0034-4885/12/1/309
155. De Gusseme A, Neves C, Willart JF, Rameau A, Descamps M (2008) Ordering and disordering of molecular solids upon mechanical milling: the case of fananserine. J Pharm Sci 97(11):5000–5012. https://doi.org/10.1002/jps.21472
156. Lepek P, Sawicki W, Wlodarski K, Wojnarowska Z, Paluch M, Guzik L (2013) Effect of amorphization method on telmisartan solubility and the tableting process. Eur J Pharm Biopharm 83(1):114–121. https://doi.org/10.1016/j.ejpb.2012.09.019
157. Willart JF, De Gusseme A, Odou G, Danede F, Descamps M (2001) Direct crystal to glass transformations of trehalose induced by milling, dehydration and annealing. Solid State Commun 119:501–505. https://doi.org/10.1016/S0038-1098(01)00283-6
158. Willart JF, Caron V, Lefort R, Danède F, Prévost D, Descamps M (2004) Athermal character of the solid state amorphization of lactose induced by ball milling. Solid State Commun 132 (10):693–696. https://doi.org/10.1016/j.ssc.2004.09.007
159. Dujardin N, Willart JF, Dudognon E et al (2008) Solid state vitrification of crystalline α and β-D-glucose by mechanical milling. Solid State Commun 148(1–2):78–82. https://doi.org/10.1016/j.ssc.2008.07.002

160. Descamps M, Aumelas A, Desprez S, Willart JF (2015) The amorphous state of pharmaceuticals obtained or transformed by milling: sub-Tg features and rejuvenation. J Non Cryst Solids 407:72–80. https://doi.org/10.1016/j.jnoncrysol.2014.08.055
161. Otsuka MM, Kaneniwa N (1983) Effect of grinding on the degree of crystallinity of cephalexin powder. Chem Pharm Bull (Tokyo) 31(12):4489–4495. https://doi.org/10.1248/cpb.31.4489
162. Shakhtshneider TP (1997) Phase transformations and stabilization of metastable states of molecular crystals under mechanical activation. Solid State Ionics 101–103:851–856. doi: https://doi.org/10.1016/S0167-2738(97)00224-5
163. Kaneniwa N, Otsuka M (1985) Effect of grinding on the transformations of polymorphs of chloramphenicol palmitate. Chem Pharm Bull 33(4):1660–1668. https://doi.org/10.1248/cpb.33.1660
164. Desprez S, Descamps M (2006) Transformations of glassy indomethacin induced by ball-milling. J Non Cryst Solids 352(42–49 Spl Iss):4480–4485. doi:https://doi.org/10.1016/j.jnoncrysol.2006.02.130
165. Vasikhovskaia VA (2016) Physical and chemical properties of starch after brittle comminution. Thesis, Novosibirsk State University. http://www.nsu.ru/xmlui/handle/nsu/10966
166. Esersky V, Savitskaya A (1992) Mechanical activation of sulfanilamides on communication. Zhurn Fiz Khim 66:3109–3114
167. Boldyrev VV (1972) Kinetic factors in mechanochemical processes in inorganic systems. Kinet Catal 13(6):1411–1421
168. Boldyrev VV (2006) Mechanochemistry and mechanical activation of solids. Russ Chem Rev 75 (3):177–199. http://iopscience.iop.org/article/10.1070/RC2006v075n03ABEH001205/meta
169. Michalchuk AAL, Tumanov IA, Drebushchak VA, Boldyreva EV (2014) Advances in elucidating mechanochemical complexities via implementation of a simple organic system. Faraday Discuss 170:311–335. https://doi.org/10.1039/C3FD00150D
170. Yoshioka M, Hancock BC, Zografi G (1994) Crystallization of indomethacin from the amorphous state below and above its glass transition temperature. J Pharm Sci 83 (12):1700–1705. https://doi.org/10.1002/jps.2600831211
171. Schammé B, Couvrat N, Malpeli P et al (2016) Transformation of an active pharmaceutical ingredient upon high-energy milling: a process-induced disorder in biclotymol. Int J Pharm 499(1–2):67–73. https://doi.org/10.1016/j.ijpharm.2015.12.032
172. Feng T, Pinal R, Carvajal MT (2008) Process induced disorder in crystalline materials: differentiating defective crystals from the amorphous form of griseofulvin. J Pharm Sci 97 (8):3207–3221. https://doi.org/10.1002/jps.21219
173. Patterson JE, James MB, Forster AH, Lancaster RW, Butler JM, Rades T (2005) The influence of thermal and mechanical preparative techniques on the amorphous state of four poorly soluble compounds. J Pharm Sci 94(9):1998–2012. https://doi.org/10.1002/jps.20424
174. Graeser KA, Strachan CJ, Gordon K, Patterson JE, Gordon KC, Rades T (2008) Physicochemical properties and stability of two differently prepared amorphous forms of simvastatin. Cryst Growth Des 8(1):128–135. https://doi.org/10.1021/cg700913m
175. Grisedale LC, Jamieson MJ, Belton PS, Barker SA, Craig M, Duncan Q (2011) Characterization and quantification of amorphous material in milled and spray-dried salbutamol sulfate: a comparison of thermal, spectroscopic, and water vapor sorption approaches. J Pharm Sci 100 (8):3114–3129. https://doi.org/10.1002/jps.22484
176. Wlodarski K, Sawicki W, Paluch KJ et al (2014) The influence of amorphization methods on the apparent solubility and dissolution rate of tadalafil. Eur J Pharm Sci 62:132–140. https://doi.org/10.1016/j.ejps.2014.05.026
177. Hancock BC, Shalaev E, Shamblin SL (2002) Polyamorphism: a pharmaceutical science perspective. J Pharm Pharmacol 54(8):1151–1152. https://doi.org/10.1211/002235702320266343
178. Turnbull D (1976) Relation of crystallization behavior to structure in amorphous systems. Ann NY Acad Sci 279:185. https://doi.org/10.1111/j.1749-6632.1976.tb39706.x

179. Andronis V, Zografi G (2000) Crystal nucleation and growth of indomethacin polymorphs from the amorphous state. J Non Cryst Solids 271(3):236–248. https://doi.org/10.1016/S0022-3093(00)00107-1
180. Politov AA, Kostrovskii VG, Boldyrev VV (2001) Conditions of preparation and crystallization of amorphous paracetamol. Russ J Phys Chem A 75(11):1903–1911. http://cat.inist.fr/?aModele=afficheN&cpsidt=13620211
181. Gaffet E, Abdellaoui M, Malhouroux-Gaffet N (1995) Formation of nanostructural materials induced by mechanical processings (overview). Mater Trans 36(2):198–209. https://doi.org/10.2320/matertrans1989.36.198
182. Caron V, Willart JF, Lefort R, Derollez P, Dande F, Descamps M (2011) Solid state amorphization kinetic of alpha lactose upon mechanical milling. Carbohydr Res 346(16):2622–2628. https://doi.org/10.1016/j.carres.2011.09.004
183. Shalaev E, Wu K, Shamblin S, Krzyzaniak JF, Descamps M (2016) Crystalline mesophases: structure, mobility, and pharmaceutical properties. Adv Drug Deliv Rev 100:194–211. https://doi.org/10.1016/j.addr.2016.04.002
184. Surovtsev NV, Adichtchev SV, Malinovsky VK et al (2012) Glycine phases formed from frozen aqueous solutions: revisited. J Chem Phys 137(6). https://doi.org/10.1063/1.4739532
185. Ogienko AG, Bogdanova EG, Trofimov NA et al (2017) Large porous particles for respiratory drug delivery. Glycine-based formulations. Eur J Pharm Sci. https://doi.org/10.1016/j.ejps.2017.05.007
186. Boldyrev VV, Shakhtshneider TP, Burleva LP, Severtsev VA (1994) Preparation of the disperse systems of sulfathiazole-polyvinylpirrolidone by mechanical activation. Drug Dev Ind Pharm 20(6):1103–1114. https://doi.org/10.3109/03639049409038355.
187. Shakhtshneider TP, Vasiltchenko MA, Politov AA, Boldyrev VV (1996) The mechanochemical preparation of solid disperse systems of ibuprofen-polyethylene glycol. Int J Pharm 130(1):25–32. https://doi.org/10.1016/0378-5173(95)04244-X
188. Shakhtshneider TP, Vasilchenko MA, Politov AA (1997) Mechanochemical preparation of drug carrier solid dispersions. J Therm Anal 48:491–501. https://doi.org/10.1007/BF01979496
189. Ivashchenko GL, Shakhtshneider TP, Boldyrev VV, Bazarnova NG, Medvedeva AS, Safronova LP (2003) Effect of mechanical activation on the physicochemical properties of piroxicam with chitosan. Mendeleev Commun 13(1):3–5. https://doi.org/10.1070/MC2003v013n01ABEH001644.
190. Drebushchak VA, Shakhtshneider TP, Apenina SA, Medvedeva AS, Safronova LP, Boldyrev VV (2006) Thermoanalytical investigation of drug-excipient interaction: Part II. Activated mixtures of piroxicam with cellulose and chitosan. J Therm Anal Calorim 86(2):303–309. https://doi.org/10.1007/s10973-005-7440-y
191. Shakhtshneider TP, Danède F, Capet F et al (2007) Grinding of drugs with pharmaceutical excipients at cryogenic temperatures. Part I. Cryogenic grinding of piroxicam-polyvinylpyrrolidone mixtures. J Therm Anal Calorim 89:699–707. https://doi.org/10.1007/s10973-006-7958-7
192. Shakhtshneider TP, Danède F, Capet F et al (2007) Grinding of drugs with pharmaceutical excipients at cryogenic temperatures. Part II. Cryogenic grinding of indomethacin-polyvinylpyrrolidone mixtures. J Therm Anal Calorim 89:709–715. https://doi.org/10.1007/s10973-006-7959-6
193. Dushkin A V, Meteleva ES, Tolstikova TG et al (2008) Mechanochemical preparation and pharmacological activities of water-soluble intermolecular complexes of arabinogalactan with medicinal agents. Russ Chem Bull 57(6):1299–1307. https://link.springer.com/article/10.1007%2Fs11172-008-0167-8?LI=true
194. Caron V, Tajber L, Corrigan OI, Healy AM (2011) A comparison of spray drying and milling in the production of amorphous dispersions of sulfathiazole/polyvinylpyrrolidone and sulfadimidine/polyvinylpyrrolidone. Mol Pharm 8(2):532–542. https://doi.org/10.1021/mp1003674

195. Shakhtshneider TP, Kuznetsova SA, Mikhailenko MA et al (2013) Effect of mechanochemical treatment on physicochemical and antitumor properties of betulin diacetate mixtures with arabinogalactan. Chem Nat Compd 49(3):470–474. https://doi.org/10.1007/s10600-013-0641-x
196. Miyazaki T, Yoshioka S, Aso Y, Kojima S (2004) Ability of polyvinylpyrrolidone and polyacrylic acid to inhibit the crystallization of amorphous acetaminophen. J Pharm Sci 93 (11):2710–2717. https://doi.org/10.1002/jps.20182
197. Balani PN, Wong SY, Ng WK, Widjaja E, Tan RBH, Chan SY (2010) Influence of polymer content on stabilizing milled amorphous salbutamol sulphate. Int J Pharm 391(1–2):125–136. https://doi.org/10.1016/j.ijpharm.2010.02.029
198. Balani PN, Ng WK, Tan RB, Chan SY (2010) Influence of excipients in comilling on mitigating milling-induced amorphization or structural disorder of crystalline pharmaceutical actives. J Pharm Sci 10(5):2462–2474. https://doi.org/10.1002/jps.21998
199. Boldyrev VV, Shakhtshneider TP, Chizhik SA (2005) On the mechanism of solubilization of drugs in the presence of poorly soluble additives. Int J Pharm 295(1-2):177–182. https://doi.org/10.1016/j.ijpharm.2005.02.011
200. Shakhtshneider TP, Myz SA, Mikhailenko MA et al (2009) Mechanochemical synthesis of nanocomposites of drugs with inorganic oxides. Mater Manuf Process 24:1064–1071. https://doi.org/10.1080/10426910902979124
201. Shakhtshneider TP, Myz SA, Dyakonova MA et al (2011) Mechanochemical preparation of organic-inorganic hybrid materials of drugs with inorganic oxides. Acta Phys Pol A 120(2). https://www.researchgate.net/profile/Rakesh_Kumar173/publication/262842074_Mechano chemical_Preparation_of_Organic-Inorganic_Hybrid_Materials_of_Drugs_with_Inorganic_Oxides/links/53db14970cf2e38c63397fbb.pdf
202. Watanabe T, Wakiyama N, Usui F, Ikeda M, Isobe T, Senna M (2001) Stability of amorphous indomethacin compounded with silica. Int J Pharm 226(1):81–91. https://doi.org/10.1016/S0378-5173(01)00776-1
203. Lobmann K, Grohganz H, Laitinen R, Strachan C, Rades T (2013) Amino acids as co-amorphous stabilizers for poorly water soluble drugs – Part 1: Preparation, stability and dissolution enhancement. Eur J Pharm Biopharm 85(3 PART B):873–881. https://doi.org/10.1016/j.ejpb.2013.03.014
204. Dudognon E, Willart JF, Caron V, Capet F, Larsson T, Descamps M (2006) Formation of budesonide/α-lactose glass solutions by ball-milling. Solid State Commun 138(2):68–71. https://doi.org/10.1016/j.ssc.2006.02.007
205. Chieng N, Aaltonen J, Saville D, Rades T (2009) Physical characterization and stability of amorphous indomethacin and ranitidine hydrochloride binary systems prepared by mechanical activation. Eur J Pharm Biopharm 71(1):47–54. https://doi.org/10.1016/j.ejpb.2008.06.022
206. Abe K, Ogawa T, Uchino T, Otsuka M, Takano-Ohmuro H, Senna M (2010) Highly-efficient amorphization of drugs by the participation of molecular complex. Trans Mater Res Soc Jpn 5 (3):717–721. 10.14723/tmrsj.35.717
207. Taylor LS, Zografi G (1997) Spectroscopic characterization of interactions between PVP and indomethacin in amorphous molecular dispersions. Pharm Res 14(12):1691–1698. https://doi.org/10.1023/A:1012167410376
208. Mikhailenko MA, Shakhtshneider TP, Eltsov IV et al (2016) Supramolecular architecture of betulin diacetate complexes with arabinogalactan from Larix sibirica. Carbohydr Polym 138:1–7. https://doi.org/10.1016/j.carbpol.2015.11.047

Chapter 5
Physical Instability: A Key Problem of Amorphous Drugs

5.1 Understanding Crystallization from Amorphous State

Physical instability of amorphous APIs is a complex problem and the major limitation which impedes the wider application of amorphous products in a common practice. There is a set of divergent factors which may cause that material under certain conditions will recrystallize losing benefits resulting from its amorphous form. It is obvious that the successful implementation of amorphous drug is strictly related to the problem how to control its crystallization ability. Therefore, the debate on the stability of amorphous drug is in a large extent identified with the discussion about its crystallization ability. One should realize that dealing with crystallization is an inherent step of each stage of work with amorphous products. Thus, through understanding of API crystallization behavior is an obligatory starting point during the rational design of amorphous drug compositions maintaining stability for the desired period of time.

Crystallization is a first-order phase transition which can be basically divided into two separated stages. During the first stage referred as nucleation the particle of new phase form nucleating centers which in the next stage grow from critical to macroscopic dimensions [1, 2]. To understand the thermodynamics and kinetics of nucleation process classical nucleation theory (CNT) can be recalled. This classical theoretical viewpoint provides qualitative description of crystallization phenomenon which, however, fails when compared with the experimental data. In principle, the values of nucleation rate are usually notably underestimated when compared to the experimental data [1]. These discrepancies result from the fundamental assumptions of CNT being comprehensively discussed in the literature (the main complain refers to the treatment of small nano-sized nucleus as a macroscopic object) [3]. According to this theory in an initial step of crystallization an assembly of a few crystalline molecules is spontaneously formed making a thin interface between a solid and a liquid phases. At the beginning, the growth of the assembly is thermodynamically unfavorable. However, when the critical size is reached the

M. Rams-Baron et al., *Amorphous Drugs*,
https://doi.org/10.1007/978-3-319-72002-9_5

process continues in an irreversible way. According to CNT the Gibbs free energy of nucleation, ΔG, is governed by two contributions: (1) a negative term, lowering the total free energy, arising from the formation of the stable crystalline phase and (2) a positive surface term taking into account the free energy costs due to creation of the liquid-crystal interface [3]. At constant pressure the free energy change, ΔG, related to nucleus formation is given by [4]:

$$\Delta G(r) = \underbrace{-\frac{4}{3}\pi r^3 \Delta G_v}_{volume\ term} + \underbrace{4\pi r^2 \gamma_s}_{surface\ term} \tag{5.1}$$

where r is the radius of nucleus, ΔG_v is the difference between the free energies of liquid and crystal per molar volume of crystalline phase (i.e. the thermodynamic driving force for crystallization) and γ_s is the interfacial free energy per unit area. The ΔG value as a function of r is schematically presented in Fig. 5.1. The maximum of ΔG(r) determines the height of the activation barrier for nucleation ΔG^* corresponding to so-called critical nucleus size r^*, which once formed, triggers the subsequent crystal growth.

$$\Delta G^* = \frac{16\pi}{3} \frac{{\gamma_s}^3}{\Delta G_v^2} \qquad r^* = \frac{2\gamma_s}{\Delta G_v} \tag{5.2}$$

The kinetics of nucleation process is given by the following steady-state nucleation rate I describing the number of nucleation events appearing per unit time in a unit volume [6, 7]:

$$I = I_0 \exp\left[\frac{-\Delta G^*}{k_B T}\right] \tag{5.3}$$

According to CNT the nucleation rate depends exponentially on the barrier height ΔG^*, k_B is a Boltzmann constant, T is an absolute temperature, and I_0 is so-called kinetic pre-exponential factor. More precisely, the pre-exponential factor can be expressed as $I_0 = \rho_s Z A_{kin}$ where ρ_s is the number of possible nucleation sites per unit volume, Z is the Zeldovich factor and A_{kin} depends on the molecular mobility of a liquid phase frequently quantified in terms of the self-diffusion coefficient [4]. Such description refers to so-called homogenous nucleation which occurs in an ideal environment i.e. far from the surface and in the absence of impurities which may accelerate the nucleation process. In practice, in most cases we are dealing with heterogeneous nucleation which occurs preferentially in particular regions (e.g. near container walls). Then, the other particles or catalyzing surfaces lower the energy costs of interface formation leading to the reduction of nucleation barrier and faster crystallization rates.

After nucleation, the second stage of crystallization occurs i.e. crystal growth. Then, the critical nucleus increases in size leading to the development of crystalline

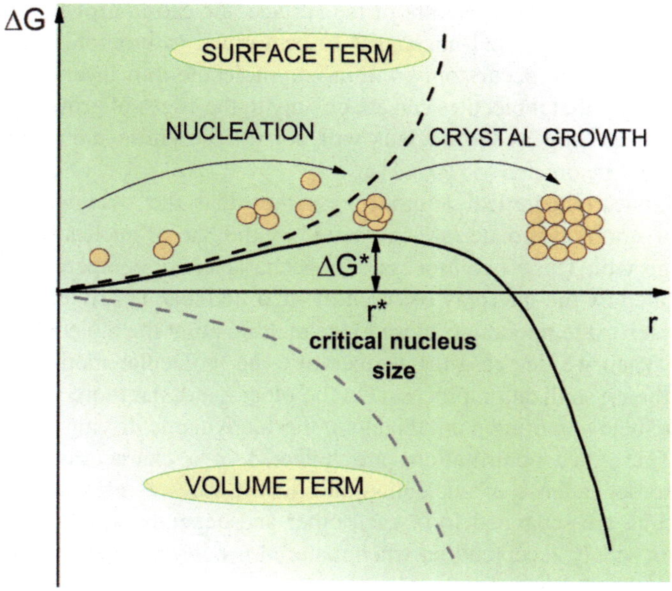

Fig. 5.1 The sketch of the nucleation free energy ΔG as a function of nucleus radius r according to CNT. The resultant value of ΔG (solid line) is a sum of two contributions i.e. positive surface term and a negative volume term. The maximum of $\Delta G(r)$ corresponds to a free energy barrier for nucleation that needs to be overcame to produce a critical nucleus and promote the crystals growth (takes place when $r > r^{*}$). Adopted from [5]

phase. According to CNT the crystal growth rate U is described by the following equation [8, 9]:

$$U = f(D/a_0)\left[1 - \exp\left(-\frac{\Delta G_c}{k_B T}\right)\right] \tag{5.4}$$

in which f is the fraction of sites at liquid-crystal interface where molecules can be added (f value is conditioned by the mechanism of crystal growth), a_0 is usually taken as a molecular diameter, ΔG_c is a thermodynamic driving force for crystal growth and D is the diffusion coefficient for molecular transport across the liquid-crystal interface.

There are several ways in which molecules can attach to the growing crystalline surface. Depending on the features of liquid-crystal interface, the three basic mechanisms of crystal growth can be distinguished i.e. (1) continuous growth model, (2) screw-dislocation model, (3) 2-d or surface nucleation growth model [10, 11]. In the case of continuous growth we are dealing with rough liquid-crystal interface with equivalent sites where molecules can join (then $f \approx 1$). For screw-dislocation model, the interface is smooth but imperfect in atomic scale and consequently, crystal growth occurs in places providing by screw-dislocations forming spiral-paths at the growing surface. Finally, the smooth and ideal (free of defects) interface is a main feature of the surface nucleation growth model. Then, the critical two-dimensional nuclei has to be formed to promote the subsequent crystal growth which occurs only within one-molecule-thin layers on the crystal surface (it means that molecules can attach only to the edges of growing layer). It is assumed that as long as the nucleus will not fill the entire monolayer, another nucleus cannot be created [8, 10, 11].

This classical theoretical approach assumes that the overall crystallization behavior of amorphous material is governed by the rate of nucleation and the rate of crystal growth. These two processes have characteristic temperature dependencies controlled by the interplay between the thermodynamic driving force and the diffusion rates. At temperatures close to the melting point the molecular mobility is very fast. When we decrease the temperature the molecular motions slow down hindering the crystallization process. On the other hand, the more we decrease the temperature below melting point the larger thermodynamic driving force for nucleation is. These two contributions are reflected in a characteristic bell-shaped temperature dependence of nucleation and growth rates as presented in Fig. 5.2. Their maxima are separated from each other and depends on the interfacial free energy. It is widely accepted that when material is cooled down below its melting temperature the ability to form a glass or undergo crystallization is determined by the degree of overlapping of temperature regions where nucleation and crystal growth are favorable. The prompt crystallization is expected in the case of materials with a prominent overlapping area while distinct separation of both regions leads to good glass forming ability. Along with the location of nucleation/growth zones with respect to each other, the applied cooling and heating rates are important as well.

Visualization presented in Fig. 5.2 is just a simplification of a very complex problem. For instance, the occurrence of several polymorphic forms of crystalline API may lead to the occurrence of more than one nucleation zone as indicated for indomethacin [12] or nilutamide [13]. In practice, the probability of nucleation or crystal growth at certain thermodynamic conditions depends on various factors. For clarity one can divide them as (1) *thermodynamic* (free energy for crystallization, configurational entropy, interfacial free energy between crystalline phase and amorphous phase), (2) *kinetic* and (3) *intermolecular*. Additionally, the influence of external conditions (such as mechanical stress, humidity, preparation method etc.) will also influence the overall crystallization behavior of amorphous API.

5.1.1 Thermodynamic Contributions

From thermodynamic perspective the crystallization tendency of amorphous drug is driven by necessity to minimalize its higher free energy. Under certain thermodynamic conditions the Gibbs free energy difference between amorphous and crystalline state acts as thermodynamic driving force for crystallization [15]. In comparison to crystalline state, the amorphous phase is characterized by the excess of free energy as well as higher enthalpy and entropy which are related to each other *via*

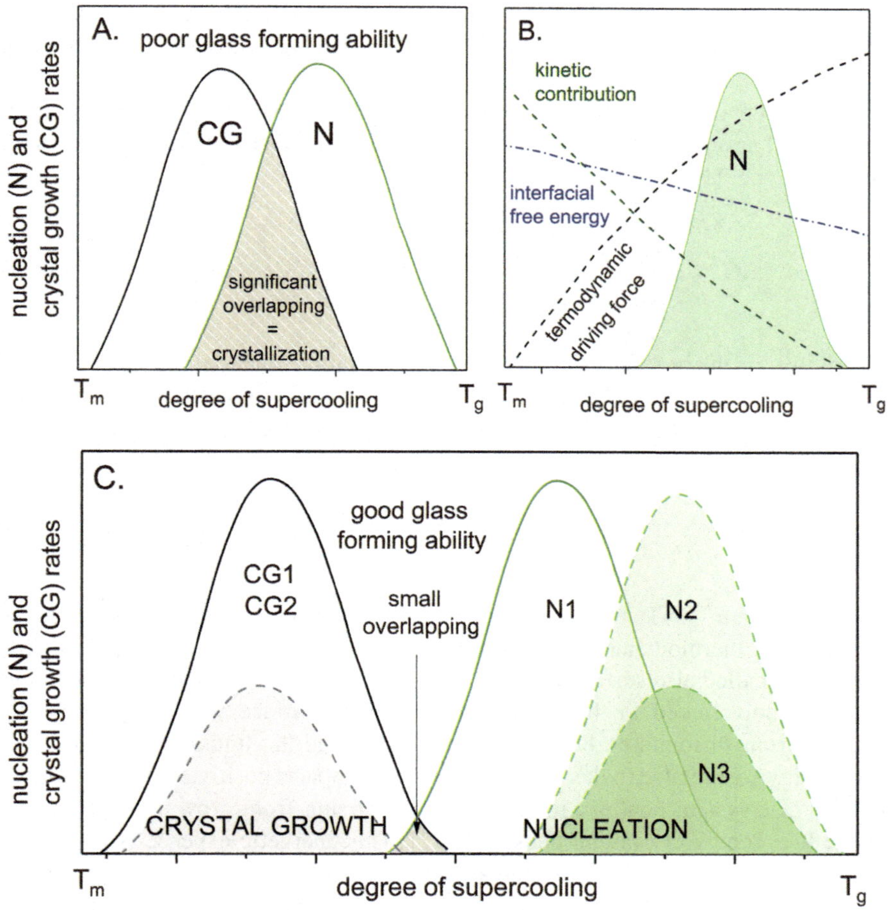

Fig. 5.2 The schematic representation of temperature dependence of nucleation and crystal growth rates (**a** and **c**). The overlapping area can be interpret as an indication of crystallization possibility on cooling (adopted from [14]). (**b**) The contribution of various parameters to nucleation rate as a function of supercooling according to CNT (adopted from [4])

the following relationship $G_{conf} = H_{conf} - TS_{conf}$ (the abbreviation used in the subscript i.e. *configurational* means that these thermodynamic values reflect the differences between amorphous and crystalline states). Higher configurational enthalpy (H_{conf}) makes recrystallization more favorable but this contribution is balanced by configurational entropy (S_{conf}) as well. Basically, the value of configurational entropy is related to the number of configurations available for investigated system (a measure of system flexibility). To enable the progress of crystallization the molecules need to self-organize themselves in order to find the optimal configurations corresponding to the regular arrangement of crystal lattice. The higher the configurational entropy, the more possible configurations to explore what should result in the deceleration of crystallization related events. Such effect will be also relevant for all these systems where amorphous API is mixed with certain excipient in order to make the reaching of crystallization favorable configurations even more demanding.

The value of ΔG cannot be directly measured but can be estimated from differential scanning calorimetry data measured for material in liquid and crystalline states, i.e. [16, 17]:

$$\Delta G = \Delta H_m(T_m - T)/T_m + \int_{T_m}^{T} \Delta C_p dT - T \int_{T_m}^{T} (\Delta C_p/T) dT \tag{5.5}$$

where ΔH_m is the heat of fusion per mole for crystalline material, T_m is the melting temperature and $\Delta C_p = \Delta C_p^l - \Delta C_p^c$ is the difference in molar heat capacities of liquid and crystal. For small undercooling $(T_m - T)$ an useful expression approximating ΔG was proposed by Hoffman [18, 19]:

$$\Delta G = \frac{\Delta H_m(T_m - T)T}{T_m^2} \tag{5.6}$$

According to CNT, the formation of stable nucleus is driven by the balance between the thermodynamic driving force for nucleation and the interfacial free energy, γ_s, called also surface tension. The growing nucleus appearing in the liquid is inseparably linked to the creation of boundary surface which separates this nucleus from surrounding liquid, and which is called the liquid-crystal interface. The energy cost related to the interface formation impacts positively the overall free energy of a system contrary to energy profits arising from growing volume of a stable low-energy crystalline phase [20–22]. This energetic expense is associated with entropy changes caused by increased ordering of liquid molecules approaching the liquid-crystal interface [23]. The interfacial free energy influences the conditions under which nucleation takes place while its direction dependence (anisotropy) impacts the final morphology of growing crystals [24, 25].

The value of interfacial free energy cannot be directly measured. Its estimation is possible only using other experimentally attainable parameters [26]. One of such approach relies on using Stefan-Skapski-Turnbull relation [11]:

$$\gamma_s = \alpha \Delta H_m / \sqrt[3]{V_m^2 N_A} \tag{5.7}$$

where α is the empirical dimensionless coefficient smaller than unity reflecting the fact that surface atoms have less neighbors than bulk atoms [27], ΔH_m is the melting enthalpy per mole, V_m is the molar volume of a crystal and N_A is the Avogadro's number. Unfortunately such indirect determination underestimates the actual interfacial free energy value by approx. 10–20% [25]. This prompted the development of theoretical approaches aimed at resolving the ambiguities associated with the experimental determination of interfacial free energy values [20, 21, 28]. Gerges and Affouard [6] applied molecular dynamic simulations (MD) and calculated the γ_s values at melting temperature for nifedipine and felodipine with uncertainties within the range of 7–8%. They found that felodipine has higher interfacial free energy and lower driving force for crystallization what explained its better resistance to crystallization in comparison to nifedipine. Furthermore, they observed that the interfacial free energy is higher for liquid-crystal interface corresponding to more stable polymorph. This was in accordance with the previous observation of Andoris and Zografi for indomethacin polymorph [12].

5.1.2 Kinetic Contributions

The kinetic term, beyond thermodynamic contributions, is considered as a second major player determining nucleation and crystal growth rates. To promote and continue crystallization the molecules have to diffuse along the liquid phase to find places where nucleation or crystal growth take place. Such transport process is frequently quantified in terms of self-diffusion coefficient. In fact the self-diffusion coefficient is commonly replaced by more accessible inverse shear viscosity η^{-1} in accordance with the Stokes-Einstein (SE) and the Debye-Stokes-Einstein (DSE) equations imposing that $D_{rot} \sim T\eta^{-1}$ and $D_{trans} \sim T\eta^{-1}$, where D_{rot} and D_{trans} are the rotational and translational diffusion coefficients, respectively. However, it should be noted that close to T_g the decoupling of diffusion and viscosity may occur resulting in a breakdown of the Stokes-Einstein relationship [29–31]. Thus, such approximation is considered as relevant in the temperature range extending from the melting point to about 1.3 T_g because in mentioned temperature range the decoupling phenomenon is usually not observed [32].

Since the diffusion-related term determines both nucleation and crystal growth rates, the higher molecular mobility will facilitate the crystallization related events. On the other hand, when a liquid is cooled down the significant increase of viscosity is observed. In the temperature range between the melting temperature and the glass transition temperature the viscosity of material usually increases 15 orders of magnitude (at T_g is typically assumed as $\eta \approx 10^{12}$ Pa·s) [33] determining the timescale of molecular motions. Below T_g the material is "trapped" in a highly viscous environment in a thermodynamically unstable glassy state and any subsequent temperature variations have limited effect upon its structure. In the supercooled liquid state, at $T > T_g$, the dynamics of material is primarily controlled by structural α-relaxation arising from cooperative, many-body motions responsible for liquid-glass transition. While below T_g secondary relaxations of different types prevail the molecular dynamics. On cooling molecular motions slow down and kinetic barrier to nucleation/crystal growth increases. Then the molecular relaxation times, the crucial parameters characterizing dynamics of relaxing species, increase in an simple activated Arrhenius-like manner or show non-Arrhenius behavior. It is illustrated when the logarithm of relaxation time $\log\tau_\alpha$ is depicted as a function of scaled temperature T_g/T on so-called Angell plot (see Fig. 5.3). This different

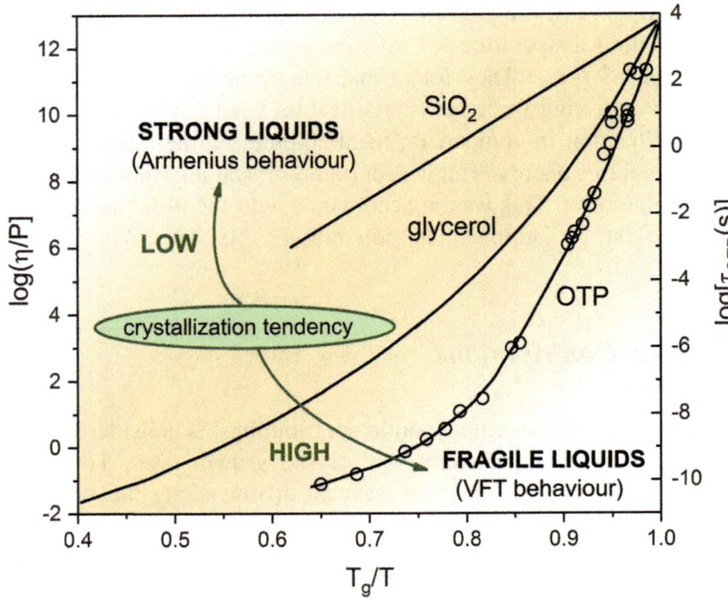

Fig. 5.3 Angell plot for various glass formers depicting possible pattern of fragility behavior. Adopted from [35]

behavior provides the basis for classifying glass-forming materials into two categories: "strong" or "fragile". The temperature sensitivity of molecular dynamics at T_g being expressed by a slope of a plot at $T_g/T = 1$ defines isobaric fragility parameter m (or steepness index) [34]:

$$m = \left.\frac{d\log\tau_\alpha}{d\left(T_g/T\right)}\right|_{T=T_g} = \frac{D\left(T_0/T_g\right)}{\left(1-\left(T_0/T_g\right)\right)^2 \ln(10)} \tag{5.8}$$

"Strong" glass formers have $m \leq 30$ while for "fragile" $m \geq 100$. The other cases are classified as intermediate strong or fragile glass formers.

To describe the non-linear temperature dependence of α-relaxation time in the supercooled liquid state the empirical Vogel-Fulcher-Tammann-Hesse (VFTH) equation is usually used [36–38]:

$$\tau_\alpha = \tau_\infty \exp(DT_0/T - T_0) \tag{5.9}$$

where τ_α is the structural relaxation time, τ_∞ denotes relaxation time at infinitely high temperature, D is a constant being considered as an indicator of fragility, and T_0 is frequently related to the theoretical Kauzmann temperature [39]. Otherwise, when the temperature dependence of logarithm of relaxation times has a linear character (like for non-cooperative secondary relaxations) the Arrhenius equation is usually used for quantification purpose:

$$\tau = \tau_{\infty} \exp(\Delta E_a / k_B T) \quad (5.10)$$

where ΔE_a is activation barrier and k_B is a Boltzmann constant.

The conceptual understanding of such behavior can be done by recalling to the Adam and Gibbs theory [40]. To do this we have to assume that liquid contains areas in which molecules can rearrange themselves independently and areas where motions occur in a cooperative manner i.e. depend on other molecules in the nearest neighborhood. On cooling the size of cooperatively rearranging regions (CRR) increases leading to the decrease of configurational entropy of a system [33]. According to the Adam and Gibbs theory the structural relaxation time (or viscosity) is inversely proportional to the changes in configurational entropy according to the following equation [41]:

$$\tau = \tau_{\infty} \exp\left(A / T S_{conf}\right) \quad (5.11)$$

where τ_{∞} is a relaxation time value in the limit of high temperature, A is a constant which reflects the activation barrier for reorientation, and S_{conf} denotes the configurational entropy of a liquid i.e. the difference in the entropy of the melt and the entropy of a glass (rather than that of crystal [42]).

Below T_g the ultraviscous regime makes the cooperative molecular motions extremely slow. At temperature corresponding to the glass transition it is commonly assumed that $\tau_{\alpha} = 100$ s, while further cooling will extend the time-scale of structural relaxations to the order of years. In the scientific jargon it has been used to state that below T_g the material is "kinetically frozen" (the time-scale of structural relaxation is too long to be probed experimentally). However, we have to remember that then the short-range non-cooperative molecular motions are still attainable and are observed as secondary relaxations. Depending on their origin we can classify them as a Johari-Goldstein i.e. JG processes and non-JG processes. The first involves the reorientations of whole molecules (intermolecular origin), while the latter are related to the local intramolecular reorientations or side groups motions. Different concepts underlying their possible mechanism have been discussed. In particular, the JG relaxations attend a lot of scientific attention due to their coupling with the α-relaxation and glass transition (they are considered as a precursors of the α-relaxation). It has been argued that the secondary relaxation modes indicates the presence of loosely-packed regions in a glassy structure in which the molecular motions take place when the rest of the structure is kinetically frozen (so-called islands of mobility [43]). The alternative concept assumes that β-relaxations involve the small angle rotational motions of all molecules that occur before the α-relaxations would take place [44]. The distinction whether we are dealing with JG or non-JG relaxation can be done on the basis of their different dynamic behavior. Readers interested in this issue will find the details in [45] and [46, 47].

According to CNT the molecular motions in general, irrespectively to their nature, will impact both stages of crystallization process (see Eqs. 5.3 and 5.4). Thus, the slowing down of molecular dynamics is considered as an effective way to provide the desired shelf-life of amorphous product. Any attempts to control the degree of molecular motions, in particular by using excipient molecules (e.g. polymers) exerting anti-plasticizing effect on API dynamics, will be inherently reflected in nucleation and crystal growth features.

5.1.3 Other Contributions

According to CNT the crystallization propensity of a material is governed by the balance of thermodynamic and kinetic contributions. In fact the problem is much more complex because other factors discussed in terms of possible correlation with amorphous state instability will influence the nucleation and growth rates as well (hydrogen bonding ability, environmental conditions etc.).

To understand the behavior of drug at the most fundamental level the nature of intermolecular attractions has to be taken into account. The weak intermolecular forces like van der Waals forces or dipolar interactions arise clearly from molecular architecture of a drug. The chemical structure imposes ability to form hydrogen-bonds. Such capability to form H-bonded structures will affect physical properties of a drug and is essential for drug-polymer formulations because the strength of interactions between both components determines the physical stability of a system. Koperwas et al. [48] show recently that changes in the attractive part of the intermolecular interaction potential correlate with crystallization ability of molecular van der Waals liquids. Their results indicated that weakening of molecular attractions by lowering the contribution of the dipole-dipole forces makes material more prone to crystallization on cooling. To strengthen this statement they performed standard molecular dynamics simulations on single-component Lennard-Jones systems in which the strength of attractive part of intermolecular potential has been modified by introducing a control parameter λ. The increase of λ parameter, related to the enhancement of dipole-dipole term of intermolecular potential, shifts the nucleation and growth rates maxima towards lower temperatures and makes the distance between both maxima more pronounced. It shows that playing with intermolecular potential by exerting impact on temperature behavior of nucleation and crystal growth rates may be decisive for physical stability of amorphous state as well.

The character of molecular interactions between drug and excipients is likewise important. Trasi et al. [14] investigated the impact of different polymers on nucleation and crystal growth rates of acetaminophen drug. Polymers selected to study were characterized by different ability to hydrogen bond formation. Among them one can distinguish polymers able to create a very strong H-bonds with acetaminophen (i.e. PVP, PVP-VA, PAA) as well as those interacting with drug in less effective manner (HPMC, Eudragit E100, HPMCAS). The strength of drug-polymer interactions was found to correlate well with the reduction of crystal growth rates. Interestingly, such a rule was not observed in the case of nucleation rates. Then, the impact of polymers was found to be very diversified and devoid of explicit rules. For example, PAA affects strongly the nucleation temperature shifting its lower limit by 60 °C but at the same time it promotes nucleation leading to the threefold increase of nucleation rate. Contrary, the HPMCAS polymer poorly impacts crystal growth but was found as the most effective inhibitor of nucleation [14].

The hydrogen bonds in the investigated system can occur between the same molecules forming homodimers and favoring crystallization from amorphous state, or between different molecules resulting in the formation of heterodimers which are expected to decrease the crystallization rate (the disruption of heterodimeric assembly must occur before the nucleation favored conformation will be formed). This concept underlines various strategies being proposed with the aim to improve the stability of amorphous API relying on binary system preparation in which API and excipient may interact with each other. Interestingly, it raises another possibility of API "self-stabilization" which may be realized in a systems with various isomeric composition. As an example etoricoxib can be given [49]. We have shown that amorphous etoricoxib exists in a dynamic equilibrium between two different tautomeric forms (as an effect of proton transfer between methyl group linked directly to the pyridine ring and pyridine nitrogen atom). Furthermore, based on observed changes in infrared response as well as on the results of DFT calculations we found that these two tautomeric forms may interact with each other forming hydrogen-bonded heterodimeric structures. In our opinion such heterodimeric composition is responsible for improved etoricoxib stability in comparison to other coxibs, i.e. celecoxib and rofecoxib. Facing with many similarities in molecular dynamics of these APIs we conclude that tautomerisation ability was the only reasonable justification of observed striking differences in their crystallization behavior. Basically, the tautomeric equilibrium is sensitive to external condition and thus is difficult to control and study. But in the case of amorphous API with tautomerisation ability this issue must be taken into account because depending on the thermodynamic conditions the tautomeric composition may be different leading potentially to unexpected changes in material properties. This issue has been poorly explored so far, however, the improved physical stability was found for some API with tautomerisation ability (e.g. glibenclamide, indapamide).

Another important factor that might affect the nucleation or crystallization rates is the presence of water. Water is a well-known plasticizer being characterized by low T_g reported as 135 K [50]. Amorphous APIs may spontaneously absorb water from their surroundings during drug production and/or storage. Water absorption will decrease the T_g of a material and enhance its molecular mobility leading to faster recrystallization. Besides it has been reported that the presence of moisture may disrupt the drug-polymer interactions [51] or even induce phase separation in drug-polymer systems [52]. In general, amorphous APIs are considered as more hygroscopic than their crystalline counterparts, and the same is related to pharmaceutical excipients in amorphous form, in particular to polymers. There are many reports indicating faster crystallization of amorphous APIs or amorphous solid dispersions stored at elevated humidity. The tendency to water uptake is conditioned by chemical structure, in particular the number of polar functional group able to form hydrogen bonds with water. Konno and Taylor [51] investigated various felodipine-polymer systems and reported that the amount of absorbed water was an increasing function of polymer concentration. But solid dispersions with PVP absorbed water more efficiently than those containing HPMC or HPMCAS. What was the most important in all cases the nucleation rate increased as a function of storage humidity.

5.1.4 Implications for Drug Stability

Crystallization and physical instability is one of the greatest challenges of developing amorphous drug formulations. Huge experimental efforts are directed into crystallization studies of various amorphous APIs with the aim to overcome the problem of their limited stability. These results may contribute to the development of appropriate protocols that would be able to predict the shelf life of amorphous product without necessity of waiting for the results of long-term stability tests. Despite the continuous progress in this field the answer to the primary question *how to predict the stability of amorphous API* still remains unresolved.

Over the years the study on relaxation dynamics of amorphous APIs appears as a key to understand the changes of molecular dynamics at different thermodynamic conditions governing nucleation and crystal growth capability of amorphous drug. Mostly they were carried out using broadband dielectric spectroscopy (BDS) giving insight into molecular dynamics over wide range of frequencies, temperatures and pressures. Research activity in the past was mostly focused on measurements using temperature as a thermodynamic variable. Recently the benefits arising from the application of pressure as a variable gained recognition. Such studies give opportunity to study molecular dynamics at conditions mimicking those prevailing during drug processing and manufacturing. Across the product lifecycle amorphous API is exposed to different external conditions. It must be realized that depending on temperature (or pressure) the amorphous material may occur in (1) the non-equilibrium glassy state (at $T < T_g$) or in (2) a thermodynamically metastable supercooled liquid state ($T_g < T < T_m$). Each of them is characterized by different timescale of corresponding molecular motions. During drug processing at high temperatures ($T > T_g$) the drug properties are governed by so-called global mobility, i.e. the structural α-relaxation associated with collective reorientations of molecules and directly related to the glass transition phenomenon. At room-temperature for drugs being in the glassy state the dynamics will be governed by local mobility (secondary relaxations). There is a strong belief that through investigation of molecular dynamics the questions crucial for drug stability assessment can be addressed:, i.e. (1) which type of molecular motions determines the crystallization rates; (2) is there a possibility to predict the structural relaxation time in a glassy state using values determined for a liquid state to further approximate the stability period. Finding an answer for these questions is of special importance because may help to resolve the long-standing challenge of pharmaceutical science which is the prediction of time-scale in which the physical stability of amorphous drug will be ensured at conditions of interest.

Is there any correlation between the timescale of crystallization and relaxation time values obtained from molecular dynamic studies?
At this moment there is no a clear answer what kind of motions, local or global, affect crystallization velocity of amorphous drug. However, the investigations have been intensively guided, and the database on the possible correlation patterns among crystallization and mobility time scales regularly expands allowing to believe that perhaps in the future, some general rules can be identified. There are several studies indicating that global mobility can be linked to physical instability and still rare examples where local motions are crucial (see Table 5.1).

In order to determine the timescale of crystallization the kinetic studies have to be performed at each temperature of interest. The greater part of researches has been focused on the study of crystallization in the supercooled liquid state. Such tests allow in a relatively fast manner verify whether amorphous API has a high tendency to recrystallize or not. However, the most valuable are crystallization studies carried out below T_g. Such studies are much more time-consuming but correspond to the actual conditions in which pharmaceuticals are stored. In general, the progress of crystallization can be experimentally monitored using different methods (e.g. X-ray diffraction methods, calorimetric, dielectric, microscopic ones). Then, to quantify the degree of crystallization as a function of time an appropriate approach has to be implemented. The most frequently used includes:

- the modified Kolmogorov-Johnson-Mehl-Avrami (KJMA) equation [53–56]:

$$\alpha = 1 - \exp\left[-\left(\frac{t - t_0}{\tau_{cr}}\right)^n\right] \tag{5.12}$$

  where α is the fraction crystallized under isothermal condition, τ_{cr} is the characteristic crystallization time accounting both nucleation and growth rates, t_0 is the induction time and n is a dimensionless parameter depending on crystal morphology and crystallization mechanism. The most popular approach to attain fitting parameters relies on analyzing the experimental data using double logarithmic scale i.e. by plotting $\ln(-\ln(1-\alpha))$ against $\ln(t - t_0)$. Then, the slope of the expected straight line corresponds to n while intercept determines $\tau_{cr} = k^{-1}$, where k is a rate constant [57]. However, in the limiting cases (i.e. for $\alpha \rightarrow 0$ and for $\alpha \rightarrow 1$) the double logarithmic function magnifies experimental errors leading to non-correct evaluation of the induction time [58].

- Avramov approach

  It allows to determine the induction time value in a more reliable manner which requires plotting the experimental data using different coordinates, namely α against $\ln(t - t_0)$ resulting in a sigmoidal-shaped curve. In order to determine the crystallization parameters, i.e. values of n, τ_{cr} and t_0, the first derivative of Eq. (5.12) with respect to $\ln(t - t_0)$ has to be calculated and plotted on the same

Table 5.1 The coupling coefficients reported for selected APIs and pharmaceutical excipients as well as solid dispersion systems

API	Coupling coefficient and examination details	Reference
Itraconazole	0.68 for τ_α and τ_{cr} determined above T_g	Bhardwaj et al. [61]
	0.94 for τ_α and τ_0 determined above T_g	
Itraconazole-PVP	0.78 τ_α and τ_{cr} determined above T_g	Bhardwaj et al. [59]
	0.99 for τ_α and τ_0 determined above T_g	
Itraconazole-HPMCAS	0.85 τ_α and τ_{cr} determined above T_g	Bhardwaj et al. [59]
	0.99 for τ_α and τ_0 determined above T_g	
Celecoxib	0.8 for τ_α and τ_{crys} time taken for 2.5% crystallization above T_g	Mehta et al. [62]
	0.3 for for τ_α and τ_{crys} time taken for 2.5% crystallization below T_g	
	1.4 for for $\tau_{\beta\text{-}JG}$ and τ_{crys} time taken for 2.5% crystallization below T_g	
Sucrose (spray-dried) (freeze-dried)	for τ_α and τ_{onset} above T_g from isothermal DSC studies 0.49 0.54	Bhugra et al. [68]
Indomethacin	0.29 for τ_α and τ_{onset} determined from isothermal microcalorimetry studies above T_g	Bhugra et al. [69]
	0.85 for τ_α and crystal growth rates above T_g	
	0.8 for τ_{crys} time taken for 2.5% crystallization	Mehta et al. [62]
	0.2 for for τ_α and τ_{crys} time taken for 2.5% crystallization below T_g	
	1.1 for for $\tau_{\beta\text{-}JG}$ and τ_{crys} time taken for 2.5% crystallization below T_g	
Felodipine	0.26 for τ_α and τ_{onset} determined from isothermal microcalorimetry studies above T_g	Bhugra et al. [69]
	0.43 for τ_α and crystal growth rates above T_g	
Flopropione	0.20 for τ_α and τ_{onset} determined from isothermal microcalorimetry studies above T_g	Bhugra et al. [69]
Nifedipine	0.41 for τ_α and crystal growth rates above T_g	Bhugra et al. [69]
	0.94 for τ_α and τ_{crys} time taken for 10% crystallization below T_g	Kothari et al. [60]
	0.62 for τ_α and τ_{crys} for10% crystallization above T_g	
Nifedipine PVPK12	1.2 for τ_α and τ_{crys} time taken for 10% crystallization below T_g	Kothari et al. [60]
	0.67 for τ_α and τ_{crys} for10% crystallization above T_g	
Ketoconazole	0.35 for τ_α and crystal growth rates above T_g	Bhugra et al. [69]
	0.49 for τ_α and τ_{cr} determined above T_g	Mistry et al. [63]
Ketoconazole 4% PVP	0.52 for τ_α and τ_{cr} determined above T_g	Mistry et al. [63]
Ketoconazole 4% PHEMA	0.60 for τ_α and τ_{cr} determined above T_g	Mistry et al. [63]

(continued)

Table 5.1 (continued)

API	Coupling coefficient and examination details	Reference
Ketoconazole 4% PAA	0.46 for τ_α and τ_{cr} determined above T_g	Mistry et al. [63]
Trehalose (freeze-dried) (spray-dried) (dehydrated)	1.23 for τ_α and τ_0 determined above T_g 1.13 for τ_α and τ_0 determined above T_g 1.18 for τ_α and τ_0 determined above T_g	Bhardwaj et al. [70]
Griseofulvin	0.94 for τ_α and τ_{crys} time taken for 0.5% crystallization below T_g	Kothari et al. [60]
	0.65 for τ_α and τ_{cr} above T_g	
Sildenafil	0.49 for τ_α and τ_{cr} determined above T_g 0.50 for τ_α and τ_0 determined above T_g	Kolodziejczyk et al. [71]
Phenobarbital	0.67 for τ_α and τ_0 determined above T_g	Caron et al. [65]
Nifedipine 5% PVP	0.50 for τ_α and τ_0 determined above T_g	Caron et al. [65]
Phenobarbital 5% PVP	0.32 for τ_α and τ_0 determined above T_g	Caron et al. [65]

PVP polyvinylpyrrolidone, *PAA* poly(acrylic acid), *PHEMA* poly-2-hydroxyethyl methacrylate, *HPMCAS* hydroxypropylmethylcellulose acetate succinate

graph (so-called Avrami-Avramov plot, see Fig. 5.4). The maximum slope of sigmoidal curve matching the maximum of its first derivative can be found for $t = \tau_{cr} + t_0$ corresponding to $\alpha \approx 0.63$. If the inflection point appears at $\alpha < 0.63$ the induction time value is different from zero. The value of n can be determined as a slope of sigmoidal function at $t = \tau_{cr} + t_0$ given by $\partial\alpha/\partial \ln(t - t_0) = 0.368n$ [57, 58].

Both the crystallization time, $\tau_{cr} = k^{-1}$, inversely proportional to the crystallization rate constant, as well as induction time, τ_0, were discussed in terms of possible correlation with relaxation times. Alternatively, other estimates of crystallization time have been considered (see Table 5.1). The relationship between the relaxation and crystallization times is usually expressed by so-called coupling coefficient *s* [31, 59–61]. It can be determined as a slope of log-log plot of studied relaxation time τ against determined crystallization time τ_{cryst} (τ_{cr}, τ_0, time taken for 2.5% of crystallization or 10% of crystallization etc.), according to the following relation where A is a constant:

$$\log\tau_{cryst} = s\log\tau + \log A \tag{5.13}$$

A coupling index equals to unity denotes that the investigated crystallization time would be fully controlled by kinetic factor. In Table 5.1 the coupling coefficients determined for several pharmaceutically relevant materials are presented, summarizing the current state-of-art in this field. Since the methodology of the studies varies greatly, together with the coupling coefficients some remarks about their experimental

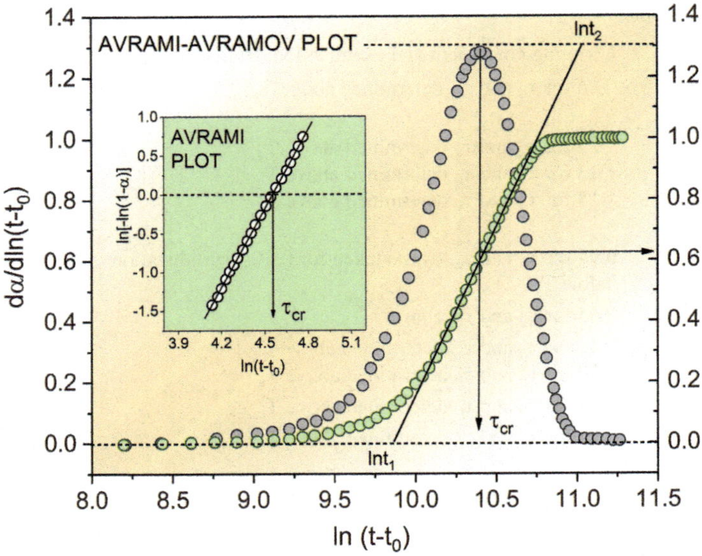

Fig. 5.4 Avrami-Avramov plot. The exemplary time evolution of crystallization fraction (e.g. normalized dielectric constant) and its first derivative versus the natural logarithm of time, analyzed according to procedure described above

determination are provided. Above T_g the correlation between the time-scale of structural relaxation and crystallization was found for several pharmaceuticals (e.g. itraconazole [61], indomethacin [62], celecoxib [62], griseofulvin, nifedipine [60], ketoconazole [63], phenobarbital [64]) and solid dispersion systems containing various polymer-API combinations (e.g. nifedipine-PVP [60], ketoconazole with PVP, PAA or PHEMA [63], itraconazole-PVP and itroconazole-HPMCAS [59]). In the case of itraconazole and its solid dispersions the evident correlation was found for structural relaxation dynamics above T_g and the crystallization induction time (then coefficient s was close to unity) while for the τ_α and τ_{cr} such coupling was less pronounced indicated the participation of other factors as well [59, 61]. The moderate coupling between τ_α and τ_{cr} above T_g was also observed for its structural analogue i.e. ketoconazole and its solid dispersions with 4% of PVP, PAA or PHEMA [63]. Interestingly the coupling coefficient obtained for pure drug and its solid dispersion was very similar indicating that the presence of polymer do not influence the strength of coupling. The same regularity was observed for nifedipine [60] and itraconazole [59] solid dispersion while as a counterexample phenobarbital and its PVP solid dispersion can be given [65]. For griseofulvin and nifedipine the strong correlation between τ_α and crystallization time-scale was found both above and below T_g indicating that global mobility is an important factor affecting physical instability both in the supercooled liquid and glassy states. For celecoxib and indomethacin the strong correlation between structural relaxation time determined above T_g and crystallization time (assessed for 2.5% crystallization) was also found ($s = 0.8$) [62]. Below T_g the strong coupling was observed between the time-scale of crystallization and secondary relaxation as was indicated by coupling coefficient greater than one for both APIs [62]. An interesting conclusions can be also done by direct comparison of crystallization and structural relaxation time-scales. Grzybowska and co-workers [66] pointed out that in glassy celecoxib the agreement between the time-scale of structural relaxation predicted at room-temperature condition and the maximal rate of celecoxib recrystallization was found (in both cases it was about 100 h). Such agreement was also reported for amorphous ezetimibe [67].

How to predict τ_α(T) in the glassy state?
Below T_g the molecular motions become very slow and their direct experimental evaluation becomes very demanding, and sometimes not possible due to exceeding the experimentally available timescale. Thus, to facilitate this process and make the assessment more effective several methods have been proposed to predict the structural relaxation times in a glassy state.

The easiest way is to transpose the dielectric spectrum collected above T_g to that registered below T_g in which only the high-frequency flank of α-process is experimentally attainable. Then, the reference dielectric loss curve, with clearly distinguishable maximum of α-process, have to be manually shifted along the frequency axis to ensure the best match with data at lower temperatures. The result of such shifting, called *master plot*, enables to predict the structural relaxation time, τ_α, from the maximum f_{max} of constructed curve as $\tau_\alpha = 1/2\pi f_{max}$ (see Fig. 5.5). This method requires, however, that the shape of the dielectric loss spectra is temperature independent. Such property called the time-temperature superposition (TTS) principle in many cases is fulfilled. In some cases variations from TTS may be due to the contribution of secondary relaxation (or excess wing). Note that at low frequency the shape of α-relaxation peak may be additionally affected by dc-conductivity contribution [72]. In spite of obvious limitations this relatively simple method is considered as a reliable approach to estimate the time scale of the structural relaxation in the glassy state in the vicinity of T_g (but its accuracy decreases with lowering temperature below T_g).

Another common approach relies on using Adam and Gibbs (AG) model [40] originally proposed to describe the equilibrium dynamics of supercooled liquids but successfully adopted to describe the temperature dependence of structural relaxation times below T_g. The following equation has been proposed to describe the glassy dynamics and can be used to predict the value of τ_α [74, 75]:

$$\tau_\alpha(T, T_f) = \tau_\infty \exp\left(\frac{DT_0}{T(1 - T_0/T_f)}\right) \tag{5.14}$$

where τ_∞, D, T_0 are the fitting parameters derived from VFT equation describing the τ_α(T) data at $T > T_g$. The parameter T_f is the fictive temperature which defines the properties of glass in terms of the equilibrium supercooled liquid having the same configurational entropy [76]. The value of T_f for a glass can be estimated using a thermodynamic parameter γ_{Cp} defined by heat capacities of liquid $C_p{}^{liq}$, glass $C_p{}^{glass}$, and crystal $C_p{}^{cryst}$, all investigated at T_g [76]:

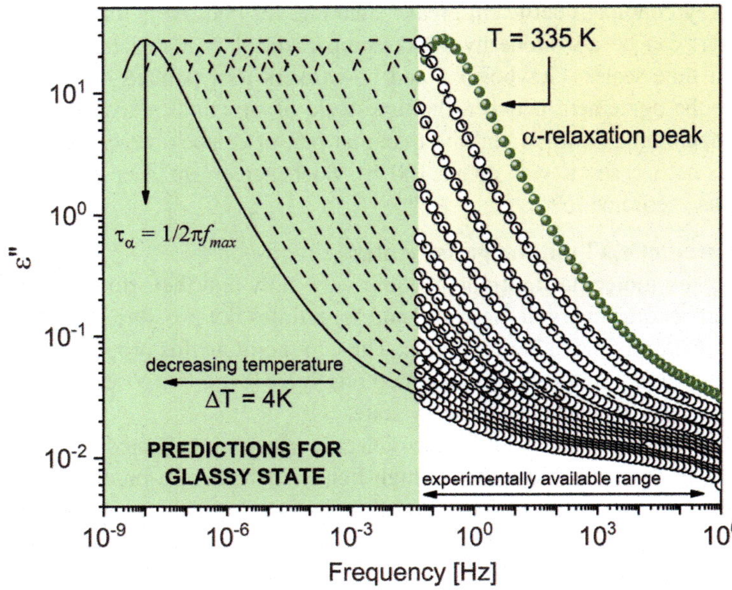

Fig. 5.5 The procedure of master plot constriction involving shifting the α-relaxation peak to dielectric spectra registered in a glassy region where maximum of α-process is experimentally unattainable. Data for bicalutamide are taken from [73]

$$\frac{1}{T_f} = \frac{\gamma_{Cp}}{T_g} + \frac{1-\gamma_{Cp}}{T}, \quad \gamma_{C_p} = \frac{C_p^{\ liq} - C_p^{\ glass}}{C_p^{\ liq} - C_p^{\ cryst}} \tag{5.15}$$

In particular cases, when $T_f = \mathrm{T}$ ($\gamma_{Cp} = 0$) the Eq. (5.14) corresponds to VFT equation while by inserting $T_f = \mathrm{T_g}$ ($\gamma_{Cp} = 1$) the Eq. (5.14) represents Arrhenius law. Both methods were used in the past to predict the values of structural relaxation times in a glassy state for various pharmaceutically relevant materials. Then, the predicted time-scale of molecular motions constituted the basis for further stability predictions at storage conditions (e.g. at room temperature). The validly of such method in approximation the time-scale of physical stability was confirmed for several pharmaceuticals (e.g. for celecoxib [66], bicalutamide [73]). Here, the predictions for ezetimibe are presented (see Fig. 5.6).

The brief description presented herein does not exhaust the topic of τ_α(T) predictions in the glassy state. Several other methods are described in details elsewhere [31, 77]. Such alternative approaches have to be employed, e.g. in the case of ionic APIs converted to amorphous form. As an example sumatriptan succinate can be given [78]. The dielectric response of conducting APIs is intrinsically dominated by dc-conductivity contribution originating from translational motion of charge carriers. As a consequence, the structural relaxation peak cannot be directly observed in the dielectric measurements and data analysis is performed

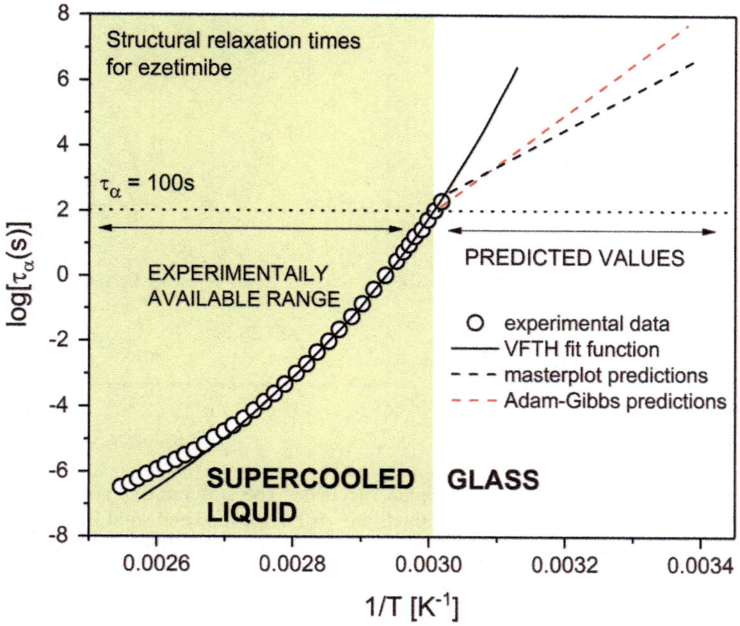

Fig. 5.6 Relaxation map for ezetimibe presented experimental τ_α values and those predicted from master plot using AG equation. Data taken from [67]

using the electric modulus formalism. Then, to determine the timescale of structural relaxation in the glass so-called aging experiment has to be performed. The *physical aging* is of particular relevance for all glass-forming systems and in general terms refers to the variation of dielectric parameters observed during sample annealing below the T_g when the glass slowly relaxes toward equilibrium. It has been shown that aging rate of ionic conductors is governed by the structural relaxation dynamics of the glass [79]. As a consequence the τ_{age} determined within the course of aging studies is a measure of τ_α in the glassy state (see data presented in Fig. 5.7). The common practice to assess τ_{age} value is to fit the $\tau_\sigma(t_{age})$ data to the following equation [80]:

$$\tau_\sigma(T, t_{age}) = A\exp\left[-\left(t_{age}/\tau_{age}\right)^\beta\right] + \tau_{\sigma,\infty}(T) \tag{5.16}$$

in which T denotes aging temperature, β is an aging stretching parameter, t_{age} and τ_{age} are aging time and relaxation time, A and $\tau_{\sigma,\infty}$ are constants depending on τ_σ values in the limit of zero and infinite times. One can see that for sumatriptan succinate the determined τ_{age} values correspond well with the α-relaxation times determined from TMDSC measurements. At room temperature the time scale of structural relaxation was found to be 24 h. This shows that in the time scale of days we can expect the drug recrystallization. Indeed, our predictions have been

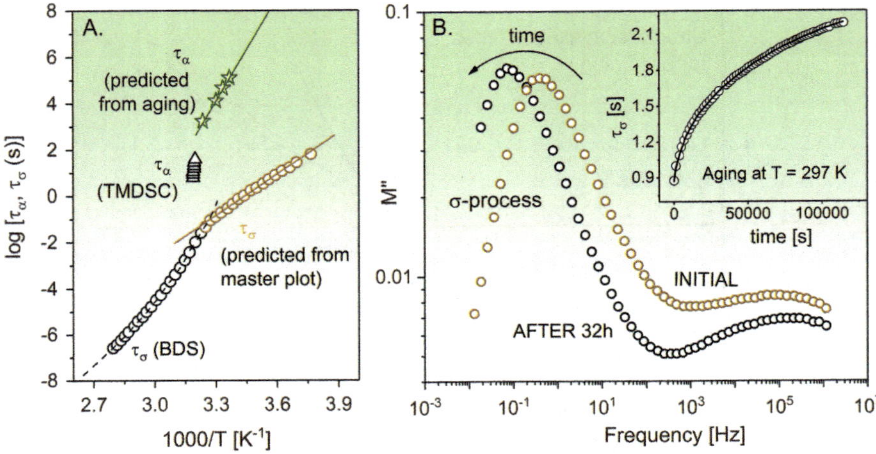

Fig. 5.7 (**a**) Relaxation map for sumatriptan succinate. The structural τ_α and conductivity τ_σ relaxation times obtained *via* different methods are visible. Dashed and solid lines correspond to VFT and Arrhenius fit functions, respectively. (**b**) Time evolution of conductivity peak during aging at T = 297 K. The inset shows the conductivity relaxation times depicted as a function of time parametrized by Eq. (5.16). Data taken from [78]

confirmed by XRD studies revealing the presence of the Bragg peaks of crystalline form of the drug after 6 days of storage at room temperature.

The results so far have indicated that the time evolution of dielectric features observed in the glassy state can provide information about the time-scale of structural relaxation regardless of the nature of investigated process [80]. Thus, another approach which may be applied to predict the values of τ_α at $T < T_g$ relies on monitoring the behavior of the secondary process during aging experiments [81]. This method is more time consuming but may help if the other approaches fails. During aging, when the glass evolves toward the equilibrium the features of β-relaxation, such as amplitude or relaxation time change with time. Casalini and Roland have shown that for amorphous polymer, polyvinylethylene, the decrease of amplitude of β-peak normalized by the initial values and analyzed at fixed frequency as a function of aging time is well described by a stretched exponential function analogical to Eq. (5.16) [81]. The aging relaxation time, τ_{age}, as well as stretching parameter determined as a fit parameters, were frequency independent and corresponded well with equilibrium data confirming that the observed changes are governed by structural relaxation dynamics. In this case the investigated β-process was classified as JG-relaxation. However, it has been suggested that the nature of β-process is not relevant here. This was indicated by results obtained for two saccharides, i.e. sucrose and trehalose, for which two secondary modes of distinct molecular origin (assigned as JG and non-JG modes) have been observed below T_g. The value of structural relaxation times in a glassy state determined according to the Casalini and Roland method were similar and independent of the intra- or intermolecular origin of investigated secondary mode [82].

5.2 How to Improve the Physical Stability of Amorphous APIs?

According to the previous discussion the improvement of stability of amorphous APIs can be achieved by appropriate modification of nucleation and crystal growth rates. It is believed that the effective suppression of these processes can extend the life-time of amorphous drug enough to meet the expectations of pharmaceutical companies. In this paragraph we will discuss the most popular strategies aimed at reducing the crystallization tendency of amorphous APIs, allowing to develop these challenging materials as an effective drug formulations.

5.2.1 Amorphous Solid Dispersions

Formulating amorphous drugs as two component systems, frustrated against crystallization, containing amorphous API and certain excipient appears as an auspicious way to enhance their physical stability. For clarification, as an additional ingredient a small molecule, another drug or polymer can be used. However, API-polymer compositions, so-called *amorphous solid dispersion* systems, have been the most actively investigated so far. Regarding the work that has been done during screening and investigation of effective amorphous drug compositions one can find a symptomatic disproportion between the efforts of scientists and a limited number of amorphous products commercially available. Meanwhile, in the field of amorphous solid dispersions a remarkable progress has been done. The summary of solid dispersion products bring into pharmaceutical market is presented in Table 5.2. The stability of API dispersed in polymer matrix has been related to the different effects, i.e.:

- solubility and miscibility of API in polymer carrier,
- slowing down of API molecular mobility due to polymer additive,
- molecular interaction between the API and polymer,
- steric hindrance imposed by the polymer additive.

5.2.1.1 Solubility and Miscibility of API in Polymer Carrier

The selection of polymer and the applied drug-polymer ratio will determine whether at particular condition the amorphous composition will be kinetically and

Table 5.2 The examples of solid dispersion products available on the market

API	Polymer	Product name	Preparation method	Year of approval
Nabilone	PVP	Cesamet™	–	1985
Itroconazole	HPMC	Sporanox®	Spray drying	1992
Tacrolimus	HPMC	Prograf™	Spray drying	1994
Griseofluvin	PEG	Gris-PEG™	Melt extrusion	2000
Lopinavir	PVP/VA	Kaletra®	Melt extrusion	2005
Etravirine	HPMC	Intelence®	Spray drying	2008
Ritonavir	PVP/PA	Novir®	Melt extrusion	2010
Ivacaftor	HPMCAS	Kalydeco	Spray drying	2010
Everolimus	HPMC	Zotress	Spray drying	2010
Itroconazole	HPMC	Onmel	Melt extrusion	2010
Fenofibrate	PEG/Poloxamer 188	Fenoglide	Spray melt	2010
Telaprevir	HMPCAS	Incivek	Spray drying	2011
Vemurafenib	HPMCAS	Zelboraf	Coprecipitation	2011
Posaconazole	HPMCAS	Noxafil	Melt extrusion	2013
Ombitasvir/Paritaprevir/Ritonavir	PVP-VA/TPGS	Viekirax	Melt extrusion	2014
Lumacaftor/Ivacaftor	HPMCAS/SLS	Orkambi	Spray-drying	2015

Data taken from [83–85]

thermodynamically stable or not. This may be illustrated by means of phase-diagram of drug-polymer system being schematically presented in Fig. 5.8. Then, the crystallization probability at certain thermodynamic conditions will be indicated by so-called solubility line and the glass transition line. The solubility line will specify conditions where the system is thermodynamically stable. It means that until the API concentration in polymer matrix will not exceed the solubility limit its recrystallization from amorphous state will not occur. The glass transition line, indicating the frontier between the dynamically distinct glass and the liquid regions, will indicate where kinetic stabilization may be expected. Finally, at certain temperature and for particular API-polymer ratio it will be possible to verify whether the system is: a thermodynamically stable liquid (1), a thermodynamically stable glass (2), a thermodynamically and kinetically unstable liquid (3), or thermodynamically unstable glass (4). These four areas correspond to the following possible scenarios of crystallization behavior: (A.) and (B.) the API-polymer composition is physically stable, (C.) the composition is physically unstable, (D.) API has tendency to crystallization which may be prevented by sufficient reduction of molecular mobility [86].

It is obvious that the knowledge about drug-polymer solubility is of great importance for rational development of solid dispersions, however, its experimental determination is demanding particular due to viscous character of polymer-based systems leading to a very slow dissolution kinetics exceeding experimentally available timescales. The several protocols have been proposed to determine the

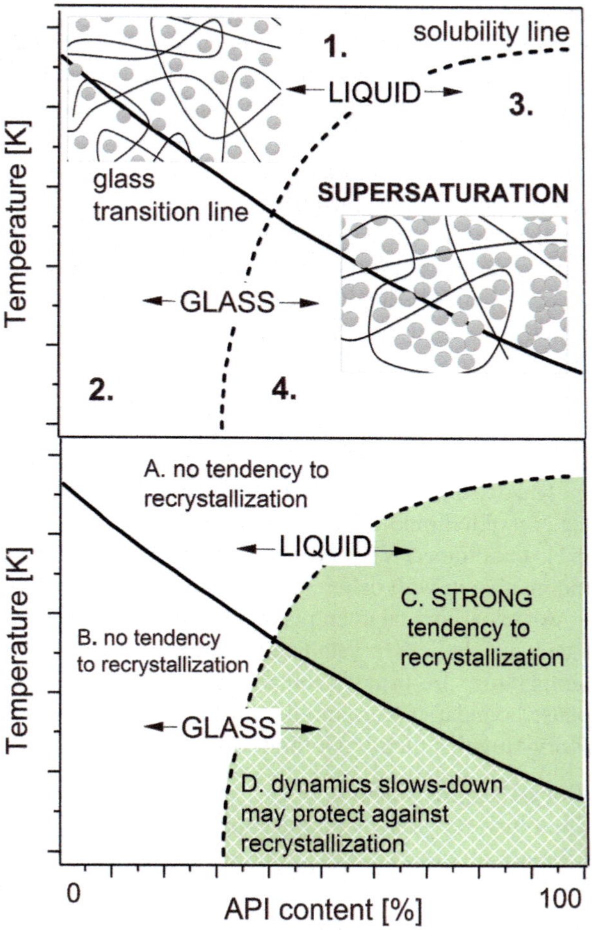

Fig. 5.8 The schematic phase-diagram of API-polymer system together with expected crystallization behavior. Depending on the temperature and API concentration we are dealing with (1) thermodynamically stable liquid, (2) thermodynamically stable glass, (3) unstable supersaturated liquid, (4) thermodynamically unstable glass. Adapted from [86]

solubility of API in a polymer carrier by means of differential scanning calorimetry. Basically, to get insight into the solubility of crystalline API in a polymer the dissolution temperatures, sometimes called the depressed melting points, have to be found for samples with different API and polymer content. Typically, the physical mixture of API and polymer is heated until the equilibrium solubility is reached. Within the course of DSC scan the dissolution process is manifested by the presence of endothermic event. The temperature at which this event comes to the end (T_{end}) is considered as corresponding to the equilibrium condition. In the literature the several modifications as well as alternative approaches can be found. Tao and co-workers [87] proposed cryomilling of mixture before DSC analysis to accelerate the determination of dissolution endpoints. Sun and co-workers suggested that the uncertainty of T_{end} assessment may be reduced by measuring DSC scans with different heating rates (0.1–10 °C/min) to subsequently determine the T_{end} value at "zero" heating rate from extrapolation. Another method requires the prior annealing of physical mixture of polymer and crystalline API at a certain temperature for a period required to achieve the equilibrium, and then scan the material with a standard heating rate (e.g. 10 °C/min) searching for a residual dissolution endotherm indicating whether the equilibrium condition was reached or not. In this way the upper and lower limit of dissolution temperature can be found [88]. Mahieu and co-workers [89] proposed an entirely different approach of dissolution curves determination. This method relies on a fact that the kinetics of demixing of supersaturated API-polymer solution is much faster than the slow kinetic of mixing, thus is faster to access experimentally. The differences in the velocity of both processes arise from the differences in molecular mobility of polymer carrier which is faster in the supersaturated condition due to strong plasticization by API molecules present in excess in working solution. The opposite effect is observed in previously discussed methods when API molecules percolate into highly viscous polymer matrix enhancing its molecular mobility which, however, due to initial conditions i.e. very high T_g of polymer, is less efficient.

The dissolution kinetics for several API-polymer systems has been investigated, e.g. for felodipine-Soluplus and felodipine-HPMCAS [90], felodipine-PVP K15 [91], felodipine-PVP/VA64 [91], sulfathiazole-PVP and sulfadimidine-PVP [92], indomethacin with different grades of PVP [89, 93, 94].

An experimental attempts to accurately determine the phase-diagrams providing basic insight into crystallization probability as a function of composition and temperature are usually complemented by appropriate models able to predict the phase boundaries at conditions inaccessible experimentally. The most helpful are a Flory-Huggins theory, Gordon-Taylor and Couchman-Karasz equations.

Flory-Huggins Theory

The Flory-Huggins theoretical approach provides a basic understanding of mixing behavior of drug-polymer systems. According to the theory the Gibbs free energy of mixing of drug-polymer system can be expressed by means of Flory-Huggins (FH) interaction parameter χ [90]:

$$\Delta G_{mix} = RT\left(\phi_{drug}\ln\phi_{drug} + \frac{\phi_{poly}}{m}\ln\phi_{poly} + \chi\phi_{drug}\phi_{poly}\right) \tag{5.17}$$

where ϕ is the volume fraction (accepts values between 0 and 1), R is the gas constant, T denotes absolute temperature, and m reflects the ratio of polymer and drug volumes, i.e. $m = [M_w^{poly}\cdot(\rho_{poly}^{-1})]/[M_w^{drug}\cdot(\rho_{drug}^{-1})]$ where M_w *is* the molecular weight and ρ is the density of the polymer and drug, respectively [90]. The mixing ability will be driven by the interplay of enthalpic and entropic contributions. For a miscible system the value of $\Delta G_{mix} = \Delta H_{mix} - T\Delta S_{mix}$ must be negative and must have positive curvature over the whole concentration range (double minima will evidence poor mixing ability) [95]. The lack of miscibility between the drug and the polymer may occur in a systems with a very weak drug-polymer interactions or those with a strong tendency to self-association. In such case the strong cohesive interactions between the individual components (drug-drug or polymer-polymer interactions) may overcompensate the adhesive drug-polymer interactions leading ultimately to the poor miscibility of a system [91]. The character of API-polymer interactions will be reflected in the value of FH interaction parameter, χ, considered as an indicator of favorable (χ negative) or unfavorable (χ positive) mixing of drug and polymer additive. The value of χ can be calculated from DSC dissolution data acquired for physical API-polymer mixtures according to the following relation [90]:

$$\left(-\frac{\Delta H_m}{R}\right)\cdot\left(\frac{1}{T_m}-\frac{1}{T_m^0}\right)-\ln\phi_{drug}-\left(1-\frac{1}{m}\right)\phi_{poly}=\chi\phi_{poly}^2 \qquad (5.18)$$

where ΔH_m is the enthalpy of melting of a pure API, T_m and $T_m{}^0$ denote melting temperatures for API dissolved in polymer matrix and pure API, respectively. The slope of the graph obtained by plotting the left side of the above equation against ϕ_{poly}^2 will give us the value of χ. By measuring the dissolution endotherms for physical mixtures containing various drug-polymer ratios we can obtain a several χ values corresponding to different T_{end} values. To take into account the temperature dependence of FH interaction parameter, assumed by theory but sometimes overlooked, the following empirical equation can be used to describe the χ values plotted vs. $1/T_{end}$:

$$\chi = A + (B/T) \qquad (5.19)$$

where constants A and B describe the entropic and enthalpic contributions to χ. Such fitting procedure allows us finally to calculate the χ value at any temperature of interest [90, 91, 96].

Gordon-Taylor (G-T) and Couchman-Karasz (C-K) Equation

The value of T_g for amorphous solid dispersion can be predicted from the data characterizing the individual components. To this date the several models have been proposed to calculate the value of T_g for binary system (Fox [97], Jenkel and Heusch [98], Kwei [99] etc.). However, the most widely used are the Gordon-Taylor (G-T) and Couchman-Karasz (C-K) equations, both based on the same equation, but with differentially defined constant k [100, 101]:

$$T_{g,mix} = \frac{w_1 T_{g,1} + k w_2 T_{g,2}}{w_1 + k w_2} \qquad (5.20)$$

$$k = \frac{\rho_1 T_{g,1}}{\rho_2 T_{g.2}} \ \text{(Gordon-Taylor)} \qquad k = \frac{\Delta C_{p2}}{\Delta C_{p1}} \ \text{(Couchman-Karasz)} \qquad (5.21)$$

where w_1, w_2 are components weight ratio, $T_{g,1}$ and $T_{g,2}$ are their glass transition temperatures, ρ_1 and ρ_2 denotes their densities, and ΔC_{p1} and ΔC_{p2} refer to the changes in specific heat capacity for each component determined at T_g. Index "1" denotes the lower T_g component i.e. API, while index "2" refers to polymer additive [102]. The G-T model is based on the assumption of the ideal volume additivity [100]. The C-K model treats the glass transition as an Ehrenfest second-order transition and assumes the characteristic continuity of mixing parameters, i.e. entropy and volume, at T_g [101]. Besides, the change in the glass transition heat capacity increment is assumed to be temperature-independent [103]. Both equations will properly predict the values of glass transition temperatures if only the components, i.e. API and polymer, will not strongly interact with each other. Otherwise the experimental data and theoretical data will disagree. Depending on the strength of drug-polymer interactions either negative and positive deviations can be expected [102]. However, there are number of examples of solid dispersions in which the predicted values of T_g correspond well with experimental data points despite the presence of specific interactions between the components [73, 104, 105].

5.2.1.2 Slowing Down of API Molecular Mobility Due to Polymer Additive

It is well recognized that the stability of amorphous API dispersed in polymer matrix is associated with the slowdown of drug molecular motions imposed by the polymer content. The impeded molecular dynamics is reflected in a higher glass transition temperature of drug-polymer system in comparison to that observed for pure API. The increase in the glass transition temperature observed after mixing API (low T_g component) and polymer (high T_g component) is frequently described as *anti-plasticizing effect*. This term has been originally applied in the polymer industry to describe small molecules, known as plasticizers, incorporated into polymer in order to reduce the T_g and improve the overall flexibility of a material [106]. Transferring this into solid dispersion systems one can say that API acts as plasticizer lowering the T_g of a polymer while the drug molecule undergoes anti-plasticization. When we look at the relaxation map in Fig. 5.9 depicting the logarithm of structural relaxation times versus the inverse of temperature, the situation when the dynamics of the API-polymer system slows down corresponds to the shift of the curve to the left in relation to the position observed for the drug alone. It means that at certain temperature the value of structural relaxation time observed for solid dispersion is longer when compared to the pure drug. At the same time the temperature corresponding to the relaxation time being equal to $\tau_\alpha = 100$ s, identified with glass transition temperature, is higher in the case of API-polymer system. One can simply say that behind the dispersion of physically unstable API in highly viscous polymer matrix there is an idea to reduce the drug mobility enough to avoid recrystallization over a pharmaceutically relevant period of time (at least 2 years). Theoretically, if we assume that the crystallization rate is governed by the dynamics only, the recrystallization will not occur as long as the molecules do not move. Thus, extending the relaxation time to the order of years e.g. at room temperature would reduce the risk of drug recrystallization during storage at these conditions. Then the material response to external perturbations, which could cause recrystallization, will be longer than the expected life-time of the drug product. To illustrate this the example of bicalutamide-PVP system can be given. Figure 5.9 shows the relaxation map for pure bicalutamide and bicalutamide dispersed in PVP carrier. One can see that the small addition of polymer (10% w/w)

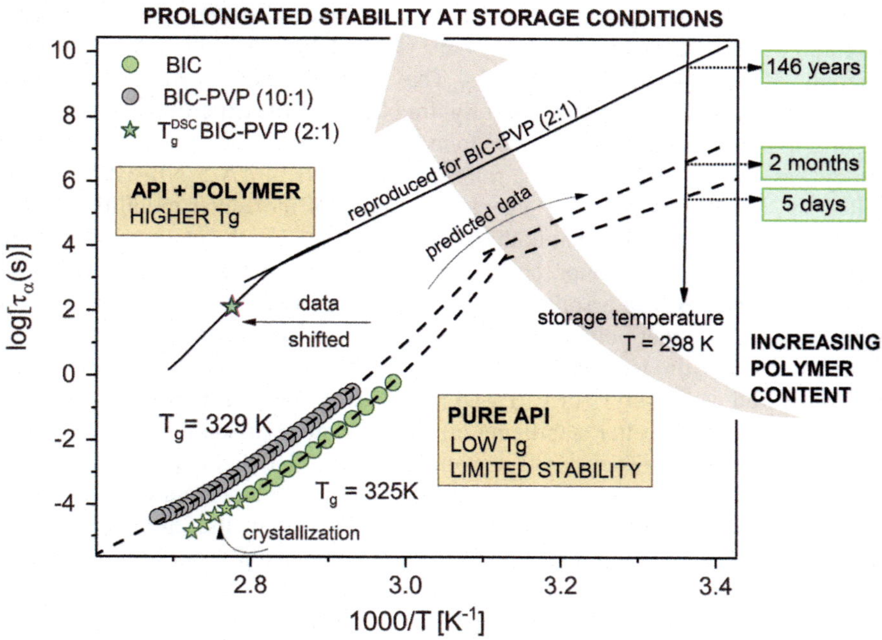

Fig. 5.9 The relaxation map of bicalutamide drug (BIC) and BIC-PVP solid dispersions containing various drug/polymer ratios illustrating slowing down of API molecular dynamics by polymer additive

extends the relaxation time observed at T = 298 K from the order of days to the order of months. The subsequent increase of polymer content led to the further spectacular increase of relaxation time value at T = 298 K which was assumed to exceed hundreds of years (value predicted from shifting the data from experimentally assessable range) [73].

5.2.1.3 Molecular Interactions Between the API and Polymer

There are many examples indicating that the observed differences in crystallization rates of pure drug and those dispersed in polymer carrier cannot be explained by taking into account only changes in T_g. The reason is that the physical stability of amorphous solid dispersion can be also enhanced by the presence of specific molecular interactions between the API and the polymer. The pattern and the strength of drug-polymer interactions can be established using the several techniques (the most common are infrared and Raman spectroscopy, solid state NMR technique). The presence of hydrogen bonds between the components was identified for several API-polymer systems, i.e. indomethacin-PVP [105, 107] nifedipine-PVP and felodipine-PVP [104], celecoxib-PVP [108], bicalutamide-PVP [73].

Kothari et al. [109] correlated the recrystallization tendency of amorphous nifedipine dispersed in various polymers (PVP, HPMCAS and PAA) with the strength of drug-polymer interactions. The strongest drug-polymer interactions were found in NIF-PVP system, contrary, for NIF-PAA the experimental indication of drug-polymer attractions was not found. Respectively, the crystallization tendency was found to decrease in the following order: NIF-PAA > NIF-HPMCAS > NIF-PVP. Besides, in NIF-PVP the slowing down of molecular mobility was the most significant indicating that both factors are engaged in the mechanism of drug stabilization. Xie and Taylor [110] reported that the physical stabilization of amorphous celecoxib in a solid dispersion was governed by these two factors as well. It was found that polymers PVP and PVP/VA were able to form stronger or more extensive hydrogen bonds with celecoxib than cellulose-based polymers. This ability as well as higher T_g of PVP and PVP/VA based systems were found to be responsible for their better stabilizing efficiency in comparison to cellulose derivatives [110]. For the dispersion of acetaminophen in PVP and PAA the similar T_g was observed. Despite of this fact the inhibition of crystallization was more effective in the case of PAA polymer what was explained by the stronger hydrogen bonds between the acetaminophen and PAA in comparison to those observed for PVP [111]. Kestur and Taylor [112] performed insightful studies of felodipine dispersed in several polymers, i.e. PVP K12, HPMCAS, poly(vinylpyrrolidone)/vinyl acetate (PVP/VA) and poly(vinyl acetate) (PVAc). The small additive of polymer (3% w/w) significantly hinders the crystal growth rates at $T > T_g$ with efficiency depending on the strength of drug-polymer interactions. An extend of hydrogen bonding between felodipine and investigated polymer can be ranked in the following order PVP K12 > PVP/VA > HPMCAS > PVAc which reflects the effectiveness of the crystal growth inhibition as well. Interestingly, no correlation between the T_g of investigated drug-polymer systems and observed crystal growth rates was found. Both T_g and T_m of drug-polymer systems were similar to those observed for pure felodipine suggesting that molecular mobility plays no role in the observed stability enhancement. Another interesting observation was related to the effect of temperature on crystal growth inhibition. At high temperatures the inhibition rate was smaller. It suggests that the stabilization *via* specific interaction may be the most effective at low temperatures. From molecular perspective it seems very reasonable since at low temperature the disruption of drug-polymer assembly is more difficult thus the formation of drug-drug dimers necessary to promote crystallization will be slower. The problem of the participation of different stabilization mechanisms at different temperatures was also investigated by Grzybowska and co-workers [113]. They showed that the overall crystallization rate of celecoxib in the supercooled liquid state was limited more efficiently by polymer PVP K30, while below T_g the small molecular weight excipient, octaacetylmaltose (acMAL), was found to be more efficient. At higher temperatures the stabilization *via* formation of heterodimeric drug-excipient structures seems to be less effective. Thus, the ability of PVP K30 to slow down the molecular mobility of celecoxib was a key in governing the crystallization rate. The acMAL with T_g similar to those observed for pure celecoxib does not influence its supercooled dynamics. Below T_g the role of strength of specific drug-excipient interaction increases. Authors speculated that then the stronger hydrogen bonds between celecoxib and acMAL in comparison to PVP K30 were responsible for better efficiency in crystallization inhibition by acMAL. To support this claim they discussed the differences in the secondary γ-relaxation behavior in celecoxib-acMAL and celecoxib-PVP K30 systems. The molecular origin of γ-process was attributed to the rotation of the phenyl ring with the sulfonamide group which may participate in hydrogen-bond formation between celecoxib and excipients. The small addition of acMAL (10% w/w) completely suppressed the γ-process indicated that all NH_2 groups of celecoxib participate in H-bond formation what prevents their further rotation. At the same time much higher amount of polymer additive was required to achieve the similar suppression effect what was interpreted as an indication of lower hydrogen-bonding potency in celecoxib-PVP K30 system. The above examples show that the intermolecular attractions between the drug and polymer leading to the presence of heterodimeric structures in the system constitute a serious impediment which slows down the progress of crystallization. Such attractions may impede the homodimers creation which already formed facilitate crystallization. As an example indomethacin (IMC) can be given as an API with strong tendency to self-association [107, 114]. Approximately 79% of the IMC molecules participate in at least one hydrogen bond, while about 40% participate in two hydrogen bonds [115]. Matsumoto and Zografi [116] suggested that the inhibition of carboxylic acid dimer formation due to the presence of hydrogen bonding interactions between indomethacin and polymers such as PVP or PVP-co-vinylacetate (PVAC) is responsible for the observed stability enhancement.

5.2.1.4 Steric Hindrance Imposed by the Polymer Additive

It is evident that the proper selection of polymer excipient will determine the overall features of drug-polymer system, in particular its stability. One can imagine the homogenous drug-polymer mixture as an entangled network formed by polymer chains in which free spaces are occupied by the drug molecules. The longer polymers will be more entangled than the short ones and their lower molecular mobility is expected. To initiate the crystallization process the drug molecules have to diffuse in polymer matrix to find each other and form the nucleus. Such process will be hindered by (1) high viscosity of a polymer matrix and (2) a spatial constrains associated with the arrangement of polymer in the space. There is a huge choice of polymers with a distinct molecular architecture (linear or branched), available in a variety of viscosity grades expressed by K values [117]. The most popular are those based on polyvinyl structure (e.g. PVP, PVPCA, Soluplus) or cellulose derivatives (e.g. HPC, HPMC, HPMCAS). Due to the adaptable carbon-carbon backbone the polyvinyl-based polymers are much more flexible in comparison to cellulose derivatives containing the rigid pyranosic ring in the structure [118]. The chemical structure of monomer influences the polymer ability to form closely and neatly packed structures and affects the free volume of the system

[119]. Li and co-workers compared the free volumes of solid dispersions prepared by spray drying technique using indomethacin and ketoconazole, and four different polymers (PVP, PVPVA, HPC, HPMCAS) [118]. They found that the rigidity of cellulose polymers is indeed reflected in their larger free volumes in comparison to those obtained for polyvinyl derivatives what clearly impact the topology of molecular packing in solid dispersions. Unfortunately, such investigations are rather rare. The general knowledge about the influence of certain molecular features of polymer carrier, such as its molecular weight, kind of side and end groups, addition of chain stiffening units, on stabilization mechanism is still lacking. The robust investigations of drug-polymer compositions with systematically varying parameters (like chain length, number of ring-based structures etc.) are necessary to probe these molecular factors independently. Such recognition is a key for a rational selection of polymer excipients for drug delivery applications.

5.2.2 Binary Drug-Drug Mixtures

The preparation of amorphous drug formulations containing two low molecular weight components, exactly two APIs, carefully selected to benefit from their dual therapeutic action, is a relatively new but dynamically developing stabilization approach. In principle such binary system can be considered as a low-molecular analogue of polymer-based amorphous solid dispersion, however, to avoid an ambiguity associated with nomenclature the term co-amorphous formulation proposed by Chieng et al. has been widely adopted to its description. The pharmacological requirements and the regulatory constrains make the selection of ingredients for drug-drug formulations a nontrivial task [120]. The main criterion is the beneficial therapeutic effect arising from the co-administration of both substances. In one of the pioneering work in this field Chieng and co-workers [121] proposed co-administration of amorphous indomethacin and ranitidine HCl. The application of ranitidine HCl was intended to reduce the gastrointestinal ailments commonly associated with the use of non-steroidal anti-inflammatory agents. This approach made a foundation for further research involving various combination of non-steroidal anti-inflammatory drugs and gastroprotective agents, such as naproxen-cimetidine [122], peroxicam-cimetidine [123] etc. Since then many other co-amorphous formulations have been investigated (the summary is presented in Table 5.3). Besides the true therapeutic justification for combined administration of two APIs, their ability to create stable amorphous systems is equally important. The stabilization mechanism in co-amorphous systems is generally related to the molecular constrains arising from their two-component nature. The anti-plasticizing effect in this case may be less important because the difference in the glass transition temperature among both components is usually not as spectacular as when polymer is used. However, when two APIs with different T_g are mixed, the resultant glass transition temperature value is composition dependent and may be predicted from the G-T or other equivalent equation. One can find examples indicating that the observed changes in T_g may be responsible for stability enhancements [124],

Table 5.3 The current state-of-art in the field of co-amorphous formulations

API-API combination	Justification of co-administration	Preparation method	Mechanism of stabilization	Reference
Indomethacin-cimetidine	Pain relief and treatment of gastrointestinal side effects of the NSAID	Solvent evaporation	Molecular interactions	Yammamutra et al. [125]
Indomethacin-cimetidine		Quench cooling, co-evaporation and ball milling	Molecular mobility	Lim et al. [129]
Naproxen-cimetidine		Ball milling	Molecular interactions	Alleso et al. [122]
Naproxen-cimetidine		Quench cooling, co-evaporation and ball milling	Molecular mobility	Lim et al. [129]
Indomethacin-ranitidine HCl		Ball milling	Molecular interactions	Chieng et al. [121]
Cimetidine-piroxicam		Solvent evaporation	Molecular interactions	Tantishaiyakul et al. [123]
Indomethacin-naproxen	Complementary painkillers	Quench cooling	Heterodimer formation	Löbmann et al. [126, 130]
Indomethacin-naproxen		Spray drying	NA	Beyer et al. [131]
Simvastatin-glipizide	Treatment of hypercholesterolemia and diabetes in metabolic disorder	Ball milling or cryomilling	Antiplasticization effect exerted by glipizide	Löbmann et al. [127]
Ezetimib-indapamid	The combined treatment of frequently coexisting hypertension and hypercholesterolemia	Quench cooling	Antiplasticization effect exerted by indapamide	Knapik et al. [128]
Atorvastatin-carvedilol Atorvastatin-glibenclamide	The beneficial efficiency due to application of statins in cardiovascular therapy and in management of type 2 diabetes mellitus in combination with anti-diabetic drugs	Solvent evaporation	Molecular interactions	Shayanfar et al. [132]

(continued)

Table 5.3 (continued)

API-API combination	Justification of co-administration	Preparation method	Mechanism of stabilization	Reference
Curcumin–artemisinin	Better antimalarial activity compared to the individual drugs	Solvent evaporation	NA	Suresh et al. [133]
Ritonavir-indomethacin	Antiretroviral medication and treatment of side-effects of ritonavir	Solvent evaporation	Antiplasticization effect exerted by ritonavir	Dengale et al. [134]
Nateglinide-metformin HCl	Synergistic therapy of oral hypoglycemia	Ball milling	Specific interactions, heterodimer formation	Wairkar et al. [135]
Ritonavir-quercetin	Reduced clearance of antiretroviral from body, anticipated improved therapeutic response	Solvent evaporation	One component in excess acts as crystallization inhibitor	Dengale et al. [136]
Talinolol-naringin	Naringin due to its glycoprotein P (P-gp) inhibiting potential may positively impacts talinalol bioavailability	Quench cooling	Molecular interactions	Teja et al. [137]
Glipizide-atorvastatin	Effective therapy for the comorbidities of hyperglycemia and hyperlipidemia	Cryomilling	Antiplasticization effect	Renuka et al. [138]
Omeprazole-amoxicillin	Bacterial infections treatment with gastric prevention	Co-grinding	Molecular interactions	Russo et al. [139]
Tranilast-diphenhydramine HCl	Concurrent treatment of allergy and inflammation	Quench cooling	Molecular interactions	Ueda et al. [140]

and counterexamples showing that composition with the highest T_g was not found as a most stable one [122]. In many co-amorphous formulations the molecular interactions between components appear to be of a paramount importance. Yamamura indicated that in amorphous precipitates of cimetidine and naproxen (20:80 mol%) the interactions between the imidazole ring of cimetidine and the COOH group of naproxen occur [125]. The molecular interactions were also found between the amide oxygen atom as well as the pyridyl nitrogen of piroxicam and H-N group of imidazole ring of cimetidine when they are coprecipitated from solvents in 1:1. mole ratio [123]. Alesso investigated co-amorphous formulations prepared by co-milling of highly unstable naproxen and cimetidine at various ratios. In this case both APIs have distinctly different T_g reported as $T_{g,1} = 6.2$ °C and $T_{g,2} = 32.1$ °C for naproxen and milled cimetidine, respectively. Thus, the cimetidine molecules exerted a strong anti-plasticization effect on naproxen as evidenced by T_g value growing with the cimetidine content. However, the formulation with the highest T_g was not found as the most stable against recrystallization. The highest stability was reported for formulation mixed in 1:1 ratio. It was explained by the favorable interaction pattern among mixture components in this particular molar concentration. Similar behavior was observed for naproxen-indomethacin [126] and indomethacin-ranitine HCl formulation [121]. Again, the highest stability was observed for 1:1 molar ratio. It suggests that compositions containing the excessive amount one of the drug are less stable due to crystallization of this part of molecules which does not participate in the formation of heterodimers. When all molecules are mixed in 1:1 ratio all of them are engaged in heterodimer formation and the stability of co-amorphous formulations seems to be improved. The further investigations have shown that the stabilization mechanism in co-amorphous formulation is a complex problem which cannot be explained only on the basis of interactions between components. In the simvastatin and glipizide the stability of co-amorphous formulation increased with the amount of glipizide in the mixture and was attributed to anti-plasticization effect exerting by higher T_g component i.e. glipizide [127]. Knapik and co-workers [128] attributed the improved physical stability of ezetimibe and indapamide mixtures to the slowing down of molecular dynamics. In this study the experimental indication for the presence of specific molecular interactions between both components was not found. On the other hand the significant differences in molecular dynamics for compositions with different drug ratio were found. Namely, (1) the glass transition temperature of co-amorphous system increased along with the increasing amount of indapamide content, while (2) the fragility of the system decrease with the indapamide content. These observation suggests that the physical stabilization of co-amorphous composition should increase with increasing content of indapamide. The XRD studies show that when the mixture is stored at 297 K the small amount of indapamide drug (8.8 wt%) is able to suppress the ezetimibe recrystallization for at least 72 days. For the therapeutic ratio (11.1–33.3 wt% of indapamide is required) the expected stability was assumed to be significantly higher [128].

5.2.3 Spatially Confined Systems

The nanoscale confinement by exerting impact on surface-to-volume ratio will strongly impose the phase behavior of a material embedded in a nanoporous carrier [141]. According to the classical nucleation theory the nucleation behavior of a material is affected by the competitive balance between the interfacial free energy and the difference in free energy between the liquid and the crystalline form. By confining in nanopores the altered interplay among these surface and volume contributions guarantees that the physical properties will differ from those observed for the bulk form. Indeed, it has been demonstrated that the carriers imposing geometrical constrains on the amorphous content can efficiently hinder its recrystallization. This stabilization effect combined with observed significant enhancement of dissolution properties make this approach truly valuable for pharmaceutical sector.

When applied pore size is small enough to exclude the possibility of critical nucleus formation the crystal growth will not occur. The crystallization will be completely suppressed when the pore diameter will not exceed the critical size, d^*, defined as follows:

$$d^* = 4\gamma T_m / [(T_m - T)\Delta H_m \rho_c] \quad (5.22)$$

where γ is the surface free energy (tension) of crystal-liquid interface, T_m and ΔH_m are the melting temperature and the enthalpy of fusion for bulk sample, and ρ_c is the density of a crystal.

Knapik et al. [142] determined the value of critical diameter for amorphous ezetimibe (see Fig. 5.10). The calculated value was further used to explain the differences in the crystallization behavior of amorphous ezetimibe embedded in two mesoporous systems Aeroperl 300 (pore diameter 30 nm) and Neusilin US2 (pore diameter 5 nm). The combined DSC, BDS and XRD study clearly indicated that the drug recrystallization was faster in Aeroperl 300. The calculated critical diameter was comparable with Neusilin US2 diameter and much lower than the pore size of Aeroperl 300. It implicates that in the latter there might be enough space to form nucleus effortlessly and trigger the crystal growth more efficiently. The better understanding of observed enhanced physical stability in Neusilin US2 host can be provided by molecular dynamic studies. In both mesoporous systems Neusilin US2 and Aeroperl 300 the spatial confinement impacted the ezetimibe dynamics. However, the changes were much better pronounced for Neusilin US2 based system. The following observations were reported: (1) the glass transition temperature increased with the pore diameter, (2) in confined systems the distribution of the α-relaxation times was broader (the larger pore size corresponds to broader α-peak), (3) above T_g besides structural α-relaxation an additional slower α′-process was found attributed to the fraction of molecules immobilized by pore walls. Finally, it has been concluded that the improved stability of ezetimibe-Neusilin US2 system results from: changes in molecular dynamics between the

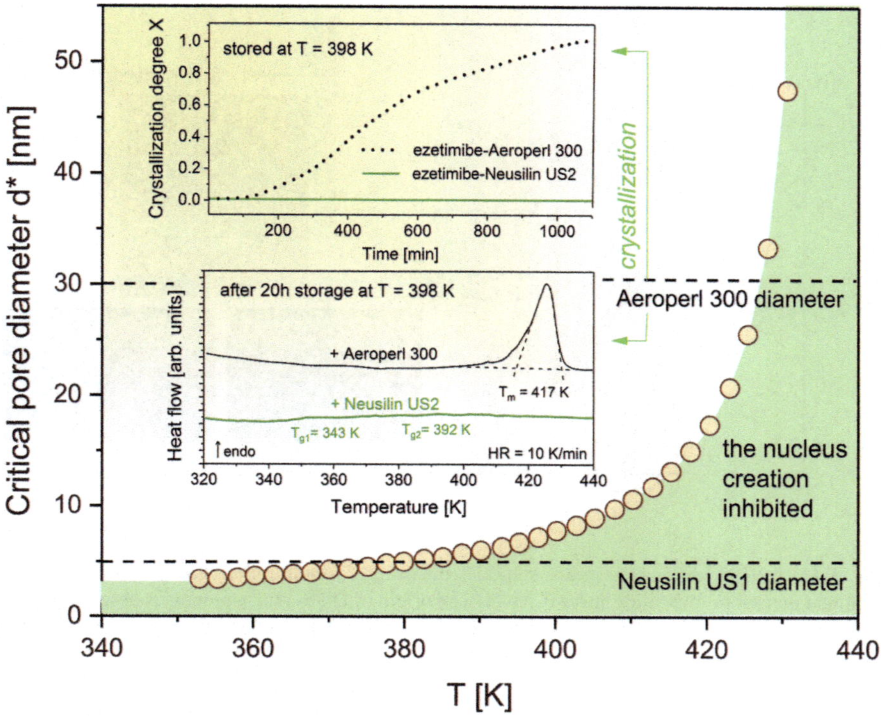

Fig. 5.10 The critical pore diameter d* as a function of temperature calculated for amorphous ezetimibe. Upper inset shows the progress of crystallization of ezetimibe in Aeroperl 300 and Neusilin US2 during storage at T = 398 K. The bottom inset shows DSC thermograms registered for both systems after t = 20 h of storage at T = 398 K. The presence of melting endotherm indicates that ezetemibe in Aeroperl 300 underwent crystallization. Data taken from [142]

bulk and confined API, the immobilization effect exerted by pore surface and the application of mesoporous carrier with pores smaller than the critical pore diameter.

Another insightful study was performed by Brás and co-workers [143] for ibuprofen confined in two mesoporous silica systems i.e. MCM-41 (pore diameter of 3.6 nm) and SBA-15 (pore diameter of 8.6 nm). The relevance of this approach in the stabilization of amorphous content was verifying by Shen and co-workers [144]. The amorphous ibuprofen trapped in the confined space limited by SBA-15 remained stable and did not convert to the crystalline form for 12 months of storage at 40 °C/75% RH. Under geometrical confinement amorphous ibuprofen revealed the complex relaxation behavior deviating from those observed for bulk counterpart (see Fig. 5.11). In comparison to the bulk sample the glass transition temperature decreased under confinement in a manner dependent on the pore size i.e. the smaller pore size corresponds to the lower T_g. In the samples based on mesoporous carriers three relaxation processes were observed in the supercooled liquid region i.e. (1) the structural α-relaxation process attributed to the motions of molecules near the pore center, (2) S-process attributed to the hindered mobility of molecules located in the

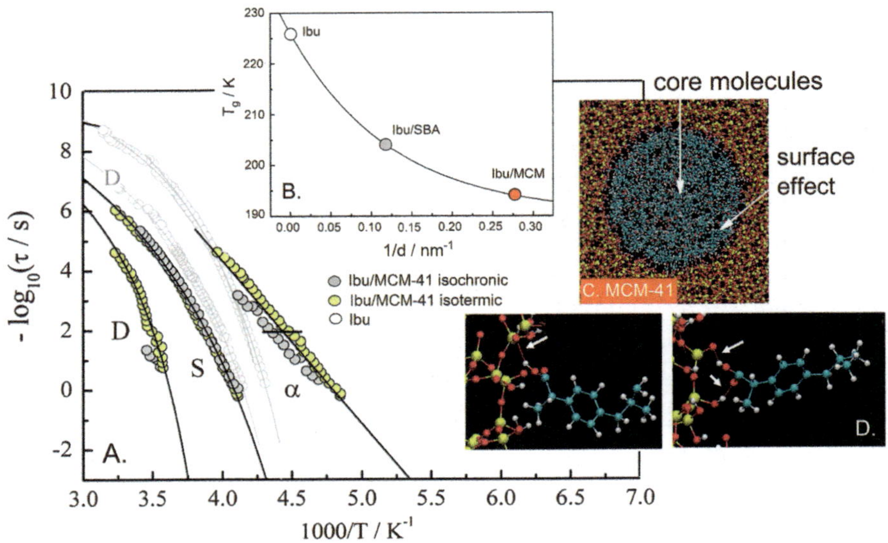

Fig. 5.11 (**a**) Relaxation times versus 1000/T measured for bulk ibuprofen (Ibu) and ibuprofen confined in mesoporous silica host MCM-41. The origin of D-, S- and α-process is described in the text. (**b**) Glass transition temperature determined from BDS studies plotted against the inverse of the pore diameter for ibuprofen in 3.6 (MCM) and 8.6 (SBA) nm pores. (**c**) The picture obtained from MD simulations (top view) performed for pore mimicking MCM-41 completely filled with ibuprofen molecules. Two distinct population of molecules can be distinguished (core molecules and those in interfacial layer). (**d**) The pattern of hydrogen bonds between the carboxylic group of ibuprofen with the silanol groups of pore carrier. Adapted with permission from Bras AR, Fonseca IM, Dionisio M, Schonhals A, Affouard F, Correia NT, Influence of Nanoscale Confinement on the Molecular Mobility of Ibuprofen. *J Phys Chem C*. 2014;118:13857–13868. Copyright 2014 American Chemical Society

vicinity of pore surface and weakly interacting with the silanol groups of the pore walls, (3) D-process attributed to the presence of hydrogen-bonded structures which impact the cis-trans conversion of the O = C–O–H group. The proper assignment of observed relaxation modes was supported by MD simulations performed for ibuprofen trapped inside the cylindrical pore of 3.6 nm diameter analogical to the investigated MCM-41 [143]. MD simulation methods can make a valuable contribution to the understanding of dynamic-stability relationship prevailing in the spatially confined systems. Yani and co-workers [145] investigated the effect of mesoporous carrier MCM-41 on the stability of amorphous ibuprofen using MD simulations. They found that the hydrogen-bonded network created by ibuprofen molecules and molecules at the surface of mesoporous carrier is crucial for stabilization because precluded the formation of drug-drug self-assemblies which promote recrystallization. This effect was correlated with the pore size. The space provided by pores with diameter larger than 20 nm was sufficient to make the strong drug-drug interactions possible and favorable leading to the less effective protection against recrystallization.

There is a plenty of reports concerning changes in molecular dynamics and/or crystallization behavior in various systems subjected to geometrical confinement and a few concerning pharmaceutical relevant materials (e.g. naproxen [146], ezetimibe [142], ibuprofen [143], fenofibrate [147]). What can be misleading, at least at the beginning, is the character of changes in T_g observed in confined materials with respect to the bulk samples (reported T_gs are lower, higher or even almost unaffected in comparison to the bulk). These changes seem to be highly material dependent. There are excellent reviews available for readers interested in this issue [148, 149]. It is widely accepted that for certain material the molecular mobility in confined conditions is governed by two contributions (1) the finite-size effect and (2) the surface contribution [150]. It is well established that in the geometrically constrained systems the character of molecular motions is strongly affected by the distance from the confining walls. Additionally, the nature of interactions between confined molecules and confining surface is equally important. Due to the surface effect two dynamically distinct population of molecules can be distinguished, first located in the core (its dynamics is accelerated with respect to the bulk conditions) and second located close to the surface (with suppressed dynamics). Thus, in a given material we will have groups of molecules corresponding to the various T_g values which will depend on a pore diameter as well. Such a picture emerged from investigations performed for confined pharmaceuticals and was confirmed by MD simulations. However, from pharmaceutical perspective the answer for the question about how it translates into crystallization inhibition is the most relevant. The prolonged stability of confined API seems to be related to: (1) molecular interactions between the molecules of confined drug and the confining surface which impedes mobility and/or suppresses drug-drug interactions, (2) the geometrical constrains excluding or hindering the nucleus formation (Fig. 5.12).

The spatial confinement of API for drug delivery purpose can be realized by the application of various silica- and silicon-based mesoporous materials. The templated silica-based materials with the uniform pore channel structure, such as MCM-41 and SBA-15, are the most commonly used. Besides, materials with randomly oriented pores are used, i.e. colloidal silicon dioxide (Aerosil®), amorphous silica gel (Syloid®), granulated silicon dioxide (Aeroperl®), magnesium alumino metasilicates (Neusilin®) and calcium silicate (Florite®).

There are several methods of API incorporation inside the porous matrix, i.e. immersion, incipient wetness impregnation, melt method [151], co-spray drying [144], solvent evaporation [142], co-grinding [152]. The efficiency of drug loading procedure, as well as drug release, depends on the character of drug-surface interactions. The native surface of mesoporous carrier is dominated by the presence of OH groups allowing for hydrogen-bonds formation. The other specific interactions (electrostatic, hydrophobic) are possible only after suitable chemical modification of the carrier surface [151]. It was shown for ibuprofen that the functionalization of mesoporous Al_2O_3 surface with hydrophobic groups Si-CH_3 (so-called silanization) resulted in the lower degree of drug loading (~20%) and faster release rate (85% over a period of 5 h). On the other hand, the application of hydrophilic groups improved drug loading capacity (up to 45%) and led to slower release rates (12–40% over a

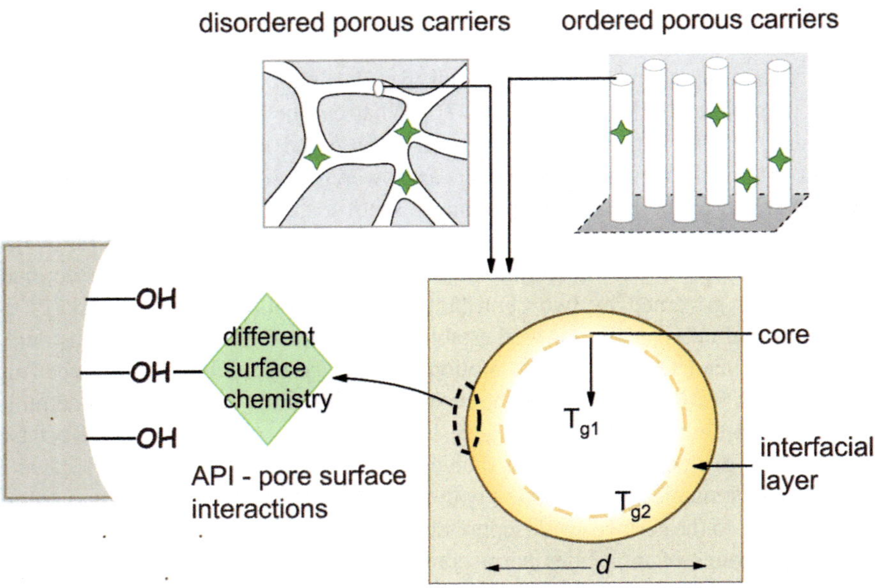

Fig. 5.12 Schematic outline of factors contributing to enhanced stability of API embedded in mesoporous carriers. The importance of pore structure, its diameter, surface chemistry, and complex mobility of confined API is indicated

period of 5 h). The highest drug loading was found for carriers with (3-aminopropyl)-triethoxysilane (APTES) modification what was attributed to the electrostatic interaction between NH_2 group of carrier surface and the COOH of ibuprofen [153]. The surface chemistry impacts the drug release rates but the pore size and its architecture are crucial as well [151]. It has been reported that the drug release rate increases with the pore size of itraconaloze in SBA-15 [154]. The dissolution rate of ibuprofen in 3D random foam-like pore network of mesoporous silica TUD-1 is faster in comparison to ordered channels of MCM-41 [155]. SBA-16 microspheres with the largest pore size of 9.0 nm and highly open and accessible pore network exhibited the fast drug release profile of poorly water-soluble indomethacin [156].

In summary, the huge advantage of mesoporous drug delivery systems arises from the diversity of available carriers which properties in a relatively easy way can be adapted to the needs of the drug being transported. The appropriate balance provided between the stability of amorphous content and the drug release ability will result in a significant improvement of drug water solubility. Besides this tunable approach allows to create amorphous drug-delivery system with modulated release.

5.3 The Stability of Amorphous Drugs Under High Pressure Condition

The procedure of screening of optimal amorphous drugs compositions is a challenging task. Doing this we have to bear in mind that during manufacturing procedures the drug is exposed to mutable external conditions such as heat, mechanical stress, and the effect of moisture [157]. All these factors might reinforce material properties and, what is crucial for amorphous pharmaceuticals, may promote recrystallization. Unintended process-induced recrystallization may adversely impact the performance of a final drug product, its dissolution properties, bioavailability etc. Thus, the potential risk of such adverse effects should be assessed at the initial stage of drug development. The investigations of API properties and drug compositions with stabilizing excipients at elevated temperature or humidity are quite common, contrary, investigations under elevated pressures seem to be much more demanding mainly due to experimental difficulties related to the high-pressure measurements. This obstacle can be eliminated by BDS spectroscopy application. Later, we will discuss how to take advantage of this method during assessment of drug properties under elevated pressures.

One can find examples of amorphous APIs with crystallization behavior that may change remarkably depending on T–p conditions. Such material appears to be stable against crystallization at ambient pressure but promptly recrystallizes after compression. As a first example etoricoxib can be given [49]. To verify the impact of compression on the crystallization behavior of etoricoxib we performed time-dependent dielectric measurements at $T = 353$ K and $p = 20$ MPa (see Fig. 5.13 right panel). At ambient pressure the drug recrystallization was not detected even at higher temperatures but after compression the crystallization process proceeded rapidly. After 26 h the drug was completely crystalline as confirmed by DSC and XRD studies. For comparison at ambient pressure at $T = 368$ K after 26 h the beginning of crystallization was still not detected (see Fig. 5.13 left panel). To understand such behavior the insight into molecular properties of etoricoxib have to be made. Supercooled etoricoxib exists as a mixture of two tautomers interacting with each other via hydrogen-bonds. We postulated that such interactions might be responsible for drug stabilization. It seems highly reasonable that reinforcing of interactions between isomers and/or variations in tautomers ratio due to compression may be responsible for observed change in its crystallization behavior under the pressure [49].

As a second example probucol can be given. The dielectric studies revealed that supercooled probucol remained amorphous when kept at $T = 333$ K and $p_1 = 0.1$ MPa but crystallized immediately after increasing the pressure to $p_2 = 10$ MPa. The procedures applied during drug fabrication do not assume keeping the material for a long time under elevated pressure. However, the example of probucol showed that even fast compression and subsequent decompression may promote crystallization. Depending on the dwelling time the crystallization rate was found to be faster when pressure was exerted in a continues manner (at $T = 333$ K 1 h was required for

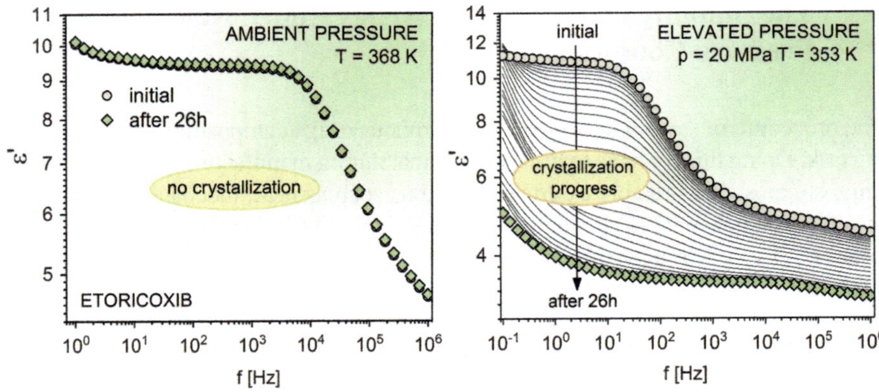

Fig. 5.13 Dielectric spectra ε′(f) registered in a time-dependent manner during keeping the sample at different T-p conditions: T = 353 K, p = 20 MPa (right panel) and T = 368 K, p = 0.1 MPa (left panel). The drop of dielectric response illustrate the progress of crystallization

complete crystallization) or slower for immediately decompressed sample (at T = 333 K 8 h were required for complete crystallization). Such prompt crystallization of API under the pressure was attributed to the changes in intermolecular distances which facilitate the homodimers formation triggering the crystallization. It was evidenced by the fact that introducing steric hindrance in a form of second API efficiently prevent pressure induced recrystallization of probucol.

To anticipate and prevent process-induced crystallization it is of particular importance to have in-depth understanding of drug and excipients properties at possible processing conditions. However, this is not an easy task. Thakral and co-workers [158] reported that the recrystallization of amorphous content in a tablet may reveal spatial heterogeneity. The study of crystallization profiles performed by two-dimensional X-ray diffractometry indicated that crystallization rate of indomethacin located in the core was an increasing function of applied pressure. Contrary, the crystallization of API at radial surface was found to be faster at lower compacting pressures what was attributed to a higher die wall friction at low-pressure conditions. Such effect should be reduced by application of soft excipients that are intended to deform preferentially during powder compaction. However, it was reported that the crystallization of indomethacin in the core was unaffected by applied magnesium stearate. The crystallization at radial surface was then limited but the observed effect was transient and after 7–14 days (depending on the type of lubrication) the crystallization degree of covered and uncovered tablets was the same.

The problem of pressure-induced crystallization is related to binary systems as well. Pressure affects the interatomic distances and may impact the strength of intermolecular attractions. It implies a risk of changes in stability of binary systems in which the intermolecular attractions are responsible for the inhibition of crystallization. Ayenew and co-workers [159] raised the important issue of pressure-induced phase separation in drug-polymer systems. They pointed out that at high pressure (from $p = 565.1$ MPa) the phase separation of amorphous naproxen-PVP K25 occurred as evidenced by the presence of two T_gs and changes in IR spectral features. The observed effect was relevant only for samples with high amount of drug (30% and 40% w/w). It was attributed to the modification of hydrogen bonds between the carbonyl group of PVP and the COOH group in naproxen due to conformational changes of PVP upon compression. They pointed out that PVP molecule can adopt two distinct conformations conditioned by the steric hindrance exerting by the planar pyrrolidone side groups [160]. Pressure-induced transition between these two rotational states may impact the PVP hydrogen-bonding ability what will be reflected in the stabilization efficiency. Such pressure-induced phase separation was also reported for itraconazole-Soluplus dispersions [161], while the opposite effect i.e. improved mixing was observed for compressed miconazole-PVP VA64 system. This systems, contrary to the precious ones, is deprived of hydrogen bonds and at ambient pressure has tendency to inhomogeneous mixing and phase-separation [162, 163]. The varied nature of effects that pressure can exert on the compressed material can be illustrated by studies of Worku and co-workers [164]. They observed that compression clearly improved drug-polymer interactions in naproxen-PVPVA64 system containing 30% w/w of naproxen. Besides, they showed in the long-term XRD stability studies that compressed samples were more stable against crystallization.

From this discussion it appears clear that pressure may influence API behavior in a highly diverse way. To recognize and understand the properties of drug formulations at elevated pressure the broadband dielectric spectroscopy can be applied. Using BDS we can monitor the response of material to the applied mechanical stress in a real time manner. The time required for pressure stabilization, at least within the pharmaceutical relevant range of pressures, does not exceed a few minutes. Besides, within the course of dielectric experiments various parameters important from tableting viewpoint, like compression or decompression rate, dwell time can be independently adjusted. BDS technique offers the unique opportunity to mimic conditions prevailing during tableting process. In such way we can gain valuable information about the outcome of compression at precisely controlled laboratory conditions. In general, during dielectric experiments we are measuring the fluctuations of dipoles in the external electric field. Readers interested in detailed recognition of dielectric measurements methodology should refer to the excellent surveys presented elsewhere [31, 77]. The pressure-induced drug recrystallization will strongly affect dielectric response decreasing the number of mobile dipoles in the processing material. Therefore, even a small fraction of crystallites will be reflected in the dielectric response. The subsequent crystallization progress will be manifested as a progressive drop of dielectric signal. The changes in dielectric permittivity can be monitored as a function of frequency, time, temperature or pressure. In the first case the dielectric loss or dielectric dispersion spectra are detected giving valuable insight into changes in molecular dynamics of investigated material in $T - p$ space. The time dependent dielectric measurements give us possibility to recognize the timescale of crystallization and resolve its kinetics at different T and p conditions. The last option includes the measurements of dielectric response at fixed frequency as a function of temperature (or pressure). Such measurements may be performed at different heating/cooling (or compression/decompression) rates giving information about the principal features of glass forming liquids concerning glass transition, crystallization or melting events. For instance, using this method we can find the limiting pressure value after which the crystallization will take place, or we can verify how long the material can be kept at particular T – p condition without the risk of crystallization. The experimental parameters can be easily modified giving possibility to verify the impact of events following each other during manufacturing cycle. For example if compression promotes the nucleation only, the crystal growth may be delayed in time and occur under favorable condition during drug storage. The properly design experiment including compression and subsequent temperature scan may help to resolve such issue during the initial drug development phase. As mentioned the dielectric BDS spectroscopy offers possibility to perform crystallization experiments at precisely controlled thermodynamic conditions. Its application is helpful during drug characterization under manufacturing conditions but also has a key importance for gaining the fundamental understanding of crystallization process. The concept of dielectric measurements at various *iso*-variants was extensively studied by Adrjanowicz and co-workers [2, 165]. It has been pointed out that performing dielectric time-dependent studies at various conditions, i.e. isothermal (T = const.), isobaric (p = const.), isochronal (τ_α = const.) one can comprehensively study the crystallization phenomenon and resolve the long-standing issue of contribution of particular factors (kinetic vs. thermodynamic) in crystallization control. For example, we can perform dielectric measurements at different T and p conditions selected in a way providing the constant value of structural relaxation times (so-called isochronal conditions). Then, keeping the impact of molecular mobility at constant level we can recognize the role of thermodynamic factor. Making some additional experiments we can further extend the studies about measurements at isochoric conditions (V = const.) or that ensuring the same contribution coming from the thermodynamic driving force (Δμ = const.).

References

1. Ruckenstein E, Djikaev YS (2005) Recent developments in the kinetic theory of nucleation. Adv Colloid Interface Sci 118(1–3):51–72. https://doi.org/10.1016/j.cis.2005.06.001
2. Adrjanowicz K, Koperwas K, Szklarz G, Tarnacka M, Paluch M (2016) Exploring the crystallization tendency of glass-forming liquid Indomethacin in the T – p plane by finding different iso-invariant points. Cryst Growth Des 16(12):7000–7010. https://doi.org/10.1021/acs.cgd.6b01215
3. Jungblut S, Dellago C (2016) Pathways to self-organization: crystallization via nucleation and growth. Eur Phys J E 39(8):1–38. https://doi.org/10.1140/epje/i2016-16077-6
4. Sosso GC, Chen J, Cox SJ et al (2016) Crystal nucleation in liquids: open questions and future challenges in molecular dynamics simulations. Chem Rev 116(12):7078–7116. https://doi.org/10.1021/acs.chemrev.5b00744
5. Garcıa-Ruiz JM (2003) Nucleation of protein crystals. J Sruct Biol 142:22–31. https://doi.org/10.1016/S1047-8477(03)00035-2

6. Gerges J, Affouard F (2015) Predictive calculation of the crystallization tendency of model pharmaceuticals in the supercooled state from molecular dynamics simulations. J Phys Chem B 119(33):10768–10783. https://doi.org/10.1021/acs.jpcb.5b05557
7. Kelton KF (1991) Crystal nucleation in liquids and glasses. Solid State Phys 45:75–177. https://doi.org/10.1016/S0081-1947(08)60144-7
8. Kirkpatrick RJ (1975) Crystal growth from the melt: a review. Am Miner 60:798–814
9. Schmelzer JWP (2008) Crystal nucleation and growth in glass-forming melts: experiment and theory. J Non Cryst Solids 354(2–9):269–278. https://doi.org/10.1016/j.jnoncrysol.2007.06.103
10. Trasi NS, Baird JA, Kestur US, Taylor LS (2014) Factors influencing crystal growth rates from undercooled liquids of pharmaceutical compounds. J Phys Chem B 118:9974–9982
11. Nascimento MLF, Zanotto ED (2010) Does viscosity describe the kinetic barrier for crystal growth from the liquidus to the glass transition? J Chem Phys 133(17):174701–174701. https://doi.org/10.1063/1.3490793
12. Andronis V, Zografi G (2000) Crystal nucleation and growth of indomethacin polymorphs from the amorphous state. J Non Cryst Solids 271(3):236–248. https://doi.org/10.1016/S0022-3093(00)00107-1
13. Trasi NS, Taylor LS (2014) Nucleation and crystal growth of amorphous nilutamide – unusual low temperature behavior. CrystEngComm 16(31):7186–7195. https://doi.org/10.1039/c4ce00118d
14. Trasi NS, Taylor LS (2012) Effect of polymers on nucleation and crystal growth of amorphous acetaminophen. CrystEngComm 14(16):5188–5197. https://doi.org/10.1039/c2ce25374g
15. Shah N, Sandhu H, Choi DS, Chokshi H, Malick AW (2014) Amorphous solid dispersions: theory and practice. Springer, New York
16. Graeser KA, Patterson JE, Zeitler JA, Gordon KC, Rades T (2009) Correlating thermodynamic and kinetic parameters with amorphous stability. Eur J Pharm Sci 37(3–4):492–498. https://doi.org/10.1016/j.ejps.2009.04.005
17. Qiu Y, Chen Y, Zhang GGZ, Liu L, Porter WR (2009) Developing solid oral dosage forms: pharmaceutical theory and practice. Elsevier, Amsterdam
18. James PF (1985) Kinetics of crystal nucleation in silicate glasses. J Non Cryst Solids 73:517–540
19. Schmelzer JWP, Abyzov AS, Fokin VM (2016) Thermodynamic aspects of pressure-induced crystallization: kauzmann pressure. Int J Appl Glas Sci 7(4):474–485. https://doi.org/10.1111/ijag.12251
20. Bai XM, Li M (2006) Calculation of solid-liquid interfacial free energy: a classical nucleation theory based approach. J Chem Phys 124(124707):1–12. https://doi.org/10.1063/1.2184315
21. Pirzadeh P, Beaudoin EN, Kusalik PG (2012) Interfacial free energy: an entropy portent to energy changes. Cryst Growth Des 12(1):124–128. https://doi.org/10.1021/cg200861e
22. Pfister C-E (2009) Interface free energy or surface tension: definition and basic properties. arXiv:0911.5232v1 [cond-mat.stat-mech] 1–20
23. Wu DT, Gránásy L, Spaepen F (2004) Nucleation and the solid–liquid interfacial free energy. MRS Bull 29(12):945–950. https://doi.org/10.1557/mrs2004.265
24. Jones DRH (1974) Review. The free energies of solid-liquid interfaces. J Mater Sci 9:1–17
25. Laird BB, Davidchack RL (2005) Direct calculation of the crystal – melt interfacial free energy via molecular dynamics. J Phys Chem B 109:17802–17812
26. Schmelzer JWP, Abyzov AS (2016) Crystallization of glass-forming liquids: specific surface energy. J Chem Phys 145(064512):1–10. https://doi.org/10.1063/1.4960342
27. Fokin VM, Zanotto ED, Yuritsyn NS, Schmelzer JWP (2006) Homogeneous crystal nucleation in silicate glasses: a 40 years perspective. J Non Cryst Solids 352(26–27):2681–2714. https://doi.org/10.1016/j.jnoncrysol.2006.02.074
28. Cheng B, Tribello GA, Ceriotti M (2015) Solid-liquid interfacial free energy out of equilibrium. Phys Rev B 92(180102):1–5. https://doi.org/10.1103/PhysRevB.92.180102
29. Tarjus G, Kivelson D (1995) Breakdown of the Stokes–Einstein relation in supercooled liquids. J Chem Phys 103(8):3071. https://doi.org/10.1063/1.470495

30. Ngai KL, Magill JH, Plazek DJ (2000) Flow, diffusion and crystallization of supercooled liquids: revisited. J Chem Phys 112(4):1887–1892. https://doi.org/10.1063/1.480752
31. Grzybowska K, Capaccioli S, Paluch M (2016) Recent developments in the experimental investigations of relaxations in pharmaceuticals by dielectric techniques at ambient and elevated pressure. Adv Drug Deliv Rev 100:158–182. https://doi.org/10.1016/j.addr.2015.12.008
32. Becker SR, Poole PH, Starr FW (2006) Fractional Stokes-Einstein and Debye-Stokes-Einstein relations in a network forming liquid. 97(055901):1–4. doi:https://doi.org/10.1103/PhysRevLett.97.055901
33. Zheng Q, Mauro JC (2017) Viscosity of glass-forming systems. J Am Ceram Soc 100:6–25. https://doi.org/10.1111/jace.14678
34. Bohmer R, Ngai KL, Angell CA, Plazek DJ (1993) Nonexponential relaxations in strong and fragile glass formers. J Chem Phys 99(5):4201. https://doi.org/10.1063/1.466117
35. Ediger MD, Angell CA, Nagel SR (1996) Supercooled liquids and glasses. J Phys Chem 100 (31):13200–13212
36. Vogel H (1921) Das temperaturabhangigkeitgesetz der viskosität von flüssigkeiten. J Phys Z 22:645–646
37. Fulcher GS (1925) Analysis of recent measurements of the viscosity of glasses. J Am Ceram Soc 8:339–355
38. Tammann G, Hesse W (1926) Die abhängigkeit der viscosität von der temperatur bie unterkühlten flüssigkeiten. Z Anorg Allg Chem 156:245–257
39. Hancock BC, Zografi G (1997) Characteristics and significance of the amorphous state in pharmaceutical systems. J Pharm Sci 86(1):1–12. https://doi.org/10.1021/js9601896
40. Adam G, Gibbs JH (1965) On the temperature dependence of cooperative relaxation properties in glass-forming liquids. J Chem Phys 43(1):139–146. https://doi.org/10.1063/1.1696442
41. Johari GP (2000) Contributions to the entropy of a glass and liquid, and the dielectric relaxation time. J Chem Phys 112:7518. https://doi.org/10.1063/1.481349
42. Masiewicz E, Grzybowski A, Grzybowska K, Pawlus S, Pionteck J, Paluch M (2015) Adam-Gibbs model in the density scaling regime and its implications for the configurational entropy scaling. Sci Rep 5(13998):1–13. https://doi.org/10.1038/srep13998
43. Johari GP (2002) Localized molecular motions of β-relaxation and its energy landscape. J Non Cryst Solids 307–310:317–325
44. Vogel M, Rossler E (2000) On the nature of slow β-process in simple glass formers: a 2H NMR Study. J Phys Chem B 104:4285–4287
45. Ngai KL, Paluch M (2010) Classification of secondary relaxation in glass-formers based on dynamic properties. J Chem Phys 120(2):857–873. https://doi.org/10.1063/1.1630295
46. Ngai KL, Capaccioli S (2008) Impact of the application of pressure on the fundamental understanding of glass. J Phys Condens Matter 20:244101. https://doi.org/10.1088/0953-8984/20/24/244101
47. Capaccioli S, Prevosto D, Lucchesi M, Rolla PA, Casalini R, Ngai KL (2005) Identifying the genuine Johari-Goldstein β-relaxation by cooling, compressing, and aging small molecular glass-formers. J Non Cryst Solids 351:2643–2651. https://doi.org/10.1016/j.jnoncrysol.2005.03.071
48. Koperwas K, Adrjanowicz K, Wojnarowska Z, Jedrzejowska A, Knapik J, Paluch M (2016) Glass-forming tendency of molecular liquids and the strength of the intermolecular attractions. Sci Rep 6:36934. https://doi.org/10.1038/srep36934
49. Rams-Baron M, Wojnarowska Z, Grzybowska K et al (2015) Toward a better understanding of the physical stability of amorphous anti-inflammatory agents: the roles of molecular mobility and molecular interaction patterns. Mol Pharm 12(10):3628–3638. https://doi.org/10.1021/acs.molpharmaceut.5b00351
50. Hancock BC, Zografi G (1994) The relationship between the glass transition temperature and the water content of amorphous pharmaceutical solids. Pharm Res 11(4):471–477

51. Konno H, Taylor LS (2008) Ability of different polymers to inhibit the crystallization of amorphous felodipine in the presence of moisture. Pharm Res 25(4):969–978. https://doi.org/10.1007/s11095-007-9331-3
52. Rumondor ACF, Marsac PJ, Stanford LA, Taylor LS (2009) Phase behavior of poly (vinylpyrrolidone) containing amorphous solid dispersions in the presence of moisture. Mol Pharm 6(5):1492–1505
53. Avrami M (1940) Kinetics of phase change. II – Transformation time relations for random distribution of nuclei. J Chem Phys 8:212–224. https://doi.org/10.1063/1.1750631
54. Avrami M (1939) Kinetics of phase change. I. General theory. J Chem Phys 7:1103–1112. https://doi.org/10.1063/1.1750380
55. Johnson W, Mehl R (1939) Reaction kinetics in processes of nucleation and growth. Trans AIME 135:416–442
56. Kolmogorov A (1937) A statistical theory for the crystallization of metals. Izv Acad Sci USSR, Ser Math 1:355–359
57. Napolitano S, Wübbenhorst M (2007) Monitoring the cold crystallization of poly(3-hydroxy butyrate) via dielectric spectroscopy. J Non Cryst Solids 353(47–51):4357–4361. https://doi.org/10.1016/j.jnoncrysol.2007.01.082
58. Avramov I, Avramova K, Rüssel C (2005) New method to analyze data on overall crystallization kinetics. J Cryst Growth 285(3):394–399. https://doi.org/10.1016/j.jcrysgro.2005.08.024
59. Bhardwaj SP, Arora KK, Kwong E, Templeton A, Clas SD, Suryanarayanan R (2014) Mechanism of amorphous itraconazole stabilization in polymer solid dispersions: role of molecular mobility. Mol Pharm 11(11):4228–4237. https://doi.org/10.1021/mp5004515
60. Kothari K, Ragoonanan V, Suryanarayanan R (2014) Influence of molecular mobility on the physical stability of amorphous pharmaceuticals in the supercooled and glassy states. Mol Pharm 11(9):3048–3055. doi:https://doi.org/10.1021/mp500229d.
61. Bhardwaj SP, Arora KK, Kwong E, Templeton A, Clas S-D, Suryanarayanan R (2013) Correlation between molecular mobility and physical stability of amorphous itraconazole. Mol Pharm 10(2):694–700. https://doi.org/10.1021/mp300487u
62. Mehta M, Ragoonanan V, McKenna GB, Suryanarayanan R (2016) Correlation between molecular mobility and physical stability in pharmaceutical glasses. Mol Pharm 13 (4):1267–1277. https://doi.org/10.1021/acs.molpharmaceut.5b00853
63. Mistry P, Suryanarayanan R (2016) Strength of drug-polymer interactions: implications for crystallization in dispersions. Cryst Growth Des 16(9):5141–5149. https://doi.org/10.1021/acs.cgd.6b00714
64. Korhonen O, Bhugra C, Pikal MJ (2008) Correlation between molecular mobility and crystal growth of amorphous phenobarbital and phenobarbital with polyvinylpyrrolidone and L-proline. J Pharm Sci 97(9):3830. http://dx.doi.org/10.1002/jps.21273
65. Caron V, Bhugra C, Pikal MJ (2010) Prediction of onset of crystallization in amorphous pharmaceutical systems: phenobarbital, nifedipine/PVP, and phenobarbital/PVP. J Pharm Sci 99(9):3887. http://dx.doi.org/10.1002/jps.22232
66. Grzybowska K, Paluch M, Grzybowski A et al (2010) Molecular dynamics and physical stability of amorphous anti-inflammatory drug: celecoxib. J Phys Chem B 114 (40):12792–12801. https://doi.org/10.1021/jp1040212
67. Knapik J, Wojnarowska Z, Grzybowska K et al (2014) Physical stability of the amorphous anticholesterol agent (Ezetimibe): the role of molecular mobility. Mol Pharm 11 (11):4280–4290. https://doi.org/10.1021/mp500498e
68. Bhugra C, Rambhatla S, Bakri A et al (2007) Prediction of the onset of crystallization of amorphous sucrose below the calorimetric glass transition temperature from correlations with mobility. J Pharm Sci 96(5):1258–1269. https://doi.org/10.1002/jps.20918
69. Bhugra C, Shmeis R, Krill SL, Pikal MJ (2008) Prediction of onset of crystallization from experimental relaxation times. II. Comparison between predicted and experimental onset times. J Pharm Sci 97(1):455–472. https://doi.org/10.1002/jps.21162

70. Bhardwaj SP, Suryanarayanan R (2012) Molecular mobility as an effective predictor of the physical stability of amorphous trehalose. Mol Pharm 9(11):3209–3217. https://doi.org/10.1021/mp300302g
71. Kolodziejczyk K, Paluch M, Grzybowska K et al (2013) Relaxation dynamics and crystallization study of sildenafil in the liquid and glassy states. Mol Pharm 10(6):2270–2282. https://doi.org/10.1021/mp300479r
72. Mehta M, McKenna GB, Suryanarayanan R (2016) Molecular mobility in glassy dispersions. J Chem Phys 144(204506):1–11. https://doi.org/10.1063/1.4950768
73. Szczurek J, Rams-Baron M, Knapik-Kowalczuk J et al (2017) Molecular dynamics, recrystallization behavior and water solubility of amorphous anticancer agent bicalutamide and its polyvinylpyrrolidone mixtures. Mol Pharm 14(4):1071–1081. https://doi.org/10.1021/acs.molpharmaceut.6b01007
74. Shamblin SL, Tang X, Chang L, Hancock BC, Pikal MJ (1999) Characterization of the time scales of molecular motion in pharmaceutically important glasses. J Phys Chem B 103 (20):4113–4121. https://doi.org/10.1021/jp983964+
75. Hodge IM (1996) Strong and fragile liquids – a brief critique. J Non Cryst Solids 202:164–172
76. Hancock BC, Shamblin SL (2001) Molecular mobility of amorphous pharmaceuticals determined using differential scanning calorimetry. Thermochim Acta 380(2):95–107. http://linkinghub.elsevier.com/retrieve/pii/S0040603101006633
77. Descamps M (2016) Disordered pharmaceutical materials. Wiley-VCH, Weinheim
78. Wojnarowska Z, Knapik J, Rams-Baron M et al (2016) Amorphous protic ionic systems as promising active pharmaceutical ingredients: the case of the sumatriptan succinate drug. Mol Pharm 13(3):1111–1122. https://doi.org/10.1021/acs.molpharmaceut.5b00911
79. Wojnarowska Z, Roland CM, Kolodziejczyk K, Swiety-Pospiech A, Grzybowska K, Paluch M (2012) Quantifying the structural dynamics of pharmaceuticals in the glassy state. J Phys Chem Lett 3(10):1238–1241. https://doi.org/10.1021/jz300349a
80. Wehn R, Lunkenheimer P, Loidl A (2007) Broadband dielectric spectroscopy and aging of glass formers. J Non-Cryst Solids 353:3862–3870. https://doi.org/10.1016/j.jnoncrysol.2007.03.023
81. Casalini R, Roland CM (2009) Aging of the secondary relaxation to probe structural relaxation in the glassy state. Phys Rev Lett 102(3):1–4. https://doi.org/10.1103/PhysRevLett.102.035701
82. Kaminski K, Adrjanowicz K, Kaminska E, Paluch M (2011) Probing of structural relaxation times in the glassy state of sucrose and trehalose based on dynamical properties of two secondary relaxation processes. Phys Rev E Stat Nonlinear Soft Matter Phys 83(6):1–8. https://doi.org/10.1103/PhysRevE.83.061502
83. Baghel S, Cathcart H, O'Reilly NJ (2016) Polymeric amorphous solid dispersions: a review of amorphization, crystallization, stabilization, solid-state characterization, and aqueous solubilization of biopharmaceutical classification system class II drugs. J Pharm Sci 105 (9):2527–2544. https://doi.org/10.1016/j.xphs.2015.10.008
84. Huang Y, Dai W-G (2014) Fundamental aspects of solid dispersion technology for poorly soluble drugs. Acta Pharm Sin B 4(1):18–25. https://doi.org/10.1016/j.apsb.2013.11.001
85. Vasconcelos T, Marques S, das Neves J, Sarmento B (2016) Amorphous solid dispersions: rational selection of a manufacturing process. Adv Drug Deliv Rev 100:85–101. https://doi.org/10.1016/j.addr.2016.01.012
86. Lehmkemper K, Kyeremateng SO, Heinzerling O, Degenhardt M, Sadowski G (2017) Long-term physical stability of PVP- and PVPVA-amorphous solid dispersions. Mol Pharm 14(1):157–171. https://doi.org/10.1021/acs.molpharmaceut.6b00763
87. Tao J, Sun Y, Zhang GGZ, Yu L (2009) Solubility of small-molecule crystals in polymers: D-Mannitol in PVP, indomethacin in PVP/VA, and nifedipine in PVP/VA. Pharm Res 26 (4):855–864. https://doi.org/10.1007/s11095-008-9784-z

88. Sun YE, Tao J, Zhang GGZ, Yu L (2010) Solubilities of crystalline drugs in polymers: an improved analytical method and comparison of solubilities of indomethacin and nifedipine in PVP, PVP/VA, and PVAc. J Pharm Sci 99(9):4023–4031. https://doi.org/10.1002/jps.22251
89. Mahieu A, Willart J-F, Dudognon E, Danède F, Descamps M (2013) A new protocol to determine the solubility of drugs into polymer matrixes. Mol Pharm 10:560–566
90. Tian Y, Booth J, Meehan E, Jones DS, Li S, Andrews GP (2013) Construction of drug-polymer thermodynamic phase diagrams using Flory-Huggins interaction theory: identifying the relevance of temperature and drug weight fraction to phase separation within solid dispersions. Mol Pharm 10:236–248. https://doi.org/10.1021/mp300386v
91. Donnelly C, Tian Y, Potter C, Jones DS, Andrews GP (2015) Probing the effects of experimental conditions on the character of drug-polymer phase diagrams constructed using Flory-Huggins theory. Pharm Res 32(1):167–179. https://doi.org/10.1007/s11095-014-1453-9
92. Caron V, Tajber L, Corrigan OI, Healy AM (2011) A comparison of spray drying and milling in the production of amorphous dispersions of sulfathiazole/polyvinylpyrrolidone and sulfadimidine/polyvinylpyrrolidone. Mol Pharm 8(2):532–542. https://doi.org/10.1021/mp1003674
93. Knopp MM, Olesen NE, Holm P, Langguth P, Holm R, Rades T (2015) Influence of polymer molecular weight on drug-polymer solubility: a comparison between experimentally determined solubility in PVP and prediction derived from solubility in monomer. J Pharm Sci 104 (9):2905–2912. https://doi.org/10.1002/jps.24410
94. Knopp MM, Olesen NE, Holm P et al (2015) Evaluation of drug-polymer solubility curves through formal statistical analysis: comparison of preparation techniques. J Pharm Sci 104 (1):44–51. https://doi.org/10.1002/jps.24207
95. Higgis JS, Lipson JEG, White RP (2010) A simple approach to polymer mixture miscibility. Philos Trans R Soc A 368:1009–1025. https://doi.org/10.1098/rsta.2009.0215
96. Lin D, Huang Y (2010) A thermal analysis method to predict the complete phase diagram of drug-polymer solid dispersions. Int J Pharm 399(1–2):109–115. https://doi.org/10.1016/j.ijpharm.2010.08.013
97. Fox T (1956) Influence of diluent and of copolymer composition on the glass temperature of a polymer system. Bull Am Phys Soc 1:123–132
98. Jenckel E, Heusch R (1953) Die erniedrigung der einfriertemperatur organischer glaser durch losungsmittel. Kolloid Z Z Polym 130:89–105
99. Kwei TK (1984) The effect of hydrogen-bonding on the glass-transition temperatures of polymer mixtures. J Polym Sci C Polym Lett 22:307–313
100. Baird JA, Taylor LS (2012) Evaluation of amorphous solid dispersion properties using thermal analysis techniques. Adv Drug Deliv Rev 64(5):396–421. https://doi.org/10.1016/j.addr.2011.07.009
101. Couchman PR, Karasz FE (1978) A classical thermodynamic discussion of the effect of composition on glass-transition temperatures. Macromolecules 11(1):116–119
102. Lu X, Weiss A (1992) Relationship between the glass transition temperature and the interaction parameter of miscible binary polymer blends. Macromolecules 25:3242–3246. https://doi.org/10.1021/ma00038a033
103. Tu W, Wang Y, Li X et al (2015) Unveiling the dependence of glass transitions on mixing thermodynamics in miscible systems. Sci Rep 5:8500. https://doi.org/10.1038/srep08500
104. Marsac PJ, Konno H, Taylor LS (2006) A comparison of the physical stability of amorphous felodipine and nifedipine systems. Pharm Res 23(10):2306–2316. https://doi.org/10.1007/s11095-006-9047-9
105. Taylor LS, Zografi G (1997) Spectroscopic characterization of interactions between PVP and indomethacin in amorphous molecular dispersions. Pharm Res 14(12):1691–1698
106. Vieira MGA, Da Silva MA, Dos Santos LO, Beppu MM (2011) Natural-based plasticizers and biopolymer films: a review. Eur Polym J 47(3):254–263. https://doi.org/10.1016/j.eurpolymj.2010.12.011

107. Yuan X, Xiang TX, Anderson BD, Munson EJ (2015) Hydrogen bonding interactions in amorphous indomethacin and its amorphous solid dispersions with poly(vinylpyrrolidone) and poly(vinylpyrrolidone-co-vinyl acetate) studied using 13C solid-state NMR. Mol Pharm 12(12):4518–4528. https://doi.org/10.1021/acs.molpharmaceut.5b00705
108. Gupta P, Thilagavathi R, Chakraborti AK, Bansal AK (2005) Role of molecular interaction in stability of celecoxib-PVP amorphous systems. Mol Pharm 2(5):384–391. https://doi.org/10.1021/mp050004g
109. Kothari K, Ragoonanan V, Suryanarayanan R (2015) The role of drug-polymer hydrogen bonding interactions on the molecular mobility and physical stability of nifedipine solid dispersions. Mol Pharm 12(1):162–170. https://doi.org/10.1021/mp5005146
110. Xie T, Taylor LS (2016) Effect of temperature and moisture on the physical stability of binary and ternary amorphous solid dispersions of celecoxib. J Pharm Sci 106:100–110. https://doi.org/10.1016/j.xphs.2016.06.017
111. Miyazaki T, Yoshioka S, Aso Y, Kojima S (2004) Ability of polyvinylpyrrolidone and polyacrylic acid to inhibit the crystallization of amorphous acetaminophen. J Pharm Sci 93 (11):2710–2717. https://doi.org/10.1002/jps.20182
112. Kestur US, Taylor LS (2010) Role of polymer chemistry in influencing crystal growth rates from amorphous felodipine. CrystEngComm 12(8):2290–2397. https://doi.org/10.1039/c001905d
113. Grzybowska K, Chmiel K, Knapik J, Grzybowski A, Jurkiewicz K, Paluch M (2017) Molecular factors governing the liquid and glassy states recrystallization of celecoxib in binary mixtures with excipients of different molecular weights. Mol Pharm 14(4):1154–1168. https://doi.org/10.1021/acs.molpharmaceut.6b01056
114. Xiang TX, Anderson BD (2013) Molecular dynamics simulation of amorphous indomethacin-poly(vinylpyrrolidone) glasses: solubility and hydrogen bonding interactions. J Pharm Sci 102(3):876–891. https://doi.org/10.1002/jps.23353
115. Xiang TX, Anderson BD (2012) Molecular dynamics simulation of amorphous indomethacine. Mol Pharm 10:102–114
116. Matsumoto T, Zografi G (1999) Physical properties of solid molecular dispersions of indomethacin with poly(vinylpyrrolidone) and poly(vinylpyrrolidone-co-vinyl-acetate) in relation to indomethacin crystallization. Pharm Res 16(11):1722–1728
117. LaFountaine JS, McGinity JW, Williams RO (2016) Challenges and strategies in thermal processing of amorphous solid dispersions: a review. AAPS PharmSciTech 17(1):43–55. https://doi.org/10.1208/s12249-015-0393-y
118. Li J, Zhao J, Tao L et al (2015) The effect of polymeric excipients on the physical properties and performance of amorphous dispersions: Part I, Free volume and glass transition. Pharm Res 32(2):500–515. https://doi.org/10.1007/s11095-014-1478-0
119. Rey L, Galy J, Sautereau H, Simon GP, Cook WD (2004) PALS free volume and mechanical properties in dimethacrylate-based thermosets. Polym Int 53(5):557–568. https://doi.org/10.1002/pi.1432
120. Chavan RB, Thipparaboina R, Kumar D, Shastri NR (2016) Co amorphous systems: a product development perspective. Int J Pharm 515(1–2):403–415. https://doi.org/10.1016/j.ijpharm.2016.10.043
121. Chieng N, Aaltonen J, Saville D, Rades T (2009) Physical characterization and stability of amorphous indomethacin and ranitidine hydrochloride binary systems prepared by mechanical activation. Eur J Pharm Biopharm 71(1):47–54. https://doi.org/10.1016/j.ejpb.2008.06.022
122. Allesø M, Chieng N, Rehder S, Rantanen J, Rades T, Aaltonen J (2009) Enhanced dissolution rate and synchronized release of drugs in binary systems through formulation: amorphous naproxen-cimetidine mixtures prepared by mechanical activation. J Control Release 136 (1):45–53. https://doi.org/10.1016/j.jconrel.2009.01.027
123. Tantishaiyakul V, Suknuntha K, Vao-Soongnern V (2010) Characterization of cimetidine-piroxicam coprecipitate interaction using experimental studies and molecular dynamic simulations. AAPS PharmSciTech 11(2):952–958. https://doi.org/10.1208/s12249-010-9461-5

124. Knapik J, Wojnarowska Z, Grzybowska K et al (2016) Molecular dynamics and physical stability of amorphous nimesulide drug and its binary drug-polymer systems. Mol Pharm 13 (6):1937–1946. https://doi.org/10.1021/acs.molpharmaceut.6b00115
125. Yamamura S, Gotoh H, Sakamoto Y, Momose Y (2000) Physicochemical properties of amorphous precipitates of cimetidine-indomethacin binary system. Eur J Pharm Biopharm 49(3):259–265. https://doi.org/10.1016/S0939-6411(00)00060-6
126. Löbmann K, Laitinen R, Grohganz H, Gordon KC, Strachan C, Rades T (2011) Coamorphous drug systems: enhanced physical stability and dissolution rate of indomethacin and naproxen. Mol Pharm 8(5):1919–1928. https://doi.org/10.1021/mp2002973
127. Löbmann K, Strachan C, Grohganz H, Rades T, Korhonen O, Laitinen R (2012) Co-amorphous simvastatin and glipizide combinations show improved physical stability without evidence of intermolecular interactions. Eur J Pharm Biopharm 81(1):159–169. https://doi.org/10.1016/j.ejpb.2012.02.004
128. Knapik J, Wojnarowska Z, Grzybowska K, Jurkiewicz K, Tajber L, Paluch M (2015) Molecular dynamics and physical stability of coamorphous ezetimib and indapamide mixtures. Mol Pharm 12(10):3610–3619. https://doi.org/10.1021/acs.molpharmaceut.5b00334
129. Lim AW, Löbmann K, Grohganz H, Rades T, Chieng N (2016) Investigation of physical properties and stability of indomethacin-cimetidine and naproxen-cimetidine co-amorphous systems prepared by quench cooling, coprecipitation and ball milling. J Pharm Pharmacol 68 (1):36–45. https://doi.org/10.1111/jphp.12494
130. Löbmann K, Laitinen R, Grohganz H, Strachan C, Rades T, Gordon KC (2013) A theoretical and spectroscopic study of co-amorphous naproxen and indomethacin. Int J Pharm 453 (1):80–87. https://doi.org/10.1016/j.ijpharm.2012.05.016
131. Beyer A, Radi L, Grohganz H, Löbmann K, Rades T, Leopold CS (2016) Preparation and recrystallization behavior of spray-dried co-amorphous naproxen-indomethacin. Eur J Pharm Biopharm 104:72–81. https://doi.org/10.1016/j.ejpb.2016.04.019
132. Shayanfar A, Jouyban A (2013) Drug-drug coamorphous systems: characterization and physicochemical properties of coamorphous atorvastatin with carvedilol and glibenclamide. J Pharm Innov 8(4):218–228. https://doi.org/10.1007/s12247-013-9162-1
133. Suresh K, Mannava MKC, Nangia A (2014) A novel curcumin–artemisinin coamorphous solid: physical properties and pharmacokinetic profile. RSC Adv 4(102):58357–58361. https://doi.org/10.1039/C4RA11935E
134. Dengale SJ, Ranjan OP, Hussen SS et al (2014) Preparation and characterization of co-amorphous ritonavir-indomethacin systems by solvent evaporation technique: improved dissolution behavior and physical stability without evidence of intermolecular interactions. Eur J Pharm Sci 62:57–64. https://doi.org/10.1016/j.ejps.2014.05.015
135. Wairkar S, Gaud R (2015) Co-amorphous combination of nateglinide-metformin hydrochloride for dissolution enhancement. AAPS PharmSciTech 17(3):673–681. https://doi.org/10.1208/s12249-015-0371-4
136. Dengale SJ, Hussen SS, Krishna BSM, Musmade PB, Gautham Shenoy G, Bhat K (2015) Fabrication, solid state characterization and bioavailability assessment of stable binary amorphous phases of ritonavir with quercetin. Eur J Pharm Biopharm 89:329–338. https://doi.org/10.1016/j.ejpb.2014.12.025
137. Teja A, Musmade PB, Khade AB, Dengale SJ (2015) Simultaneous improvement of solubility and permeability by fabricating binary glassy materials of talinolol with naringin: solid state characterization, in-vivo in-situ evaluation. Eur J Pharm Sci 78:234–244. https://doi.org/10.1016/j.ejps.2015.08.002
138. Renuka, Singh SK, Gulati M, Narang R (2015) Stable amorphous binary systems of glipizide and atorvastatin powders with enhanced dissolution profiles: formulation and characterization. Pharm Dev Technol 22(1):13–25. https://doi.org/10.3109/10837450.2015.1125921
139. Russo MG, Sancho MI, Silva LMA et al (2016) Looking for the interactions between omeprazole and amoxicillin in a disordered phase. An experimental and theoretical study.

Spectrochim Acta A Mol Biomol Spectrosc 156:70–77. https://doi.org/10.1016/j.saa.2015.11.021
140. Ueda H, Kadota K, Imono M, Ito T, Kunita A, Tozuka Y (2017) Co-amorphous formation induced by combination of tranilast and diphenhydramine hydrochloride. J Pharm Sci 106 (1):123–128. https://doi.org/10.1016/j.xphs.2016.07.009
141. Hamilton BD, Ha J, Hillmyer MA, Ward MD (2012) Manipulating crystal growth and polymorphism by confinement in nanoscale crystallization chambers. Acc Chem Res 45 (3):414–423. https://doi.org/10.1021/ar200147v
142. Knapik J, Wojnarowska Z, Grzybowska K, Jurkiewicz K, Stankiewicz A, Paluch M (2016) Stabilization of the amorphous ezetimibe drug by confining its dimension. Mol Pharm 13 (4):1308–1316. https://doi.org/10.1021/acs.molpharmaceut.5b00903
143. Bras AR, Fonseca IM, Dionisio M, Schonhals A, Affouard F, Correia NT (2014) Influence of nanoscale confinement on the molecular mobility of ibuprofen. J Phys Chem C 118:13857–13868
144. Shen S-C, Ng WK, Chia L, Dong Y-C, Tan RBH (2010) Stabilized amorphous state of ibuprofen by co-spray drying with mesoporous SBA-15 to enhance dissolution properties. J Pharm Sci 99 (4):1997–2007. http://dx.doi.org/10.1002/jps.21967
145. Yani Y, Chow PS, Tan RBH (2016) Pore size effect on the stabilization of amorphous drug in a mesoporous material: insights from molecular simulation. Microporous Mesoporous Mater 221:117–122. https://doi.org/10.1016/j.micromeso.2015.09.029
146. Cordeiro T, Santos AFM, Nunes G et al (2016) Accessing the physical state and molecular mobility of naproxen confined to nanoporous silica matrixes. J Phys Chem C 120 (26):14390–14401. https://doi.org/10.1021/acs.jpcc.6b04078
147. Dwyer LM, Michaelis VK, O'Mahony M, Griffin RG, Myerson AS (2015) Confined crystallization of fenofibrate in nanoporous silica. CrystEngComm 17:7922–7929. https://doi.org/10.1039/C5CE01148E
148. Richert R (2011) Dynamics of nanoconfined supercooled liquids. Annu Rev Phys Chem 62:65–84. https://doi.org/10.1146/annurev-physchem-032210-103343
149. Alcoutlabi M, McKenna GB (2005) Effects of confinement on material behaviour at the nanometre size scale. J Phys Condens Matter 17(15):R461–R524. https://doi.org/10.1088/0953-8984/17/15/R01
150. Morineau D, Xia Y, Alba-Simionesco C (2002) Finite-size and surface effects on the glass transition of liquid toluene confined in cylindrical mesopores. J Chem Phys 117 (19):8966–8972. https://doi.org/10.1063/1.1514664
151. Xu W, Riikonen J, Lehto VP (2013) Mesoporous systems for poorly soluble drugs. Int J Pharm 453(1):181–197. https://doi.org/10.1016/j.ijpharm.2012.09.008
152. Watanabe T, Wakiyama N, Usui F, Ikeda M, Isobe T, Senna M (2001) Stability of amorphous indomethacin compounded with silica. Int J Pharm 226(1–2):81–91. https://doi.org/10.1016/S0378-5173(01)00776-1
153. Kapoor S, Hegde R, Bhattacharyya AJ (2009) Influence of surface chemistry of mesoporous alumina with wide pore distribution on controlled drug release. J Control Release 140 (1):34–39. https://doi.org/10.1016/j.jconrel.2009.07.015
154. Mellaerts R, Aerts CA, Van Humbeeck J, Augustijns P, Van den Mooter G, Martens JA (2007) Enhanced release of itraconazole from ordered mesoporous SBA-15 silica materials. Chem Commun (13):1375–1377. doi:https://doi.org/10.1039/B616746b
155. Heikkilla T, Salonen J, Tuura J et al (2007) Mesoporous silica material TUD-1 as a drug delivery system. Int J Pharm 331(1):133–138. https://doi.org/10.1016/j.ijpharm.2006.09.019
156. Hu Y, Wang J, Zhi Z, Jiang T, Wang S (2011) Facile synthesis of 3D cubic mesoporous silica microspheres with a controllable pore size and their application for improved delivery of a water-insoluble drug. J Colloid Interface Sci 363(1):410–417. https://doi.org/10.1016/j.jcis.2011.07.022
157. Priemel PA, Grohganz H, Rades T (2016) Unintended and in situ amorphisation of pharmaceuticals. Adv Drug Deliv Rev 100:126–132. https://doi.org/10.1016/j.addr.2015.12.014

158. Thakral NK, Mohapatra S, Stephenson GA, Suryanarayanan R (2015) Compression-induced crystallization of amorphous indomethacin in tablets – characterization of spatial heterogeneity by two-dimensional X-ray diffractometry. Mol Pharm 12:253–263. https://doi.org/10.1021/mp5005788
159. Ayenew Z, Paudel A, Van Den Mooter G (2012) Can compression induce demixing in amorphous solid dispersions? A case study of naproxen-PVP K25. Eur J Pharm Biopharm 81(1):207–213. https://doi.org/10.1016/j.ejpb.2012.01.007
160. Tonelli AE (1982) Conformational characteristics of poly(N-vinyl pyrrolidone). Polymer (Guildf) 23:676–680
161. Singh A, Bharati A, Frederiks P et al (2016) Effect of compression on the molecular arrangement of itraconazole-Soluplus solid dispersions: induction of liquid crystals or exacerbation of phase separation? Mol Pharm 13:1879–1893. https://doi.org/10.1021/acs.molpharmaceut.6b00046
162. Singh A, Van HJ, Van Den Mooter G (2014) A new twist in the old story-can compression induce mixing of phase separated solid dispersions? A case study of spray-dried miconazole-PVP VA64 solid dispersions. Pharm Res 31:3191–3200. https://doi.org/10.1007/s11095-014-1411-6
163. Singh A, De Bisschop C, Schut H, Van Humbeeck J, Van Den Mooter G (2015) Compression effects on the phase behaviour of miconazol-poly(1-vinylpyrrolidone-co-vinyl acetate) solid dispersions-role of pressure, dwell time, and preparation method. J Pharm Sci 104:3366–3376. https://doi.org/10.1002/jps.24540
164. Worku ZA, Aarts J, Van Den Mooter G (2014) Influence of compression forces on the structural stability of naproxen/PVP-VA 64 solid dispersions. Mol Pharm 11:1102–1108. https://doi.org/10.1021/mp5001313
165. Adrjanowicz K, Koperwas K, Tarnacka M et al (2016) Changing the tendency of glass-forming liquid to crystallize by moving along different isolines in the T – p phase diagram. Cryst Growth Des 16(11):6263–6268. https://doi.org/10.1021/acs.cgd.6b00798

Chapter 6
Amorphous Drug Formulation

6.1 Introduction

Modern drug discovery tools have biased towards compounds with poor aqueous solubility even though the importance of drug-like properties has long been recognized [1], and the trend continues to deteriorate [2]. The poor solubility of these drug candidates imposes great challenges to pharmaceutical scientists and engineers who are ultimately responsible for developing a bioavailable drug product to support the clinical programs and commercialization, if successful. A poorly water soluble drug candidate not only may lengthen the formulation development phase, increase the resource demands, delay the clinical trials due to insufficient in vivo exposure, but also may impact the ultimate success of the entire program due to suboptimal bioavailability [3].

Oral solid dosage forms undergo two basic processes once landing in the gastrointestinal (GI) tract: dissolution of the drug molecules in the GI fluid (i.e. in vivo dissolution), followed by permeation across the GI membrane into the blood stream (i.e. absorption). These two processes may be treated based on the principles of simple diffusion,

$$\frac{dX_{solid}}{dt} = -\frac{DA}{h}(C_s - C_{GI}) = -\frac{DA}{h}\left(C_s - \frac{X_{dissolv}}{V_{GI}}\right) \tag{6.1}$$

$$\frac{dX_{dissolv}}{dt} = \frac{dX_{solid}}{dt} - k_{absorb}(X_{dissolv} - X_{blood}) \approx \frac{dX_{solid}}{dt} - k_{absorb}X_{dissolv} \tag{6.2}$$

$$\frac{dX_{blood}}{dt} \approx k_{absorb}X_{dissolv} \tag{6.3}$$

where X is the drug amount, C is drug concentration, D is the diffusivity of drug molecule in the GI fluid, A is the available surface area of the solid, h is the thickness of the diffusion layer, V is the volume, and k is the apparent rate constant. The subscripts, *solid*, *dissolv*, *s*, *GI*, *absorb*, *blood* designate solid, dissolved

M. Rams-Baron et al., *Amorphous Drugs*,
https://doi.org/10.1007/978-3-319-72002-9_6

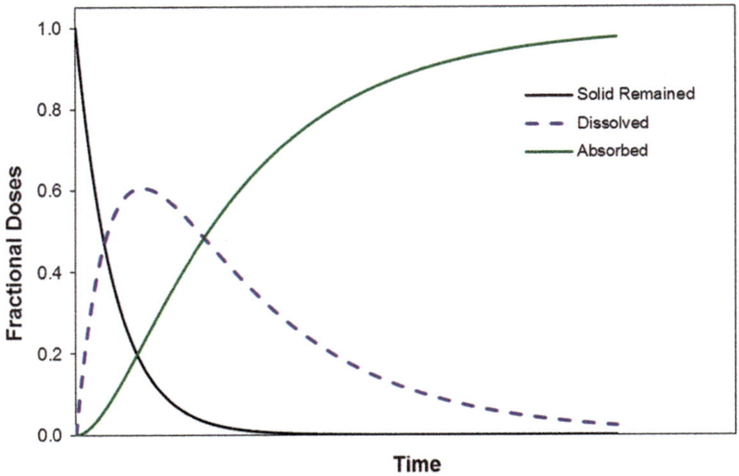

Fig. 6.1 Schematic drug amount-time profiles of in vivo dissolution and absorption

(in solution), saturation, gastrointestinal fluid, absorption, and blood, respectively. Equation (6.1) is a variant of the Noyes-Whitney equation.

According to these equations, rate of oral absorption, dX_{blood}/dt, is determined by the dissolved drug amount, $X_{dissolv}$, or drug concentration in the GI fluid, C_{GI}. The higher the dissolved drug amount in the GI tract, the faster is the rate of drug absorption. The amount of dissolved drug in GI fluid is not a constant, but is rather a quantity dynamically depending on the rate of generation (dissolution) and the rate of elimination (absorption). Schematic drug amount-time profiles of in vivo dissolution and absorption are shown in Fig. 6.1.

The diffusivity, D, is a property of the drug molecule in the dissolution medium. The thickness of the diffusion layer, h, is a function of the hydrodynamics in the GI depending on the GI physiological conditions such as s gastric emptying and GI motility. In general, neither of these two parameters can be manipulated via formulation approaches. However, two other parameters, the solubility, C_s, and the available surface area for dissolution, A, may be readily modified by formulations, which are the foundations of various enabling formulation technologies for poorly water-soluble drugs.

The general requirement of enabling formulations is to improve dissolved drug concentration in the GI tract, either by enhancing the dissolution rate via increased surface area, or improved solubility, or by eliminating the dissolution step altogether (such as complexation, and simple solubilization). Significant advances have been achieved in formulating poorly water-soluble drugs over the last two decades, leading to the commercialization of various delivery technologies such as self-emulsifying drug delivery systems (SEDDS) [4–6], nanoparticles [7–9], and amorphous solid dispersions (ASDs) [10–12]. Particularly, the amorphous solid dispersion approach has gained significant advances in both basic research and commercial product approvals. Compared to solution or semi-solid based enabling technologies, ASD is more attractive not only because it can increase the pharmacokinetic exposure of otherwise poorly absorbed drugs, but also because the final product can be delivered to patients as tablets or capsules, which may provide greater chemical stability and improved patient convenience (e.g. Kaletra™ tablets [13]).

6.2 Understanding Oral Bioavailability Enhancement

Amorphous form has higher free energy than the corresponding crystalline counterparts, which is the foundation for higher aqueous solubility and enhanced oral bioavailability. However, utilization of an amorphous form does not necessarily lead to enhancement in oral bioavailability. The ultimate outcome of oral absorption of an amorphous drug formulation depends on many factors, with the dissolution behavior being the most important. Solubility advantages of the amorphous forms have been dealt with in Chap. 3 of this book. This section discusses the behavior of amorphous form and amorphous solid dispersions during dissolution and the implications for oral absorption. The aim is to lay down a foundation for ASD formulation design/development in achieving the ultimate goal of oral bioavailability enhancement.

6.2.1 The Rediscovery of Liquid-Liquid Phase Separation (LLPS)

Since the recognition of the potential advantages of using the metastable amorphous forms of pharmaceuticals, scientists have delved into the understanding their solubility advantages. Earlier reports indicated solubility enhancement of amorphous forms ranged from under 2-folds to about 20-folds [14–17]. Hancock et al. [18] conducted the first systematic evaluation on this topic and compared the measured with the predicted amorphous/crystalline solubility ratios based on the simple concept of free energy differences. While the predicted solubility ratio ranged from slightly over unity to hundreds, the experimentally determined solubility ratios from 12 drugs were only within the range of 1.1- to 24-folds. This discrepancy was certainly beyond the experimental errors and raised an important question: why so big differences? The most probable cause was ascribed to crystallization from the supersaturated solution during the solubility measurement, as shown in Fig. 6.2 for indomethacin [18].

The metastable nature of the supersaturated solution, created by the dissolution of an amorphous form, causes inherent challenges to measure itself, i.e., the true solubility advantage of an amorphous pharmaceutical. A more serious implication is that the solubility advantage would be negated, partially or entirely, if crystallization occurs during in vivo dissolution in GI fluid over the time frames relevant to oral absorption.

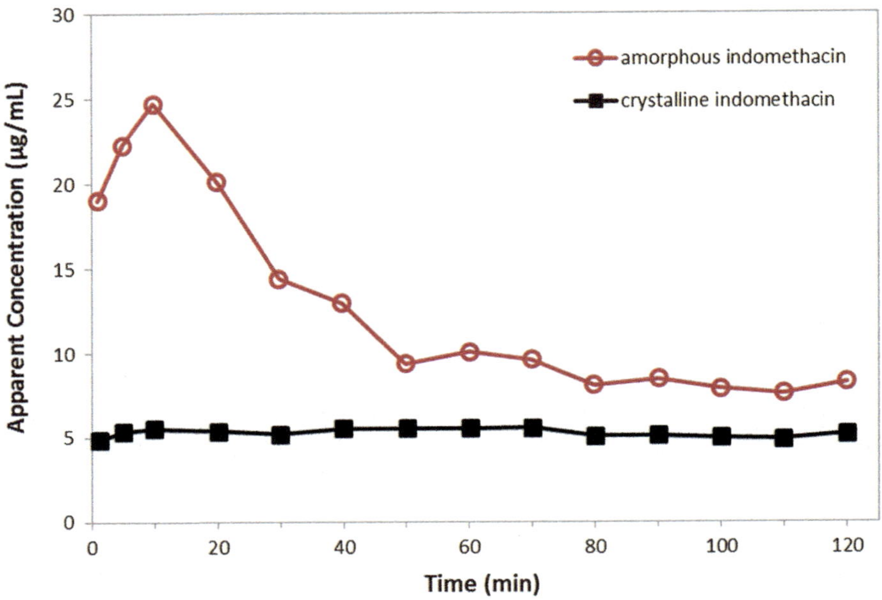

Fig. 6.2 Apparent concentration-time profiles for amorphous and crystalline (γ) form of indomethancin during dissolution measurement. Data source from [18]

In an attempt to understand the dissolution behavior of amorphous pharmaceuticals, the work from Alonzo et al. has shed some significant light on the subject. In the example of felodipine [19], the neat amorphous form did not provide solubility advantages at all, as both crystalline and amorphous forms resulted in equivalent dissolution profiles (Fig. 6.3). Raman spectroscopy indicated that rapid crystallization of amorphous felodipine occurred once in contact with the dissolution medium. Crystallization in the bulk amorphous material led to its similar dissolution profile as the crystalline form. Further explorations showed that, while amorphous felodipine alone did not provide solubility advantage, adding a small amount of polymer in the dissolution medium helped to realize the solubility advantages of the amorphous form. Figure 6.4 compares the apparent dissolution profiles of amorphous felodipine in the presence very small amount of polymers, 0.1% (w/w) PVP (polyvinylpyrrolidone), HPMC (hypromellose, or hydroxypropyl methylcellulose), HPMCAS (hypromellose acetate succinate), respectively, that were pre-dissolved in the dissolution media. All these polymers promoted the supersaturation over the crystalline solubility, with HPMCAS being the highest, followed by HPMC and PVP. These polymers were confirmed not to increase the equilibrium solubility of the crystalline form at the relevant concentrations. However, the presence of the small amount of polymer somehow made it possible to reveal the true solubility advantages of amorphous felodipine during dissolution and this is purely a kinetic phenomenon.

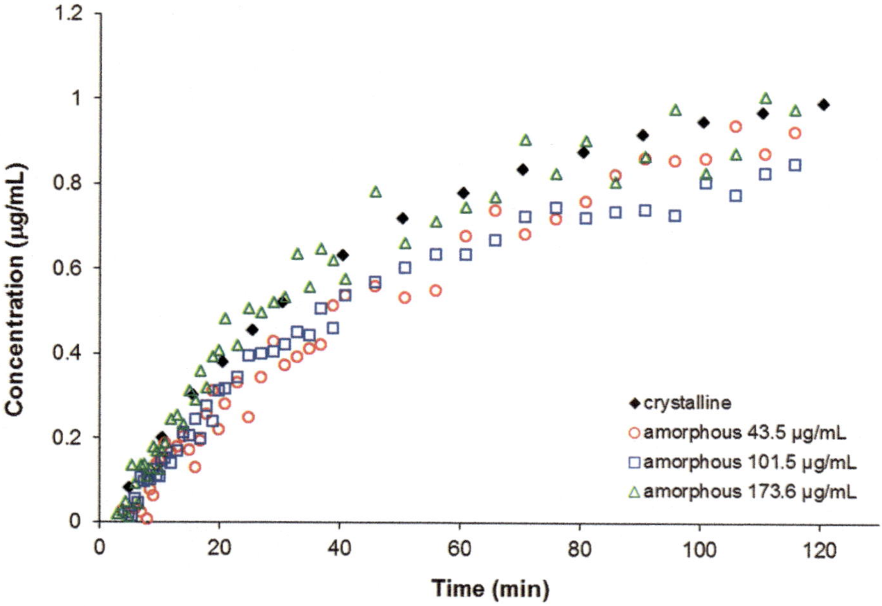

Fig. 6.3 Apparent dissolution profiles of crystalline and amorphous felodipine neat drug substance. The amount of solids added into the dissolution media is indicated in the legend. Data source from [19]

The role of these polymers to sustain the drug concentration of amorphous felodipine was related to the crystallization inhibiting ability of the polymers. HPMC was shown to inhibit both nucleation and crystal growth of supersaturated felodipine solution, even at very low concentrations (1–3.5 ppm). An example is shown in Fig. 6.5 [20]. PVP, HPMC, and HPMCAS all can impact the crystallization kinetics of felodipine. Therefore, by incorporating appropriate crystallization-inhibiting polymers, the solubility advantages of amorphous forms are now better revealed and utilized for oral bioavailability enhancement. It is not surprising since similar phenomena have been observed by others [21–34], both in solution and in solid state [35, 36]. While the exact mechanism is not fully understood [37], recent investigations have concluded that the hydrophobicity of the polymer has a significant impact [38], potentially because these systems were all poorly water-soluble drugs.

However, there were some "clouds" over the above observations. Careful readers may recognize that in the presences of 0.1% HPMCAS, the apparent drug concentration during dissolution of amorphous felodipine reached ~16 µg/mL, well above the predict solubility of 9.1 µg/mL at 37 °C, whereas the highest drug concentration was right at the amorphous solubility in the presence of 0.1% HPMC (Fig. 6.4). The initial thought was that perhaps an apparent concentration above the "equilibrium" amorphous solubility might have been reached for kinetic reasons, that is, solution could be "supersaturated" with respect to the amorphous

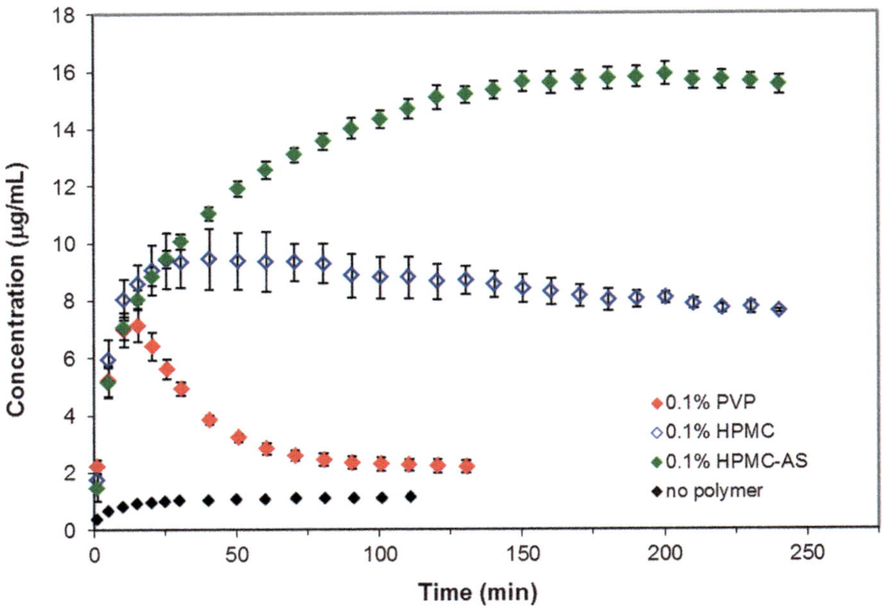

Fig. 6.4 Apparent dissolution profiles of neat amorphous felodipine in the presence of 0.1% polymers (w/w) in pH 6.8 phosphate buffer. Data source from [19]

solubility under some circumstances, just like supersaturation against the crystalline solubility.

Similar apparent supersaturation with respect to amorphous solubility was also noticed when studying the dissolution of amorphous solid dispersions. Figure 6.6 shows the apparent concentration-time profiles of 50/50 (w/w) and 90/10 (w/w) HPMC/felodipine amorphous solid dispersions during dissolution. While the dissolution of 50/50 HPMC/felodipine ASD generated drug concentration right at the amorphous solubility, the 90/10 HPMC/felodipine ASD resulted in a drug concentration approaching to ~26 µg/mL, far exceeding the amorphous solubility. Similar observations were made with the PVP/felodipine [39] and the HPMCAS/felodipine (unpublished data) ASD systems when the polymer/drug ratio is high (e.g. 90/10).

It would be exciting if higher concentration can be reached beyond the solubility of amorphous forms, which implies significantly upside potentials for oral bioavailability enhancement. However, further examination indicated these higher-than-amorphous-solubility concentrations actually resulted from an artifact of the experimental techniques, where a UV/Vis fiber-optics probe was used to monitor the absorbance and to determine drug solution concentration. Nanoparticles of amorphous felodipine were confirmed in these solutions where the higher-than-amorphous-solubility drug concentrations were observed. These amorphous felodipine nanoparticles also contributed to the UV/Vis absorption, supported by the Mie theory-based calculations, resulting in higher calculated drug concentration than that of the truly dissolved felodipine molecules [40]. Visual observations indicated that these nanoparticle

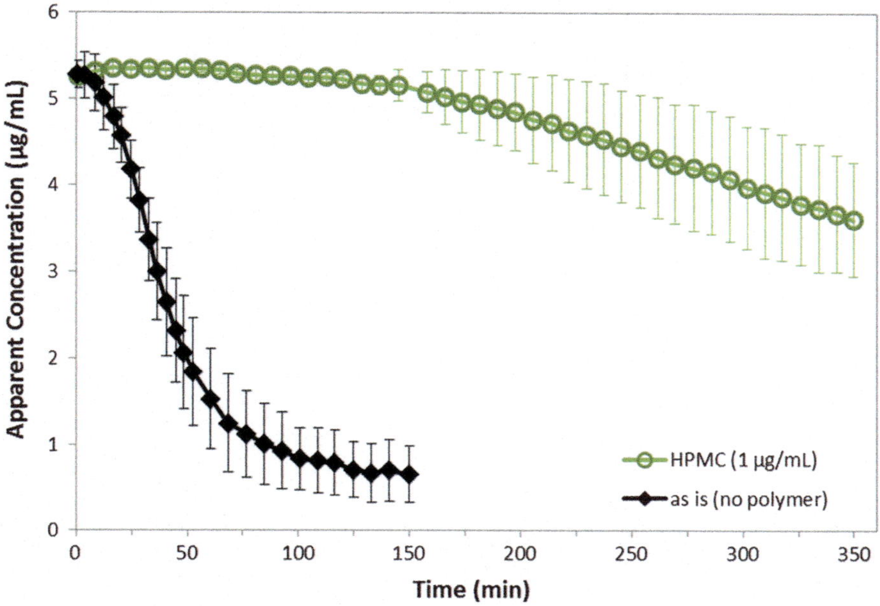

Fig. 6.5 Apparent concentration profiles of supersaturated felodipine aqueous solution at a supersaturation of ten in the absence and presence of 1 μg/mL HPMC at 25 °C. Data source from [20]

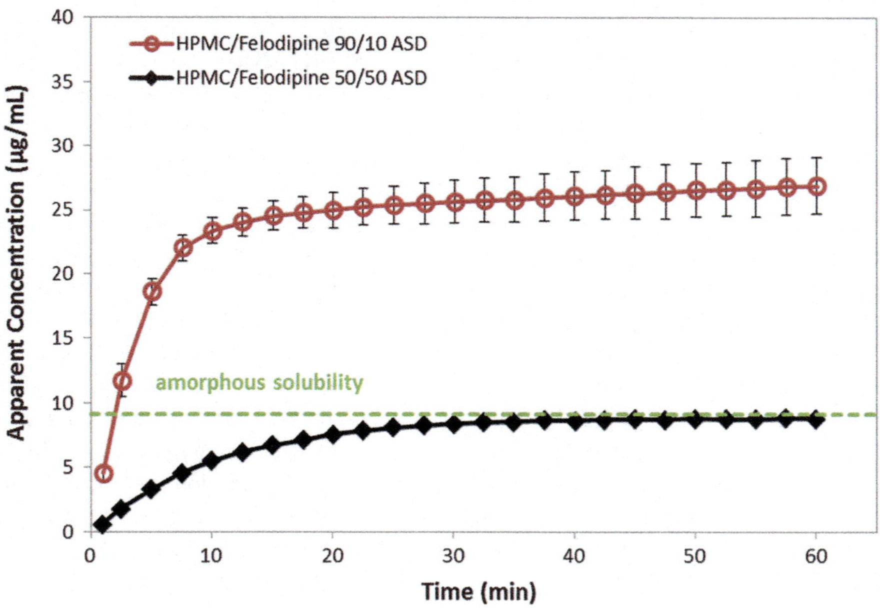

Fig. 6.6 Apparent concentration-time profiles of HPMC/felodipine amorphous solid dispersions during dissolution in 50 mM pH 6.8 phosphate buffer at 37 °C. Data source from [39]

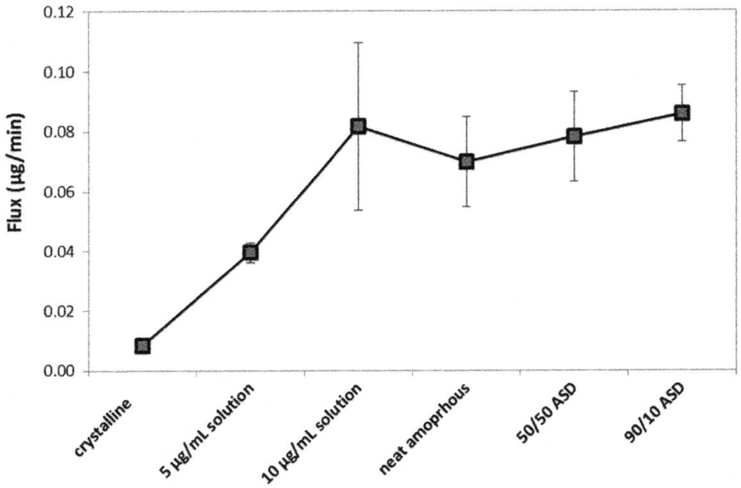

Fig. 6.7 Mean flux when various felodipine systems were introduced at the donor side of a side-by-side diffusion chamber. ASDs were composed of HPMC/felodipine. Data source from [39]

containing solutions were translucent with tint of colloidal appearance. Dynamic light scattering (DLS) experiment revealed the presences of large number of mostly submicron particles, which increased with increasing drug/solvent ratio. These findings were further confirmed by diffusion experiments. Using a side-by-side diffusion chamber with a permeable membrane that only allows the passage of the solvent and free drug molecules (MW cutoff 6–8 kDa), the 90/10 HPMC/felodipne ASD in the donor chamber generated similar mass flux to the 50/50 HPMC/felodipine ASD, both of which were similar to a 10 µg/mL felodipine solution [39]. The results indicated that both amorphous solid dispersions generated similar drug concentration at the amorphous solubility limit (Fig. 6.7). No concentration beyond the amorphous solubility was achieved by amorphous solid dispersions.

Further diffusion experiments were conducted by Raina et al. on felodipine and nifedipine systems in 50 mM phosphate pH 6.8 buffer [41]. In these diffusion experiments, various drug concentrations were introduced into the donor chamber by diluting a concentrated stock solution in methanol. Pre-dissolved HPMC were present in both donor and receptor chambers at concentration to effectively inhibit the crystallization of the drug (if any). Drug concentration in the receptor chamber was monitored using µDISS UV/Vis dip probe. The flux profiles were shown in Fig. 6.8a, b for felodipine and nifedipine, respectively. In both cases, the membrane flux initially increased linearly with drug concentration in the donor compartment, but plateaued after the donor compartment reached approximately the amorphous solubility (~11 µg/mL for felodipine and ~72 µg/mL for nifedipine). No further increase in membrane flux was observed above the corresponding amorphous solubility within measurement error, indicating that the free drug concentration reached the same highest level for each case, and that further increase was not achieved.

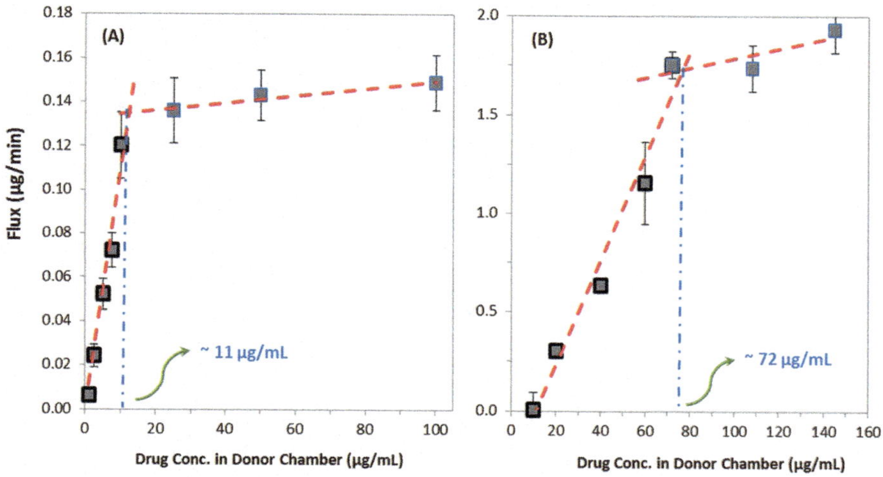

Fig. 6.8 Flux as a function of drug concentration in the donor chamber for felodipine (**a**) and nifedipine (**b**) at 37 °C. Data source from [41]

By that time it seemed that any potential drug concentration above the amorphous solubility resulted in instantaneous phase separation and that no metastable zone against amorphous solubility was observed on the experimental time scale, unlike the crystallization from supersaturated solutions. These observations suggest that new liquid (drug-rich) phase can be formed rather promptly and that the amorphous solubility is the potentially attainable ceiling of drug concentration.

In a separate study on the pH-dilution behavior of a number of weakly basic drugs by Hsieh et al. [42], amorphous drug precipitates were also observed in equilibrium with the corresponding amorphous solubilities when the drug molecules didn't crystallize upon titrating from high solubility at low pH to low solubility at high pH. These observations, along with those described above, led to the rediscovery of the so-called liquid-liquid phase separation [41–44], or LLPS, that were not uncommon for various organics-solvent systems [45–47].

The LLPS phenomenon may be understood from the perspective of free-energy diagram of mixing of two small molecules (Fig. 6.9). For binary systems where the two components are not infinitely miscible, a miscibility gap exists, i.e., thermodynamically, there are two minimums in the free energy landscape, each corresponding to a stable phase. In the case of water/drug system, they correspond to a drug scarce phase (drug solubilized in aqueous phase) and a drug-rich phase (water solubilized in liquid or amorphous drug phase). Even though the overall mixing is thermodynamically favorable for all drug/water compositions ($\Delta G_{mix} < 0$), diluting one component with the other, no matter which direction, all leads to one of the two stable phases at the free energy minimums. Once exceeding the compositions of the free energy minimum (i.e. from $W_{d,1}$ and above, or from $W_{d,2}$ and below), the overall free energy gain decreases in the composition range of $W_{d,1} - W_{d,2}$, and a local maximum exists. The free energy curve in this composition range can be divided into three segments,

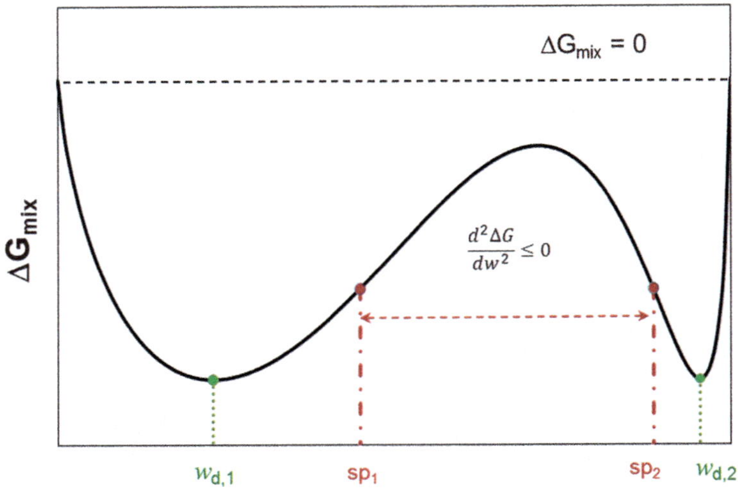

Fig. 6.9 Free energy schematic of mixing for a water-amorphous drug binary system. W_d is drug concentration, $W_{d,1}$ and $W_{d,2}$ corresponds to the compositions of the two immiscible phases, and sp_1 and sp_2 correspond to the spinodal compositions

corresponding to: $W_{d,1} - SP_1$, $SP_1 - SP_2$, and $SP_2 - W_{d,2}$. In composition range of $SP_1 - SP_2$, the second derivatives of the free energy change is negative, meaning that any microscopic fluctuation in the composition will lead to a decrease in the overall free energy and therefore the solution composition will spontaneously decompose into the two stable phases corresponding two minimums at $W_{d,1}$ and $W_{d,2}$, respectively. This phenomenon is called the spinodal decomposition and is the underlying kinetic cause for the rapid liquid-liquid phase separation observed during dissolution of amorphous forms and amorphous solid dispersions in absence of crystallization. The remaining two segments, $W_{d,1} - SP_1$ and $SP_2 - W_{d,2}$, correspond to the metastable zone because the second derivative of the free energy change is positive where minor composition fluctuation in these regions leads to an increase in the overall free energy. During dissolution of an amorphous solid dispersion, as water molecules floods into the molecularly dispersed matrix, the amorphous drug molecules there are diluted, resulting in transient drug concentrations moving from right to left of the phase diagram, i.e. towards the miscibility limit (i.e. $W_{d,1}$). As long as this transient drug concentration sweeps through the spinodal zone, the liquid-liquid phase separation would then occur. Similar argument may be applied to dissolution of neat amorphous drugs.

The rediscovery of LLPS has significant implications for oral drug absorption [48]. Based on the various experimental observations and the arguments from the phase diagram, LLPS is expected to occur widely, not only for amorphous solid dispersion systems. Once a transient supersaturated solution is formed above the amorphous solubility limit in the spinodal zone, LLPS is to occur, in the absence of crystallization. Therefore, the rediscovery of this phenomenon reveals a common mechanism of oral bioavailability enhancement behind many enabling formulation

Table 6.1 Reported equilibrium solubility and LLPS concentrations for various compounds at 37 °C

Compound	pH of medium[a]	Equilibrium solubility (μg/mL)	Predicted amorphous solubility (μg/mL)[b]	LLPS concentration (UV detection) (μg/mL)	LLPS concentration (DLS detection) (μg/mL)
Ritonavir	6.8	1.3 ± 0.20	20.6 ± 0.3	18.8 ± 0.07	18.6 ± 0.1
Ritonavir (in H_2O)	7.4	2.4 ± 0.03	39.2 ± 0.6	37.2 ± 0.9	39.8 ± 0.3
Efavirenz	6.8	8.2 ± 0.20	19.8 ± 0.7	18.4 ± 0.8	17.0 ± 0.03
Loratadine	6.8	1.6 ± 0.10	6.8 ± 0.4	7.6 ± 0.1	7.5 ± 0.4
Ketoconazole	10.0	3.7 ± 0.10	56.6 ± 1.5	54.4 ± 0.5	54.5 ± 0.7
Indomethacin	2.0	3.0	26.7 ± 1.0	30.4 ± 0.5	31.8 ± 0.8
Felodipine	6.8	0.94 ± 0.08	8.5 ± 0.2	9.8 ± 0.4	9.4 ± 0.4
Clotrimazole	10.0	0.4 ± 0.02	4.0 ± 0.2	5.2 ± 0.1	5.0 ± 0.5
Clozapine	10.0	8.8 ± 0.10	133.3 ± 1.8	135.8 ± 0.6	136.9 ± 1.7

[a]With water saturation
[b]Medium: 100 mM phosphate buffer, except indicated otherwise. Data source from [44]

technologies that are based on supersaturation, including amorphous solid dispersions, self-emulsifying drug delivery systems (SEDDS), cosolvent-based solubilization techniques, and some complexation methods. This liquid-liquid phase separation may also occur even with the conventional formulation approaches, including formulating with highly soluble pharmaceutical salt forms and weakly basic drugs [49]. A prerequisite seems to be the generation of local transient concentration that exceeds the amorphous solubility during in vivo drug release from the dosage forms. Table 6.1 lists the LLPS concentrations for a number of pharmaceutical compounds [44].

6.2.2 *Formation of Amorphous Drug Nanoparticles*

We have concluded that, in the absence of crystallization, supersaturation-based formulations often undergo liquid-liquid phase separation during in vitro and in vivo dissolution when a transient local drug concentration sweeps through the spinodal zone. These LLPS phenomena are believed to be quite common to various enabling formulation technologies and have thus important implications for oral drug absorption.

One remaining significant question is, given that enabling formulations (via supersaturation) undergo LLPS, why they often differ in oral bioavailability?

It is not difficult to see the important role played by the amorphous drug precipitates (strictly speaking, it is a liquid phase where a small amount of water is absorbed into the amorphous drug) upon the occurrence of the LLPS. At the beginning of LLPS during in vivo dissolution, the drug concentration equals to the amorphous solubility. The initial drug absorption is at a maximally achievable rate:

$$\frac{dX_{blood}}{dt} = k_{absorb}C_{GI} = k_{absorb}C_S^a$$

where C_S^a is the amorphous solubility, and k_{absorb} is the apparent rate constant of aborption.

When the kinetics of drug dissolution from the amorphous drug precipitates is sufficiently fast, drug concentration in the GI fluid can be maintained at the amorphous solubility, and the drug absorption rate will be sustained at the maximum rate. However, if dissolution rate is slow, drug concentration in the GI fluid will be at a lower level commensurate to the dissolution rate from these precipitates, and the oral absorption rate will be lower. The dissolution rate from the amorphous drug precipitate is mainly impacted by the particle size (available dissolution surface area) because the solubility is the same. Therefore, smaller particle size of the amorphous drug precipitate favors rapid dissolution and hence the oral absorption. But how small can be achieved, and, how to control their particle sizes under the dynamic in vivo environment?

Milky suspensions have long been observed in the dissolution of copovidone-based hot melt extrusion tablets such as Kaletra® and Novir®, however these observations were not published until much later [50, 51]. Photon correlation spectroscopy (PCS) characterization of ritonavir extrudate suspended in water (extrudate/water ratio correspond to 0.5 mg/mL ritonavir) resulted in nano-suspension with mean particle sizes in the range of 40–60 nm. These particles were observed to grow and might also cause a macroscopic precipitation over the experimental time frame of 6–24 h, resulting from particulate aggregations [50]. Asymmetrical flow field-flow fractionation (AsFlFFF) with online multi-angle light-scattering (MALS) particle size analysis of the nano- and micro-suspensions from ritonavir/lopinavir co-extrudate revealed three distinctly different types of colloidal to nanoparticulate assemblies in the aqueous dispersions in the order of elution: colloidal polymer with marginal amounts of APIs, API-rich nanoparticles, and nanoparticulate assemblies assigned to sorbitan monolaurate and/or hydrophilic fumed silica [51]. Similar findings were also encountered with the ASD of ABT-102 [52].

The above observations are consistent with those obtained during dissolution of amorphous solid dispersions containing high percentages of polymers, such as those from 90/10 HPMC/felodipine amorphous solid dispersion [39, 40]. In all these experiments, nano to submicron particles were observed during dissolution of amorphous solid dispersions that resembles colloidal appearance. These nano and submicron particles were also observed to grow in size with time, accompanied by reduction of total number of particles, indicating the occurrences of particulate aggregation.

From the point of view of oral drug absorption, formation of smaller amorphous drug nanoparticles is preferred because they dissolve sufficiently fast to replenish the depletion of drug concentration in the GI fluid during the dynamic dissolution/

absorption processes. It is also important that these particles be kinetically stable over the time frame relevant to the oral absorption. Even though LLPS occurs during in vivo dissolution of many enabling formulations, the difference in their formations and the behaviors of the colloidal/nano suspensions are believed to be responsible for their differences in oral bioavailability.

Harmon et al. reported a fine investigation on the dissolution-induced nanoparticle formation from copovidone-based anacetrapib amorphous solid dispersions containing 20% drug and 0–10% TPGS (*D*-α-tocopheryl polyethylene glycol 1000 succinate, or vitamin E TPGS) surfactant [53]. During the powder dissolution of the ASD (total drug concentration equals to 100 μg/mL of drug), the authors noticed that the amorphous drug nanoparticle (detected in the filtrate after passing through a 1-μm membrane filter) released rapidly and completely at the TPGS level of 3.5% and above (Fig. 6.10). However, no drug nanoparticle was observed at 0% and 1% TPGS levels. The rate of nanoparticle formation was slower and in a transition state at the 2% TPGS level.

Further probing of these systems indicated that polymer release was almost instantaneous, and within 1 min virtually all polymers were released regardless the level of TPGS in the ASD (Fig. 6.11). On the contrary, releases of the API and the TPGS surfactant were observed to be synchronized with each other and the rates depended on the surfactant levels (Fig. 6.12). SDS replacement experiment also confirmed that ~73%, 68%, and 55% of the TPGS were associated with the amorphous drug nanoparticles produced from the ASDs containing 2%, 5%, or 10% TPGS, respectively. These findings indicated that the surfactant played an important role in generating the drug nanoparticles. With other experimental observations, the authors proposed a scenario called "hydrophobic capture" during dissolution of the copovidone-based ASDs, referring to the amorphous drug aggregate formation due to hydrophobic drug-drug interactions when hydrophilic polymers rapidly leave the dispersion matrix during dissolution. The role of surfactant is to prevent an otherwise rapid, local drug domain aggregation event, thus allows the amorphous drug domains to escape the hydrophobic capture and diffuse into bulk solution as more stable nanoparticles. The size of the drug nanoparticles decreased with increasing level of the surfactant TPGS, consistent with the concept of particle surface coverage by surfactant.

Findings from the dissolution of anancetrapib/copovidone/TPGS ASD system can provide significant insight into the mechanisms on how amorphous drug nanoparticles are generated during dissolution, even though potential of this hydrophobic capture may be less in systems where less hydrophilic polymer (e.g. HPMCAS) is used. Reasonable factors affecting the hydrophobic capture may include the nature of the polymer (balance between hydrophilicity and hydrophobicity), nature and strength of the drug/polymer interaction, strength of drug/drug interaction, the strength of the water/drug and water/polymer interactions, and strength of interaction between the surfactant and the surface of the amorphous drug nanoparticles. In addition, the drug loading impacts the closeness of the drug molecules during dissolution after polymer escapes, thus can impact the hydrophobic capture. A lower drug loading favors the formation of drug nanoparticles, as confirmed by the 98/2 copovidone/anacestrapib

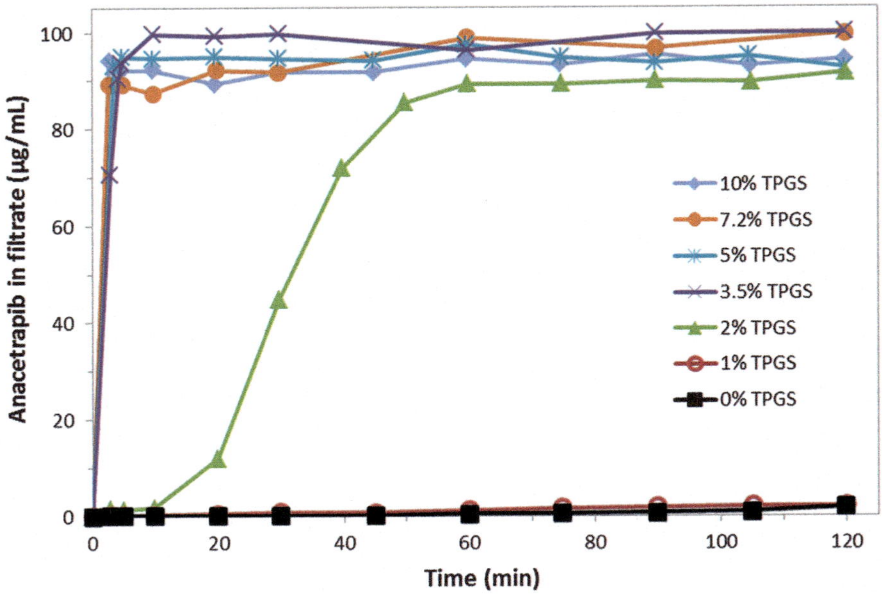

Fig. 6.10 Recovered anacetrapib from 1-μm membrane filtrate after anacetrapib/copovidone ASD containing various levels of TPGS were dispersed in water. Data source from [53]

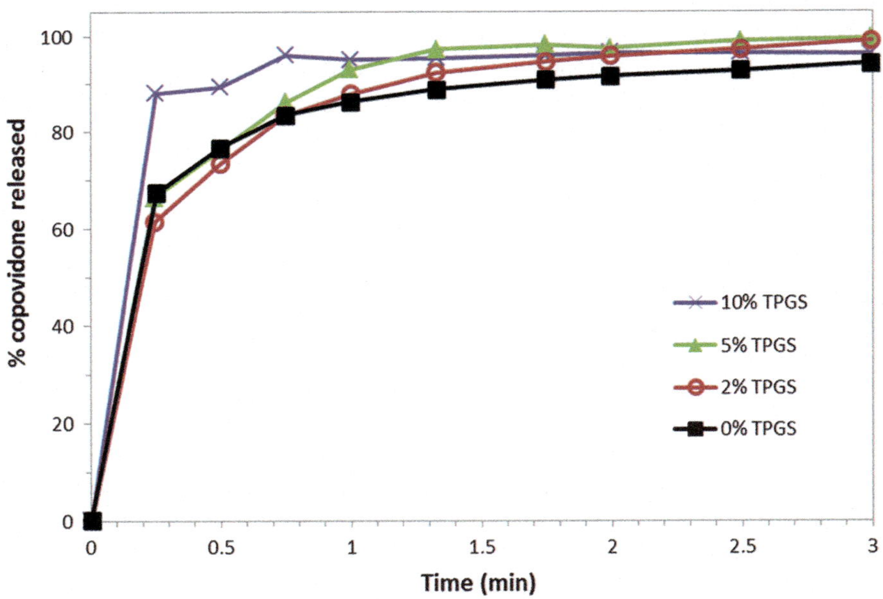

Fig. 6.11 Recovered copovidone from 1-μm membrane filtrate after anacetrapib/copovidone ASD containing various levels of TPGS were dispersed in water. Data source from [53]

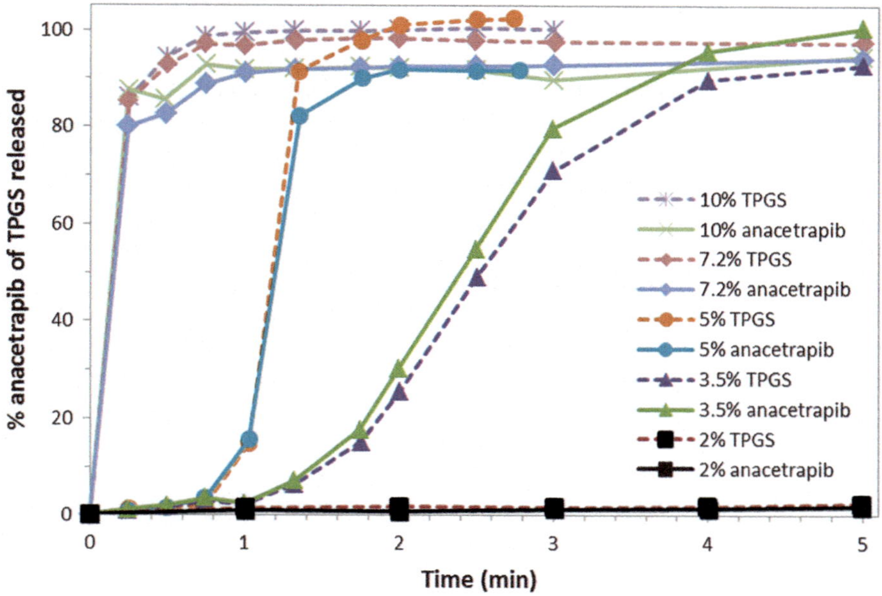

Fig. 6.12 Release of API and surfactant from 1-μm membrane filtrate after anacetrapib/copovidone ASD containing various levels of TPGS were dispersed in water. Data source from [53]

ASD, in which drug nanoparticles were generated without the need of any surfactant. This observation is consistent with ours, where ASDs of high polymer/drug ratio generated colloidal particles during dissolution [19, 39].

All these observations point to the important roles played by the relative and competitive strength of the intramolecular and intermolecular interactions among the four components during dissolution of amorphous solid dispersions: drug, polymer, surfactant, and water. Most polymers used in ASDs are hydrophilic in nature therefore dissolve quickly into bulk water, leaving mostly the amorphous drug molecules behind. Water interacts with polymers much more favorably, hence, as water molecules floods into the ASD matrix, they tend to pull the polymer molecules out into solution and push the drug molecules together. Without some external help such as the surfactant, the hydrophobic nature of drug-drug interaction often causes the formation of a drug-rich phase (amorphous drug phase separation) before drug molecules are released into the bulk. The amorphous drug phase separation upon contacting with dissolution media is consistent with the concept of hydrophobic capture. Once the amorphous phase separation occurs, amorphous drug nanoparticles become difficult to generate. The dissolution from the bulk amorphous drug phase is much slower than from amorphous drug nanoparticles, and the oral absorption will be compromised.

6.2.3 The Keys

This section has described the rediscovery of the liquid-liquid phase separation, which is a common phenomenon during dissolution of many enabling formulation technologies beyond the amorphous solid dispersion approach. LLPS has significant implications for oral absorption of drugs. The amorphous solubility provides a basis for oral bioavailability enhancement, but it also represents a practical upper limit that can be achieved by any enabling formulation. In order to fully utilize the advantage of amorphous solubility, the particle size of amorphous drug precipitate upon LLPS is important. Small and stable amorphous drug nanoparticle formation is necessary to push the limit of oral bioavailability enhancement. The drug loading, polymer, and surfactant all can play a role in how amorphous drug nanoparticles are formed and maintained in a fashion that is maximally beneficial to oral absorption. It is noted that crystallization of the drug molecules in the bulk ASD, or from the solution, or amorphous drug phase separation in the bulk ASD, upon contacting with the dissolution medium, disfavors the generation of LLPS and amorphous drug nanoparticles.

6.3 The Major Manufacturing Technologies for Amorphous Solid Dispersions

A good drug formulation cannot be developed without considering the manufacturing methods. Many unit operations in conventional drug product manufacturing, such as blending, milling, compression, capsule filling, coating, are also used in the production of amorphous solid dispersions. Most of these are treated in a conventional way and can be found in many textbooks. The mechanical properties of ASDs may require some special attention and will be deferred in a later part. This section introduces the two key manufacturing processes used to produce amorphous solid dispersions on commercial scales, namely, hot melt extrusion (HME) and spray drying (SD).

6.3.1 Hot Melt Extrusion (HME)

Hot melt extrusion is a landmark manufacturing process for amorphous solid dispersions. The technology itself has long been utilized in the polymer industry [54] and its pharmaceutical applications started to emerge in the 1970s. Since the introduction of the first commercial melt extrusion product by AbbVie, this technology has gained widespread recognitions in the manufacturing of ASDs and has resulted in the approval of many commercial products [55–63]. The proprietary

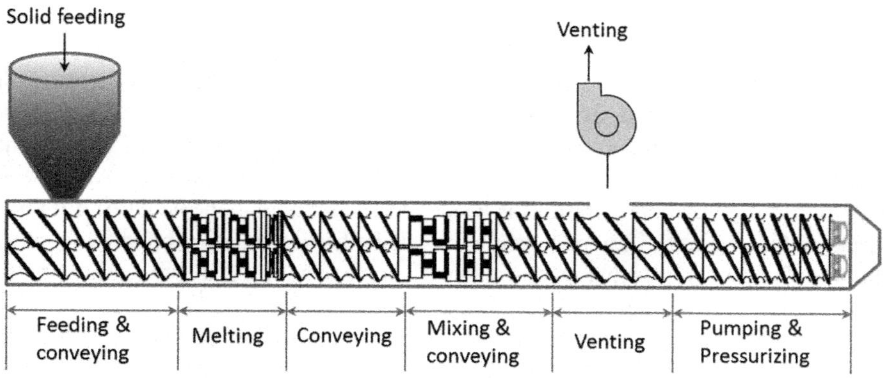

Fig. 6.13 Schematic of a twin-screw extruder and typical processing zones for HME

Meltrex™ technology underlines the success of many of AbbVie's commercial ASD products including Kaletra®, Norvir®, Viekira™, and Venclexta™ tablets.

While single screw extruders are still available, the most typical pharmaceutical HME processes utilize a co-rotating twin screw extruder (TSE). Twin-screw extruder provides higher shear, improved conveying and mixing, easier material feeding and overall a more reproducible and robust amorphous solid dispersion manufacturing process. It represents the current state-of-the-art manufacturing process of ASDs via the melting method.

A twin screw extruder consists of a barrel or a set of modular barrels that can be heated and cooled and separately controlled, enclosing a set of twin screws which convey, compound, melt, mix materials inside the barrel, and subsequently force the melt mixture through a die mounted at the end of the barrel. During extrusion process, the solid drug/polymer/excipient mixture is plasticated under the induced shear force and applied heat to form a homogenous melt before the terminal extrusion step. A passive or active (e.g. by applying vacuum) venting step may be necessary prior to the final extrusion step. In order to facilitate downstream processing, a strand pelletizer with cutter or a calendar may be used to produce glassy pellets/lentils of suitable sizes, which are often further cooled on an air-cooled conveyor belt.

Figure 6.13 represents a schematic of twin-screw extruder. The modular designs of both screw and barrels are commonly utilized in the design of pharmaceutical HME TSEs, which allows the length of the overall screw and the configurations of the screw to be changed as needed, thus provides flexibility of HME process development to accommodate the varying properties of different drug molecules and formulations. Modular barrels can be flanged together or bolted together. The entire extruder may be divided into different processing zones (e.g. feeding, conveying, liquid addition, melting, mixing, venting, pumping), each consisting one or more barrels. Different temperatures can be applied from one barrel to another to facilitate the processing goal of each zone.

6.3.1.1 Screw Configurations

The two screws have identical lengths and matching screw configurations, each consisting of a shaft and segmented screw elements in the same order. Each screw element can have different designs and lengths. They can be combined in different ways, thus creating a wide array of possible screw configurations. Again, the flexibility in screw configurations is needed to accommodate the needs of different drug molecules and different formulations. It is the combination of screw configurations/barrel controls that work together to achieve appropriate processing objectives such as conveying, melting, mixing, and extruding.

A screw element may be specified by inside screw diameter (D_i), outside diameter (D_o), pitch (S), flight distance (axial and perpendicular), perpendicular flight width (e), helix angle (φ) and flight clearance. A schematic of screw element is shown in Fig. 6.14.

Different parameters may be used to characterize a screw element. The outer diameter-to-inner diameter ratio (D_o/D_i) of a screw element is related to the free volume, which increases with increasing D_o/D_i. The pitch of a screw, S, is an important parameter and is usually expressed as pitch-to-diameter (S/D) ratio, such as 1D, 3/2D, 2D, etc. A standard element has a pitch of 1 S/D, corresponding to a helix angle of ~17.7°. Given the relationship $tan\varphi = S/\pi D$, the helix angle, φ, increases as S/D increases. High pitch (1.5–2D) screws provide better conveying speed and are usually used for powder feeding. Low pitch screws ($<1D$) have lower conveying speed but compress material better and are often found in the pressure build-up zone and are used to pump melt. Screws of moderate pitches (1–1.5D) are often used to transport melt or to compress powder after feeding. Figure 6.15 shows conveying screws of various pitches.

Kneading blocks are created by combining multiple single kneading discs (also called mixing elements) with different offset angles (30°, 45°, 60° or 90°) between the adjacent discs. The kneading discs are often made of self-cleaning profiles. The offset angle determines the conveying and mixing properties. Increasing offset angles, conveying capability decreases but the mixing capability increases (which is similar to decrease the helix angle for the conveying screws). When a 90° offset is utilized, the block then has pure mixing but no conveying capability so it is also called a neutral kneading block. The common kneading discs have a width of 1/4 S/D. Increase the disc width (e.g. 1/2 S/D) introduces higher shear while a narrower disc (e.g. 1/8 S/D) increases chopping and promote dispersive mixing. Kneading block can be made right-handed or left-handed. While right-handed block conveys material forwards, the left-handed one creates back flow and is called reversing kneading block. Figure 6.16 shows mixing blocks of various designs.

There are two types of mixing in twin-screw extruder: distributive and dispersive. Distributive mixing refers to spatial rearrangement of species and does not reduce the domain sizes (e.g. distributing de-agglomerated particulates throughout space). On the contrary, dispersive mixing refers to the process that reduces the

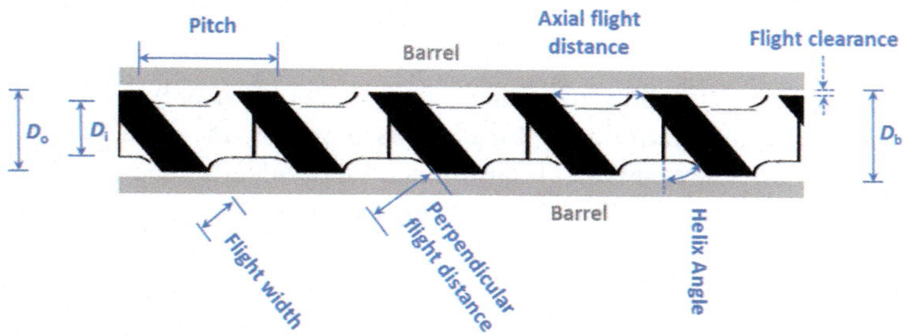

Fig. 6.14 A schematic of a screw element

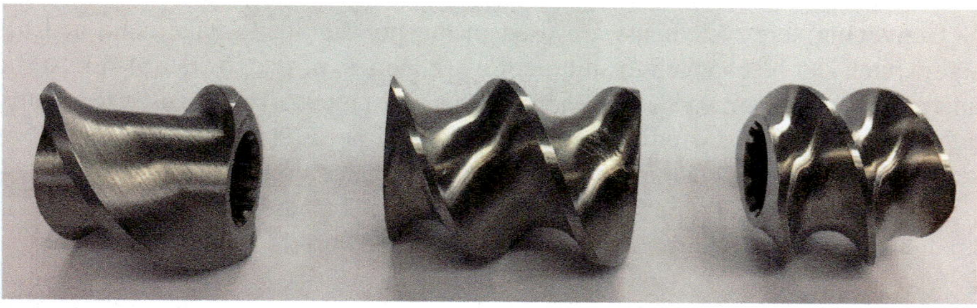

Fig. 6.15 Example conveying screws (from left to right: 2 S/D, 1.5 S/D, and 1 S/D)

Fig. 6.16 Example mixing blocks consisting of five mixing discs (left, 45° offset, 1/5 S/D disc; middle: 90° offset, 1/5 S/D disc; right, 45° offset, 3/10 S/D disc)

domain sizes via shear-induced stresses (e.g. droplet deformation and break-up). Dispersive mixing requires high flow stresses, such as high viscosity under high shear, in order to provide the dispersive forces to overcome the cohesive forces of the agglomerates or immiscible droplets; whereas distributive mixing is dictated only by the flow-generated strain and does not require high stresses. The conveying screws have more characteristics of distributive mixing while the kneading blocks (particularly the neutral or reverse flowing ones) have more characteristics of dispersive mixing.

6.3.1.2 Feeding/Conveying

During HME, the active pharmaceutical ingredient (API), polymer, and/or other excipients are usually introduced together as a pre-blend via a gravimetric or volumetric feeder. They can also be introduced individually, and in this case, via gravimetric feeders to accurately control the formulation composition. As indicated previously, twin screw extruder has excellent conveying capacity, greater than the commonly used feeding rate (i.e. starve fed). As a result, TSE is under fill except at the melting and mixing blocks. This feature is advantageous because the screw speed becomes independent of the feeding system and can be controlled independently to optimize the extrusion process without impact the feeding rate, unlike the single-screw extruder.

Conveying screw elements are used in the powder feed intake zone and the screw pitch can be changed to suit the flow properties of the powder blend. At the adjacent downstream, screw of somewhat decreased pitches may be used to slightly compress the powder. Additional feed ports may be used to introduce liquid or semi-liquid materials such as surfactant and plasticizer along the screw.

Low temperature is used for the feeding barrel but the barrel temperature is gradually increased before the material is conveyed into the melting zone.

6.3.1.3 Melting

The melting zone consists of a combination of kneading blocks of various designs. As discussed, the offset angle between the adjacent kneading discs determines the conveying/mixing capability. Mixing capability increases as this offset angle increase, at the expense of the conveying capability. Increasing the disc width increases the shear but will also increase the melt temperature. A short section of reverse-flowing kneading block may also be added to increase back mixing and residence time in this zone to facilitate melting.

Melting process is impacted by a number of factors, including degree of filling (higher is better), pellet size (smaller is better), melt viscosity (higher is better), kneading discs (narrow is better), and the residence time (longer is better).

It should be noted that the energy required for melting comes primarily from the mechanical shear, and may not by the barrel heating. Therefore, the screw configurations in the melting zone and the screw speed greatly impact the melting process, while the barrel temperature may have negligible effect. Indeed, during very aggressive melting process, the barrel may actually serve as a heat sink. It is therefore very likely that the temperature of the melt near the screw surface is significantly higher than the barrel temperature in the melting zone.

6.3.1.4 Mixing of Melt

The melt formed in the melting zone is conveyed to a mixing zone, which again, consists of a kneading block or combination of kneading blocks. The purpose is to create a homogeneous melt mixture via dispersive mixing, and to form amorphous solid dispersion at the molecular level (glassy solution). It should be noted that the API/polymer/surfactant mixture coming out from the melting zone may not melt fully and may contain some residual crystals. This additional mixing zone provide further opportunity for those residual crystals to melt the before extrusion step.

The mixing efficiency is dependent upon the screw speed, throughput, viscosity of the melt and the screw geometry. The greater the screw speed and the smaller the throughput, the better is the mixing performance. Dispersive mixing is promoted by neutral or reverse kneading blocks. However, care need to be taken to avoid potential chemical degradation. It is a common approach to introduce multiple small mixing sections to achieve the desired mixing performance in the end and mitigate potential chemical degradation.

6.3.1.5 Devolatization

The fully mixed melt is further transported to the devolatization zone to remove air and residual moisture/solvent. Bubbles may be formed if these gases are not removed sufficiently before extrusion. Venting is generally achieved by opening the top barrel of the devlolatization zone. Conveying screws of slightly increased pitches over the previous section are often used in this zone to decrease the fill level so that the melt is not drawn into the vent stream located at the top barrel section. Typically, vacuum is applied to facilitate the degassing process.

6.3.1.6 Extrusion/Discharge

Extrusion zone is located at the end of the extruder in the discharge zone. The aim is to build up the pressure require for extrusion with as little energy as possible. Discharge feed screws are used for this purpose to ensure full fill level and stable melt flow through the die.

The critical attributes of melt extrudate are the residual crystallinity, homogeneity, and degradation products. These may be impacted by the formulation compositions, melt viscosity, barrel temperature, screw speed, and feed rate.

6.3.2 Spray Drying

Spray drying is another common technology for manufacturing amorphous solid dispersions [12, 64–67]. Similar to solvent coating, the formulation is first dissolved in a solvent or solvent mixture. The solution is then pumped through a nozzle into a drying chamber. The droplet are atomized at the exit of the nozzle tip, and dried by a stream of hot air. In typical coating process, spray drying of the coating liquid should be minimized to facilitate the adsorption of the partially dried droplets onto the coating surface. In spray drying, the droplets need to be dried sufficiently before they may land on a surface and cause sticking. Finally, the dried material is separated from the drying medium using a cyclone and collected in a collection device. The solvent vapor is then removed by a condenser and the remaining air is recycled to the spraying dryer. A secondary drying of the collected SD particles may be needed to further reduce the level of the residual solvent. Figure 6.17 shows a schematic of spray-drying process.

The characteristics of spray dried materials are determined by the formulation, the droplet size, and the drying kinetics. Atomization controls the size of the spray droplets, which results in a large increase of product surface area. The droplet size is impacted by formulation composition (drug, polymer, and other components), the solvent used, the solid concentration of the spray solution, spray solution viscosity and surface tension, nozzle design and geometry, spray rate, and other spray parameters (e.g. pressure, temperature). The drying kinetics is impacted by the mass and heat transfer between the carrier gas and the droplet surface, the mass and heat transfer inside the droplet, droplet size and the characteristics of the formulation of the droplet (composition, viscosity and surface properties).

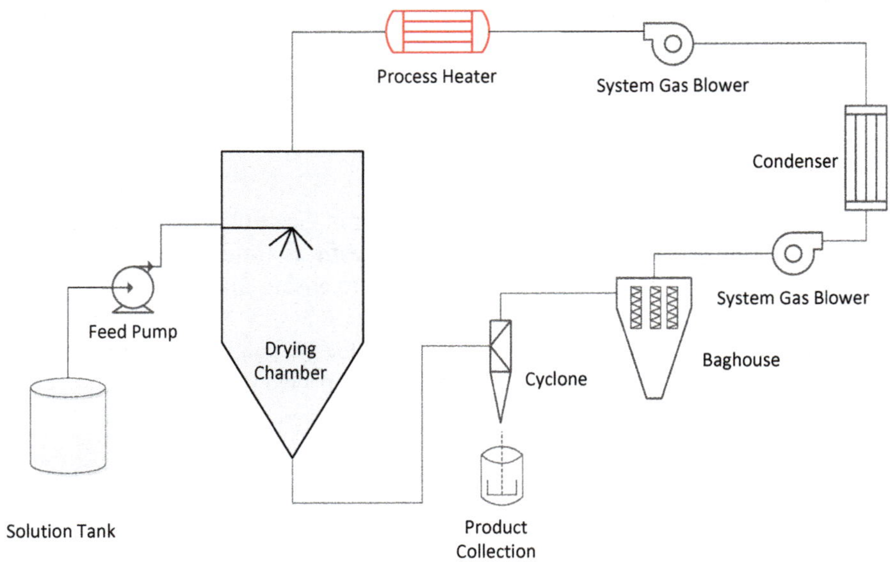

Fig. 6.17 Schematic representation of spray drying process. Modified from [68]

Like a coating process, many process parameters can impact attributes of spray-dried materials [12, 69]. Important process parameters include the inlet/outlet temperature, drying air flow rate, solution feed rate, humidity or solvent activity, spraying conditions (nozzle design and atomization), solution solid content, viscosity, and surface tension. These process parameters often convolute with each other, therefore, an empirical design of experiment (DOE) approach would tend to be cumbersome, laborious, and ultimately unreliable to achieve process understanding, design and scale-up. However, characteristics may be derived from these parameters to achieve better understanding and simpler methodologies for design and transfer of the spray drying process. Here we describe the thermodynamics and particle drying kinetics during spray drying.

6.3.2.1 Thermodynamics of Spray Drying

The thermodynamics of spraying drying process may be modeled based on the principles of conservations of heat and mass transfer, similar to what is generally applied in the coating process.

Conservation of Heat:
inlet air enthalpy + atomization air enthalpy + spray liquid enthalpy = exhaust enthalpy

$$h_{in}m_{in} + h_A m_A + h_{solv.}(1 - x_{solid})m_{soln} = h_{out}m_{out} \tag{6.4}$$

$$m_{in} + m_A = m_{out} \tag{6.5}$$

Here h is the specific enthalpy (kJ/kg), m is the mass flow rate (kg/h), and x is the solid content (w/w) of the spray solution. The subscripts, *in*, *A*, *solv*, *solid*, *soln*, *out*, designates parameters for the inlet, atomization, solvent, solid, solution, and exhaust, respectively.

The enthalpy of the air/solvent mixture is a function of both the temperature and the solvent partial pressures. Therefore, the exhaust temperature, T_{out}, can be calculated based on the enthalpy h_{out}, or the conditions at the inlet, atomization, and solution spray, as below.

$$h_{out} = \frac{h_{in}m_{in} + h_A m_A + h_{solv}(1 - x_{solid})m_{soln}}{m_{in} + m_A} \tag{6.6}$$

The above equation may be simplified [68] using heat capacity, C_p, and heat of vaporization, ΔH_{vap}, as:

$$T_{out} = \frac{m_{in}T_{in} \times C_{p,sa}(T,\omega) + m_A T_A \times C_{p,sa}(T,\omega) - \Delta H_{vap}(1 - x_{solid})m_{soln}}{(m_{in} + m_A) \times C_{p,sa}(T,\omega)} \tag{6.7}$$

The subscript, *sa*, designates parameters for solvent/air mixture. The heat capacity of the solvated air is a function of temperature and composition, ω, which is the solvent ratio of the air, expressed as mass of solvent per unit mass of the dry air (kg/kg_{da}),

$$\omega = \frac{M_{solvent}}{M_{da}} \tag{6.8}$$

The subscript, *da*, designates parameters for dry air. It is a common approach that the inlet temperature, T_{in}, is varied in order to achieve a desired outlet temperature, which can be obtained by reversing Eq. (6.7).

Conservation of Mass (solvent):
The total organic solvent at the outlet comes from the inlet, the atomization, and the spray solution, i.e.,

$$\omega_{in}m_{in} + \omega_A m_A + (1 - x_{solid})m_{soln} = \omega_{out}m_{out} \tag{6.9}$$

Therefore, the solvent ratio at the outlet is expressed as,

$$\omega_{out} = \frac{\omega_{in}m_{in} + \omega_A m_A + (1 - x_{solid})m_{soln}}{m_{in} + m_A} \tag{6.10}$$

The mass flow rate is related to volumetric flow rate (m^3/h), Q, by:

$$m = \frac{Q}{\nu(1 + \omega)} \tag{6.11}$$

Here the specific volume of air/solvent mixture, ν, expressed as cubic meter per kg of dry air (m^3/kg_{da}), is the volume of air/solvent mixture divided by the mass of the dry air:

$$\nu = \frac{V}{M_{da}} = \frac{V}{28.9645 n_{da}} = \frac{1000RT}{(1 + \omega)P}\left(\frac{1}{MW_a} + \frac{\omega}{MW_{solv}}\right) \tag{6.12}$$

where n, R, T, P, and MW are the number of moles, gas constant, temperature, pressure, and molecular weight (g/mol), respectively.

Similarly to relative humidity (RH), relative solvent saturation, RS, may be defined as the ratio between the vapor pressure of the solvent in the air/solvent mixture to the saturated solvent vapor pressure at the that temperature. Then the relative solvent saturation at the outlet, RS_{out}, may be calculated as:

$$\begin{aligned}\%RS_{out} &= 100 \times \frac{P_{chamber} w_{out}}{\left(\frac{MW_{solv}}{MW_{da}} + \omega_{out}\right) p^*_{solv}} \\ &= 100 \times \frac{P_{chamber}\left[\omega_{in} m_{in} + \omega_A m_A + (1 - x_{solid}) m_{soln}\right]}{\frac{MW_{solv}}{MW_{da}}(m_{in} + m_A) + \omega_{in} m_{in} + \omega_A m_A + (1 - x_{solid}) m_{soln}}\end{aligned} \quad (6.13)$$

where p^*_{solv} is the saturated vapor pressure of the solvent at the outlet temperature, T_{out}.

The above treatments assume ideal gas properties of the air/solvent mixture and each of the individual components, as well as the complete removal of the solvent from spray droplets. This treatment allows one to calculate the expected exhaust temperature and the corresponding solvent content in the exhaust air stream under thermodynamic equilibrium conditions. It is useful to determine whether the inlet air can provide sufficient drying capacity needed by the process. It is also used to determine the inlet conditions in order to accommodate a desired outlet condition. If significant residual solvent persists in the spray-dried particles, the solid content in Eq. (6.13) may need to be adjusted to reflect an effective increase in the solid content.

It should be noted that spray drying is nearly an adiabatic process in that all enthalpy comes from the inlet air, atomization air, the solution spray and the heat loss can be neglected. The Mollier diagram (or psychrometric chart in case of air/moisture) is a handy tool for process engineers. Figure 6.18 shows a psychrometric chart that may be useful for spray-drying of aqueous solutions. Due to the nature of the isenthalpic process, spray drying conditions moves along an isenthalpic line, as demonstrated by the red line in Fig. 6.18. Therefore, one can quickly read out the exhaust conditions when the inlet condition is known in conjunction with a targeted parameter value at the exhaust (e.g. temperature, or solvent saturation), and vice versa.

6.3.2.2 Droplet Drying Kinetics During Spray Drying

The drying kinetics of the sprayed droplets determines the morphology of the spray dried material, which may impact the downstream process (flow, compaction) as well as product performance (i.e. phase separation, stability, oral absorption). While the process is complex and a complete model has yet to be developed, some qualitative and semi-quantitative descriptions of the process are available [67, 70–73].

Droplet drying and particle development may be divided into three stages: stage 1 droplet, stage 2 wet particle, and stage 3 dry particle, which reflect different kinetic controlling mechanisms of the drying process (Fig. 6.19).

Before stage 1 is initiated, the droplet undergoes rapid heating without much mass change. In stage 1, the rate of solvent evaporation is nearly constant (so as the droplet temperature), reflecting the fact that the evaporation kinetics is controlled by the rate of heat transfer from the drying stream to the droplet. Evaporation results

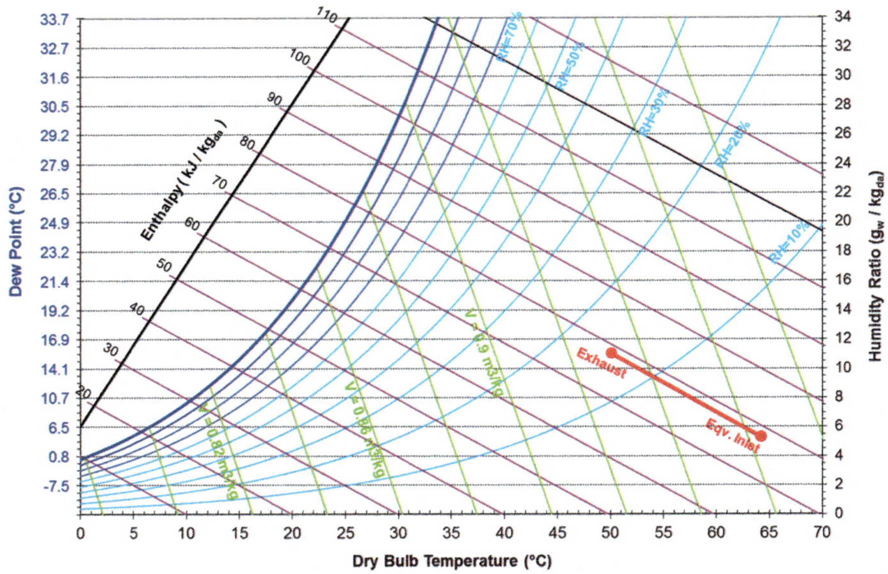

Fig. 6.18 Example of a psychrometric chart for spray drying of aqueous solutions

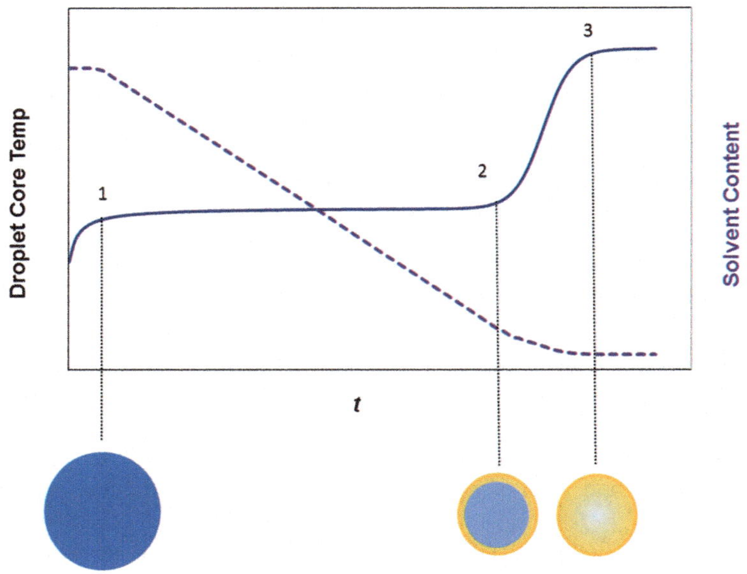

Fig. 6.19 A schematic of droplet drying. Modified from [67]

in a receding droplet surface, thus concentrating the solutes (drug and polymer) at the surface and creating a concentration gradient between the droplet surface and the core. As a result, the solute molecules move inward and the solvent moves outwards. However the diffusion rate of the solutes will slow down and may no longer be able to keep up with the rate of evaporation at some later time point. A shell/crust starts to form on the droplet surface (i.e. "skinning") and retards the outward movement of the solvent, marking the beginning of the stage 2. The rate of evaporation drops with time during stage 2 and the drying kinetics is controlled by the solvent mass transfer across the shell. Nevertheless, heat continues to supply from the inlet air and the heat transfer exceeds that is needed by the mass transfer, resulting in an increase in the temperature of the wet particles. Finally, as the shell becomes thicker and thicker, mass transfer may be insignificant at the end, resulting in minimum solvent level that can be achieved during stage 2. However, as heat is continuously provided to the particle, the resulted temperature increase leads to a build-up of internal pressure inside the particle. Depending on the strength, thickness of the shell and extent of pressure build-up, the particle can explode, inflate or crack, releasing the remaining solvent in a burst manner, which marks the stage 3 of the drying and the formation of dry particle.

6.4 Considerations for Drug Product Development

The previous sections have discussed the underlying mechanism of oral bioavailability enhancement via LLPS and the generation of amorphous drug nanoparticles, and basic characteristics of the manufacturing processes of amorphous solid dispersions, both of which are important in the development ASD products. This section discusses other various considerations necessary for a rational drug product development.

6.4.1 Pharmaceutical Properties

The chemical structure, physicochemical and biopharmaceutical properties are the starting point to understand a drug molecule, to envision how different barriers may be encountered during manufacturing, storage, in vivo dissolution, and oral absorption of the drug, and to formulate strategies to address these challenges [74–76].

Hydrophobicity (e.g. octanol/water partition coefficient, $\log P_{o/w}$) is a property related to the permeability and biopharmaceutical dispositions of the drug molecule in the body (e.g. elimination via excretion and/or metabolism). Hydrophilic molecules (low permeability) tend to be eliminated unchanged from the body into urine and bile; whereas hydrophobic (high permeability) molecules tend to be extensively metabolized in the body, according to the Biopharmaceutical Drug Disposition Classification System (BDDCS) [77–79]. Solubility is directly related to drug dissolution and absorption kinetics, and together with the permeability, determines the Biopharmaceutical Classification System (BCS) of the molecule [80] upon which many modern regulatory considerations are based. While many poor water soluble drugs are BCS II (low solubility/high permeability) molecules, with increasing trend in molecular sizes, more and more drug candidates appear to fall into the BCS IV (low solubility/low permeability) or board line BCS II/IV category as passive diffusion via the gut membrane becomes more difficult with the increasing sizes.

The acidity constant, p*K*a, for weakly acidic and weakly basic compounds, is important in our understanding of the solubility, dissolution, and absorption behavior in the GI tract. For example, weak acids have higher solubility at higher pH GI regions coinciding with the site of absorption (e.g. small intestine), thus favors absorption. Weak bases, on the contrary, are solubilized in the stomach where pH is lower. The solution of the weak bases in stomach becomes supersaturated in GI regions where absorption takes place and may precipitate as amorphous particles in the absence of crystallization, i.e., LLPS. These precipitates, if controlled properly (e.g. nanoparticle size), may provide adequate dissolution rate to replenish drug concentration in the GI fluid thus facilitate oral absorption. These two scenarios both can promote oral absorption, albeit in different ways. It is worth to note that a suitable salt of many poorly water-soluble acidic or basic drug molecules may adequately enhance oral bioavailability. Therefore a conventional formulation and manufacturing process can be developed, eliminating the need to resort to the ASD technologies which are more complex and resources demanding.

Chemical stability in both the solution and solid state are important factors for drug development and can impact the shelf-life of a drug product. However, chemical degradation tends to occur less frequently in solid dosage forms of poorly water soluble drugs. A main reason is that degradations in solid dosage forms are frequently modulated by moisture and the very low aqueous solubility for today's insoluble compounds plays favorably to the chemical stability of these drug products [75]. However, it does not necessarily indicate that chemical instability is no longer an issue. Rather than exhibiting during the storage of a final product, chemical degradation may arise as a significant issue during the manufacturing process, where the high molecular mobility in a solution or in a melt liquid state, resulted from the process of manufacturing ASDs under much harsher conditions than normal product storage, can create degradation issues.

Solid state properties of both the crystalline and amorphous forms are important for developing amorphous solid dispersions. The crystallinity and the melting parameters (melting temperature, melting enthalpy and melting entropy) are clearly related to the strength of interaction between the drug molecules and the processing conditions during manufacturing of ASDs (e.g. melt extrusion). The free energy differences between the amorphous and the crystalline form is related to the solubility advantage of the amorphous form and determines the theoretical enhancement in bioavailability of ASD. The glass transition temperature and the change in heat capacity are related to the glass formation ability of the molecule and the crystallization propensity of the glass, all of which have implications to developing amorphous solid dispersions.

6.4.2 Physical Stability of Amorphous Forms and Amorphous Solid Dispersions

The amorphous form is of higher free energy so the conversion to a crystalline state is inherently favorable and is a thermodynamically spontaneous process. Hence, the kinetic stability of the amorphous forms and amorphous solid dispersions need to be addressed both scientifically and practically to comply with regulation requirements. Chapter 5 has dealt with the subject of physical stability. Below we present a brief discussion within the context for formulation development.

Physical stability of amorphous forms has been a concern and the focus of research since the amorphous emerged as an alternative approach to crystalline forms in the pharmaceutical field. Molecular mobility has long been recognized [10] as a primary factor impacting the stability of amorphous phases. The simple fact is that crystallization proceeds at a much faster rate at above T_g than below. However, physical stability is not determined by molecular mobility alone. For example, amorphous forms of small molecules such as acetaminophen and griseofulvin are very unstable even when the they are stored below their respective glass transition temperatures [81, 82], whereas the amorphous forms of a number of flexible, large drug molecules such as ritonavir and lopinavir, are very kinetically stable even above the glass transition temperature [81]. It is worth to note that molecular mobility is comparable at the glass transition temperatures. These observations cannot be explained by molecular mobility alone. It rather indicates that the physical stability of amorphous forms also has origins intrinsic to the molecular structures. One significant finding of the authors is the recognition of the configurational entropy as a thermodynamic factor in determining the intrinsic physical stability of an amorphous molecule [83, 84]. The higher the configurational entropy, the larger number of conformations (translational and rotational) that a molecule can experience, thus the lower the probability a molecule will assume the proper conformation in order to crystallize into the specific configurations required to pack into a crystal lattice. From experience, people have well recognized that large, flexible molecules are in general more difficult to crystallize in supersaturated solutions. In particular, crystallization of proteins have been found quite challenging, primary due to the huge number of available conformations the molecule can have, consistent to the our observation.

The molecular mobility and the configurational entropy in combination may provide a quick risk estimate on the crystallization tendency of an amorphous form. A "kinetically stable" amorphous form is expected if it has low mobility and high entropy (large, flexible molecules). On the contrary, physical stability might be of concern if an amorphous form has high mobility and low configurational entropy (small, rigid molecules). Moderate physical instability may be associated with the intermediate molecules, i.e., either low mobility/low entropy or high mobility/high entropy combination.

The above evaluation concerns only the physical stability of a pure amorphous form. However, it also provides significant insights to the physical stability of

ASDs. Amorphous solid dispersion, whether it is a single phase glassy solution (molecular dispersion), or it is phase separated, has a lower free energy and is more stable than the pure amorphous phase, as long as there is some extent of miscibility. Therefore, a kinetically stable amorphous form translates into a kinetically more stable amorphous solid dispersion in a straightforward manner. Hence, physical stability of amorphous solid dispersion of this class of molecules represents the lowest risk and no special attention may be required on this subject. A "kinetically unstable" amorphous form could still lead to kinetically unstable amorphous solid dispersion. Therefore this class of amorphous system represents the highest risk in terms of physical stability of ASD development and requires careful attention and mitigation. For molecules that fall in the middle, i.e., those having low mobility/low entropy or high mobility/high entropy, some efforts of stabilization may be required to address the physical stability risk. It is important to recognize, however, that low entropy molecules require a different strategy from high mobility molecules, in order to address the physical instability more effectively. Decreasing molecular mobility such as by increasing the glass transition temperature of the matrix is more effective in improving the physical stability of molecules that has high mobility. For molecules with low entropy, theoretically we should target to increase the entropy so crystallization tendency is reduced. However, entropy is more of an intrinsic property of the molecule and it is not readily modifiable. Introducing strong drug-polymer interactions helps to lock certain conformations of the drug molecule in the matrix, thereby increase the energy barrier for the drug molecule to assume the specific conformation needed for crystal packing. This approach essentially imparts part of the conformations of the polymer to the drug molecules thus increases the configurational entropy and enhance the physical stability of the amorphous drug molecules. It can be deemed as rendering drug/polymer complex as a single entity. The stronger is this interaction, the more effective is the stabilization. Interaction of drug with other components of the formulation than the polymer may also be utilized in a similar suit. It is noted that stabilization is often achieved by lowering molecular mobility (e.g. antiplastization) and at the same time enhancing interaction between drug and formulation components [85, 86].

6.4.3 Biopharmaceutical Considerations

Physicochemical properties including solubility, permeability, p*K*a need to be considered together with the dose strength and biopharmaceutical properties of the drug molecule. While BCS class II drugs may be absorbed throughout small and large intestine, BCS IV or board line BCS II/IV drug molecules are generally not well absorbed in the lower GI tract (e.g. colon) thus may have a limited absorption window that is limited in the upper GI tract (e.g. duodenum, jejunum, ileum). Molecules that are substrate for efflux transporters (e.g. P-glycoprotein, or P-gp), or metabolizing enzymes (e.g. CYP 3A and other cytochrome P450 enzymes) have intrinsic tendencies to limit their concentration in the body and therefore present significant challenges to oral bioavailability. Even if the molecule has good permeability in the colon, the limited fluid volume in the region can still present a barrier for good drug absorption. It is even more challenging to develop a once-daily sustained release dosage regimen for molecules not well absorbed in the lower GI tract. For many extremely poorly water soluble drugs, often better rewards may be realized if the efforts are focusing on improving oral absorption in the upper GI tract.

Food effect can be significant for most poorly water soluble drug molecules. The reasons can vary, however. Food may affect the bile flow, splanchnic blood flow, GI pH, gastric emptying, and cause physical/chemical interactions with the drug. Food can significantly alter the physiology of the GI tract. During the fasting phase, the stomach first goes through a quiet period of 30–60 min, then a period of irregular contraction, and ends up with a house keeper contraction, where the high amplitude contractions empty the entire stomach content into small intestine. During fed state, the stomach undergoes regular and frequent contractions to facilitate the grinding of food contents. Large solid particles >2 mm are retained for further processing but smaller ones are emptied when the pylorus opens up briefly in response to the peristaltic wave. Water is generally emptied in minutes. The nature of food contents also strongly influences the stomach emptying time and fats may stay in stomach significantly longer than proteins and carbohydrates. In addition, the stomach pH is generally much lower (~2) in the fasted state than in the fed state (pH 4–6). Fed state also increases the bile flow and splanchnic blood flow. The volume of intestinal fluids may increase two- to threefolds following a meal and the levels of phospholipids and bile salts in the gut also may increase four- to fivefolds. Obviously, the above factors can impact the oral absorption in various ways. For example, the solubility of weakly acidic and weakly basic drug molecules is certainly impacted by the gastric pH. Many poorly soluble drugs are mainly absorbed in the upper GI regions, therefore, the components of a fatty meal may help to solubilize the hydrophobic drugs in the stomach and release the drug molecule to the small intestine gradually over a longer time period, thus significantly increase the oral bioavailability. On the contrary, the entire content from a dosage form is quickly flushed into the small intestine during fasting state, and the fraction not dissolved in the upper GI are moved into the lower bowl region, where it may not be well absorbed. Digestive products of food contents will also impact drug solubilization and oral absorption. The triglycerides in the fatty meals form mono- and di-glycerides when partially digested. These partially digested fats form various mixed micelles in the presence of bile salts, and can contribute to drug solubilization and oral absorption [87].

The BDDCS has also been used to evaluate the potential impacts of uptake and efflux transporters on the oral availability of a drug molecule and the potential food effect [88]. Based on BDDCS, food effects may be predicted if intestinal transporters are involved in drug absorption, in which the high fat meals may provide an inhibitory effect. BDDCS predicts that high fat meals will have no significant effect on Class 1 drugs. Despite many of them are transporter substrates, the effects are negated due to their high gut permeability and high intestinal fluid solubility. Class 2 compounds, which are highly metabolized and therefore are often dual substrates of enzymes and transporters, are predicted to increase in bioavailability due in part to transporter inhibition. Class 3 compounds, which are poorly metabolized and poorly permeable, often rely on uptake transporters, and transporter inhibition results in a reduction in oral absorption after a meal. Of course, all these are general rules and in reality, food effect can be very complex.

6.4.4 Process Considerations

Manufacturing of amorphous solid dispersions may be divided into upstream (pre-ASD), ASD, and downstream (post-ASD) stages. Depending on the selected ASD technology, different process trains may be required for the upstream and downstream processes. No doubt, the focus is on the ASD stage.

A wide range of methods of preparing amorphous forms have been reported in literature [11, 12], including melting, solvent evaporation (simple evaporation, rotary evaporation, spray drying), cryo-milling, compaction, vapor deposition, freeze-drying and supercritical fluids (SCF). However, to obtain stable amorphous forms, it is necessary to eradicate all long range orders in the original crystal packing, otherwise this order may serve as the nuclei and harm the physical stability of the prepared amorphous forms during storage. Additionally, the nuclei may induce crystallization during in vivo dissolution so negatively impact the oral bioavailability. Mechanical trituration methods such as milling/compaction are inefficient to eradicate the long range order entirely and are of low efficiency. These methods are therefore not suitable for manufacturing amorphous solid dispersions. Vapor deposition, while it has been used to prepare most stable amorphous forms, it has low throughput, is costly, and will remain as a laboratory method in the foreseeable future. Freeze-drying requires good aqueous solubility so is not suitable for poorly water-soluble drugs. SCFs require special and costly equipment and may not be widely applicable. Both the melting and the solvent evaporation methods completely destroy the crystalline lattice and can be operated on both laboratory and commercial scales. They are the methods of choice for manufacturing amorphous solid dispersions, even though each has its pros and cons. For compounds that are not feasible to both spray drying due to limited solubility in volatile solvent and HME due to high melting temperature or melt decomposition, a technique called solvent-controlled precipitation or micro-precipitated bulk powder (MBP) may be used to manufacture ASD [89]. However, it may have significant limitations and is beyond the scope of this discussion.

The solvent evaporation method requires the use of an organic solvent or solvent system to solubilize the drug, the carrier polymer, and other formulation components such as surfactant, and then to remove the solvent during the process of ASD formation (e.g. spray-drying). A drawback of spray-drying is the need of a special facility that can handle organic solvents, which has more stringent requirements on

HVAC, occupational safety, and environmental regulations, and is more costly to operate. In order to ensure patient safety, residual solvent need to meet applicable regulatory limits. Based on the intrinsic toxicity of each solvent, ICH Q3C (R6) has set limits on the permitted residual solvent levels in drug product according to its class [90]. Class 1 solvents are known to cause unacceptable toxicities and should be avoided completely in the manufacturing process, unless strong benefit/risk justification is available. Class 2 solvents are associated with less severe toxicity therefore they has a lower limit. Class 3 solvents are less toxic solvents should be used wherever practical, and a limit of permissible daily exposure (PDE) of 50 mg/day or 5000 ppm can be used without further justification. To comply with these limits, solvent used in spray drying process are those belonging to Class 2 and preferably Class 3 solvents having relatively low boing points and high volatility (high vapor pressure). Solvent of high boiling point may be used as a cosolvent at low percentages to improve solubility, if necessary. However their removal is more difficult. Even with these considerations, finding an adequate solvent system for ASDs can sometime be challenging, considering the solubility of the drug, solubility of the polymer, toxicity and the easiness in the solvent removal. Sometimes heating the spray solution to a higher temperature before spray may be necessary to ensure a smooth process. Table 6.2 lists some of the solvents commonly used for spray drying of amorphous solid dispersions.

Phase miscibility during the spray drying process is also a consideration for solvent selection. The spray solution consists of at least a solvent, a drug, and a polymer. Therefore it is at least a ternary system. Flory-Huggins lattice theory can be extended to treat a ternary system and the free energy of mixing, ΔG_{mix}, can be written as,

$$\frac{\Delta G_{mix}}{nRT} = \phi_s \ln \phi_d + \phi_d \ln \phi_d + \frac{\phi_p}{m} \ln \phi_p + \chi_{sd}\phi_s\phi_d + \chi_{sp}\phi_s\phi_p + \chi_{dp}\phi_d\phi_p \quad (6.14)$$

where ΔG, ϕ, R, T, χ, m, and n are the free energy, the volume fraction, universal gas constant, the absolute temperature, the number of lattice sites occupied by the polymer, and the total number of lattice sites, respectively. The subscripts s, d and p designate the solvent, the drug and the polymer respectively.

Based on the investigations of systems consisting of one polymer and two small molecules by Scott [91] and Tompa [92], a ternary system will be less prone to phase separation and more compatible when the solvent-drug interaction is comparable with the solvent-polymer interaction, i.e. $\chi_{sd} \approx \chi_{sp}$, even when the drug-polymer have positive interaction parameters ($0 < \chi_{dp} < 2$). Therefore, as a general rule, polymer/solvent selection for spray solution should be based on matching solvent-drug and solvent-polymer interactions, as promoted by Li et al. [93], even though a favorable drug-polymer interaction ($\chi_{dp} < o$) allows more freedom on this requirement. Experimentally, Davis et al. has developed and expanded a miniaturized solvent casting (MSC) method as a predictive tool for ASD stability [94].

The melting method, in the form of hot melt extrusion, provides a robust, scalable manufacturing process for amorphous solid dispersions. It avoids the use organic

Table 6.2 Some commonly used solvent for spray-drying

Solvent	Boiling point (°C)	Dielectric constant	Vapor pressure (kPa, 20 °C)	ΔH_{vap} (kJ/mol)	Water solubility (g/100 g)	ICH class	ICH limit (ppm)
Acetone	56.2	20.7	24	29.1	Miscible	3	5000
Chloroform	61.7	4.81	20.8	35.0	0.795	2	60
Methanol	64.6	32.6	12.8	35.3	Miscible	2	3000
Methylene chloride	39.8	9.08	47.5	28.0	1.32	2	600
Isopropanol	82.6	18.2	4.4	45.7	Miscible	3	5000
Ethanol	78.5	24.6	5.9	38.7	Miscible	3	5000
Dimethyl formamide	153	36.7	0.4	46.7	Miscible	2	880
DMSO	189	47	0.06	52.9	25.3	3	5000
Ethyl acetate	77	6	9.7	31.9	8.7	3	5000
Butyl acetate	126.1	5.07	1.13	40.5	0.68	3	5000
Water	100	78.5	1.75	40.7	–		–
Tetrahydrofuran	66	7.52	20	26.9	Miscible	2	720

solvent therefore eliminates the need of special solvent-handling facility and better assures product safety and compliance with drug and environmental regulations. However, hot melt extrusion may not be suitable for all drug molecules. The API crystal lattices need to be wiped out entirely (via melting/dissolution) during the short duration (on the order of minute) of the extrusion process. Not all drug molecules can survive this harsh environment of high temperature and high shear rate. Drugs of high melting temperatures (e.g. >220 °C) can be particularly challenging, which is especially true for drugs that undergo degradations upon melting. Ideally, APIs with low-to-moderate melting temperature (150–200 °C or lower) and reasonably stable in the melt are good candidates for HME. HME process is also limited by the glass transition temperature and chemical stability of the carrier polymers, even if drug molecule is stable. Many polymers degrade above 200 °C. Some polymers useful for forming amorphous solid dispersions, such as HPMC and other cellulose derivatives, are not very friendly to the HME process because of chemical degradation. Polymers commonly used in HME include poly(vinylpyrrolidone-co-vinyl acetate) or copovidone, PVP, poly(ethylene-oxide) or PEO, poly(ethylene glycol) or PEG, poly(vinyl alcohol) or PVA, poly(propylene oxide-*block*-ethylene oxide-*block*-propylene oxide) (various Poloxamer®), Eudragit® polymers (copolymers of methacrylic acid and various esters of acrylic monomers), and polyvinyl caprolactam-polyvinyl acetate-polyethylene glycol graft copolymer (Soluplus®). Some new extrudable grades have just emerged, which include a low viscosity grade of HPMC. Drug loading of API in the polymeric carrier is another challenge for HME.

Different ASD manufacturing technologies also dictate different upstream and downstream processes. For spray-drying, a solution of API, polymer, and/or surfactant is needed, which can be straightforward if solubility is decent. However it can be challenging if the solvent can only provide limited solubility to API and polymer. Solubility generally increases with temperature, therefore heating of the suspension before spraying may be an option. The ASD particles obtained from spray-drying is often fluffy, low density, poor flowing and often require a densification step prior to tablet compaction or capsule filling. Roller compaction (a dry granulation) is often utilized for the densification purpose. It is worth to note that dry granulation in general is not a very efficient granulation process therefore the granule properties may not be very optimal (such as wide particle size distribution, loss of compactibility, variable flow properties). In particular, the poor flowing nature of the spray-dried ASD particles can cause challenges to the feeding of the powder from bin to the roller compactor, thus causing more variability in the granule properties and may sometimes causing problems even to the roller compaction operation itself due to inadequate feeding. If this situation is encountered, the spray drying conditions need to be tweaked to modify the particle properties. The reproducibility of ASD particle morphology and PSD are very important to enable a consistent downstream roller compaction process. For HME process, a blending step is generally required at the upstream for the solid components. Liquid components such as surfactants are usually introduced directly into the extruder between the solid intake and the melting zone. Because API particle size impacts the melting/dissolution kinetics, API particles are usually small for HME formulations and a PSD specification may be required to ensure a robust HME process. The small API particle size often leads to poorer flow of the formulated blends. However twin screw extruders have good tolerability on powder flow and it rarely causes a problem during feeding. For the downstream process, the extrudates are generally milled, and blended with additional formulation components, before tablet compaction or capsule filling. Direct shaping of HME has also been explored. However many challenges still exist and it is very difficult to obtain defect-free final dosage form via direct shaping.

Mechanical properties of an amorphous form can differ from those of the crystalline counterparts. Material properties can be treated at three levels [95]. The constituent molecules (chemical species) underpin the first level of material properties. The second level involves the molecular order or the lack thereof (i.e., structure in the condensed states such as amorphous and crystalline structures including polymorphs). The third level is more complex and involves particle morphology, particle size, and other particle characteristics such as moisture sorption, pore structure (e.g., granule structure), and consolidation, etc. These three levels may be understood in a similar way to the primary, secondary, tertiary structures of proteins. While amorphous and crystalline forms have the same chemical structure, the difference in molecular order, particle morphology and PSD can impact their mechanical properties.

Only sparse attention has been paid to the mechanical properties of amorphous pharmaceuticals. Kopp et al. [96] studied the mechanical strength of different polymorphs of phenobarbital and its amorphous form prepared by a melt–solidification–aging technique. The amorphous form and crystalline form III of phenobarbital gave the toughest discs and would be therefore the most suitable to manufacture coherent tablets. Hancock et al. [97] compared the mechanical properties of crystalline and amorphous form of a Pfizer compound, 3-[(4-*O*-[4,6-bis (fluorophenylcarbomyl)]-β-D-glucopyranolsyl-)β-D-flucopyranosyl]oxy-(3β, 5α, 25R)-spirostan-12-one. In this study, the amorphous form was obtained by spray drying and the particle size distribution was controlled to be similar to that of the crystalline form. Nevertheless, the particle morphologies were very different: the crystalline consisted of agglomerates of elongated prisms whereas the amorphous comprised more equant but less uniform primary particles. As a result of the less discriminating aspect ratio, the amorphous form had slightly better flow properties. Both forms were compacted to the same solid fraction of 0.85. While they both had similar tensile strengths, the amorphous form required 30% less compression stress to achieve compacts of similar porosity, indicating better compressibility. Both crystalline and amorphous compacts were classified as very "brittle" by tensile testing of compacts with an introduced "controlled flaw". Tablets made from the amorphous were even more brittle. These observations were consistent with the brittle nature of many "glassy" materials at temperature well below their T_g [98]. Notably, the indentation hardness of this amorphous drug is about 30% higher than the crystalline form, opposite to the case of amorphous and crystalline phenobarbital. The probable explanation lies at the T_g: while this amorphous form has a high T_g of 142 °C, amorphous phenobarbital has a T_g of ~43 °C. Amorphous materials are expected to behave more brittle at temperature far below glass transition but will become more ductile at temperatures closer to T_g.

While the mechanical properties of neat amorphous drug substance may be important, those of the carrier polymers can be dominating because drug loading is relatively low (typically less than 30%) in amorphous solid dispersions and the majority is comprised by the polymer. There were some systematic studies on the mechanical properties of polymers after being treated with different processes for ASD manufacturing [99, 100]. Iyer et al. characterized the mechanical properties of HPMCAS and copovidone before and after hot melt extrusion and/or spray-drying. The results are reproduced in Table 6.3, which include the brittle fracture index, BFI, compression pressure, CP, dynamic bond index, DBI, quasistatic bond index, QSBI, reduced modulus of elasticity, E′, dynamic hardness, H_d, dynamic hardness, H_{qs}, strain index, SI: tensile strength, TS, intrinsic tensile strength, TSI, compromised tensile strength, TS_o, and viscoelasticity index, VI. The compression pressure to achieve the same solid fraction (SF) of 0.85 was 24% and 61% higher for HPMCAS after spray drying and hot melt extrusion, respectively, indicating decreased tabletability, even though spray-dried HPMCAS resulted in higher tensile strength (improved compactability). For both HPMCAS and copovidone, most of the relevant mechanical properties decreased upon HME, indicating that HME could impact reworkability by reducing deformation of materials and by increasing density during the extrusion process. The mechanical properties of polyvinyl alcohol was examined similarly [100]. Hot melt extrusion processing reduced tabletability and compressibility of PVA, similar to the findings with HPMCAS and copovidone. The impact of hot melt extrusion process was also examined with Soluplus®, Kollidon® VA 64 and

Table 6.3 Mechanical properties of excipients impacted by HME or spray-drying process

Property	MCC PH101	Lactose monohydrate	DCP-A	HPMCAS	Copovidone	Spray-dried HPMCAS	Melt-extruded HPMCAS	Melt-extruded copovidone
True density	1.556	1.537	2.830	1.291	1.202	1.296	1.277	1.203
SF	0.85	0.85	0.85[a]	0.85	0.85	0.85	0.85	0.85
CP, MPa	112 (4.6%)	123 (5.4%)	3530	55 (4.6%)	85 (3.9%)	68 (11%)	89 (9.2%)	77 (2.6%)
H_d, MPa	130 (3.8%)	156 (8.3%)	7900	74.4 (10.3%)	101 (3.4%)	80.7 (1%)	174 (14%)	100 (9.7%)
H_{qs}, MPa	40.5 (2.9%)	59.3 (8.7%)	1060	32 (5%)	33 (13%)	20 (2%)	17.5 (14%)	16.4 (5%)
TS, MPa	5.7 (5.2%)	1.76 (0.3%)	59	2.8 (24%)	2.6 (1.1%)	3.4 (3%)	0.35 (8.7%)	0.79 (8.9%)
TS_o, MPa	4.7 (0.8%)	1.07 (4.1%)	42	1.1 (15%)	1.7 (16%)	2.1 (3.7%)	0.2 (1.7%)	0.6 (7.4%)
E', GPa	8.1 (17%)	11.3 (11%)	1710	3.0 (32%)	6.3 (17%)	2.9 (0.1%)	8.3 (0.2%)	5.2 (0.02%)
DBI = TS/H_d	0.044	0.011	0.007	0.038	0.026	0.043	0.002	0.008
QSBI = TS/H_{qs}	0.14	0.03	0.06	0.09	0.08	0.17	0.02	0.05
BFI = $(0.5(TS/TS_o) - 1)$	0.11	0.32	0.2	0.82	0.26	0.31	0.37	0.11
SI = H_d/E'	0.016	0.014	0.0046	0.025	0.016	0.028	0.021	0.019
VI = H_d/H_{qs}	3.22	2.63	7.4	2.32	3.02	3.98	9.95	6.11
Yield strength, P_y, MPa (out of die)	255.7	584.7	1307	130.8	133.4	275.4	357.8	221.1
Intrinsic tensile strength (TS_I)	13.9	18.3			8.28	9.86	1.84	2.68

Microcrystalline cellulose (MCC), lactose and dibasic calcium phosphate anhydrous (DCP) are included for comparison purpose. Values given in parenthesis are the percentage error of the estimate

[a]Values are extrapolated to SF = 0.85. Data source from [99]

Eudragit® EPO with similar conclusions. It may be concluded that HME process shifts material properties towards more elastically deformation, increases elastic recovery, and weakens the bonding strengths. Very few reports detailed compaction studies on amorphous solid dispersions [101–104]. However, the overall non-ideal mechanical properties of amorphous solid dispersion may be improved by incorporating materials that undergo plastic deformation such as microcrystalline cellulose [99], which is similar to the standard approach to improve the compaction properties of other conventional formulations.

6.5 Rational Formulation Design and Product Development

Rational formulation development starts with understanding the basic physicochemical, biopharmaceutical, pharmacokinetic (PK), and pharmacodynamics (PD) properties of the drug molecule, the delivery technology, the dosage form design, and the manufacturing processes. Each drug substance possesses inherent properties that may require some specific considerations during the ASD formulation design and the manufacturing technology selection. In addition to the standard physicochemical properties of crystalline forms, amorphous drug formulation development requires the understanding of the basic properties of the amorphous form: glass transition temperature, glass formation ability, thermodynamics, crystallization tendency both in solid state and in solution including the assessment of molecular mobility and configurational entropy, to name a few. Polymers are then screened based on various properties to accommodate the formulation and process, and their capability to inhibit crystallization and to promote liquid-liquid phase separation. Further screening of drug loading, polymer, and surfactant are then performed to optimize the formation of amorphous drug nanoparticles during dissolution and to maximize oral bioavailability. Preliminary processes are then developed to manufacture the candidate formulation on small scale to support product characterization, early analytical method development, in vivo animal PK screening, and human PK studies. Formulation decision is generally made by considering a multitude of factors, including the human PK outcomes, the formulation design and formulation properties, the manufacturing process, and in-house capability. Once a formulation is selected, product characterization continues and issues emerged during the development will need to be addressed and mitigated. Meanwhile, the manufacturing process may be scaled up, characterized, and optimized to support the various phases of the clinical programs and to support the final registration of the product. Product and process characterization frequently intertwines during the entire product development cycle, where the formulators, engineers, and analysts work collaboratively to ensure a reproducible manufacturing process is developed to deliver a final product with consistent quality attributes and robust in vivo performance.

6.5.1 Selecting the Carrier Polymer

The carrier polymer usually dominates in the ASD formulation. It greatly impacts the overall properties of the formulation (glass transition temperature, mobility, physical/chemical stability, in vitro and in vivo dissolution), oral bioavailability and the manufacturability of the drug product. Polymer is selected based on the intended ASD manufacturing technology, thermal properties (T_g, stability), solubilization capacity, impacts on ASD stability, and ability to inhibit crystallization, promote LLPS and generate amorphous drug nanoparticles, and enhance oral bioavailability.

Hot melt extrusion requires suitable stability of the polymer at the temperature of extrusion. The processing temperature window of HME formulation resides between T_g of the formulation matrix and degradation temperature (T_d) of the polymer or the drug molecule. The regular grades of HPMC, due to their high molecular weight, high T_g, and the narrow windows between T_g and T_d, lead to significant molecular weight reduction during HME and are not considered as extrudable. However, extrudable low molecular weight grades of HPMC have emerged, such as the Affinisol™ HPMC HME from Dow and the Benecel HMPC from Asland. Conventionally, copovidone is a primary polymer used for HME applications. It has a T_g of 106 °C and T_d of 230 °C, therefore affords a wide processing temperature window. Extrusion temperatures up to 180–190 °C have been successfully applied for copovidone. Copovidone is the carrier polymer used for a number of Meltrex® products including Kaletra®, Novir®, Viekira™, and Venclexta™ tablets. The drawback of copovidone is that the extrudate are usually brittle. Polymers and copolymers based on methacrylic acid and derivatives (such as various Eudragit® polymers) have also been used in manufacturing HME formulations [105]. These polymers may have improved thermoplastic properties. Polyvinyl alcohol (PVA, partially hydrolyzed from PVAc) has been studied for HME applications and provide some benefits as improved drug loading [61]. In addition to polymer degradation, chemical incompatibility between the polymer and drug molecules during HME process is also a factor to consider. Interested readers may refer to Lang et al. [106] and the cited references there for a more in depth accounting of polymer and polymer/drug degradations.

Contrary to HME formulation, the polymer in a spray-drying formulation has less constraint on degradation temperature with respect to temperature window of processing. However, the solubility of the polymer in suitable solvent system needs to be considered together with drug solubility in that solvent system.

The glass transition temperature of the ASD matrix is important factor in formulation design. We have discussed that T_g is an indicator for molecular mobility under typical storage conditions (15–30 °C). The molecular relaxation time (i.e. the reciprocal of mobility) at the glass transition temperature, τ_g, is comparable across different amorphous systems and is on the order of 100 s. Molecular relaxation time, τ, is a function of temperature and the strength parameter, D, of the amorphous system. Above T_g, τ has a non-Arrhenius temperature dependence as described by the VTF equation [107–109]. Below T_g, τ has a Arrhenius temperature dependence based on the of AGV equation [110–113] via the concept of fictive temperature, T_f, and its approximation to T_g for freshly prepared glasses.

$$\tau(T > T_g) = \tau_0 \exp\left(\frac{DT_0}{T - T_0}\right) \tag{6.15}$$

$$\tau(T < T_g) = \tau_0 \exp\left(\frac{DT_0}{T - (T/T_f)T_0}\right) = \tau_0 \exp\left[\frac{T_g}{T} \cdot \ln\left(\frac{\tau_g}{\tau_0}\right)\right] \tag{6.16}$$

In the above equation, τ_0, is a constant that equals to 10^{-14} s. Equation (6.16) predicts that molecular mobility in glasses is similar at a similar temperature represented at the T/T_g scale. A simpler empirical rule is the so-called $T_g - 50$, that is, an amorphous system is considered to be physically stable when stored at 50 °C below its glass transition temperature. This rule is based on the extrapolation of the VTF equation to temperature below T_g (i.e. the ideal equilibrium glass), and is different from the behavior of a real glass. Nevertheless, the larger the difference between T_g and the storage temperature, the slower is the molecular mobility and the better is the stability (both physical and chemical). The rule of $T_g - 50$ indicates that a T_g of at least 75–80 °C of the ASD is preferred for products intended for ambient storage.

The glass transition of a homogeneous blend depends on the individual components and the blend composition, and is often described by the Gordon-Taylor equation [114, 115]:

$$T_g = \frac{w_1 T_{g1} + K_G w_2 T_{g2}}{w_1 + K_G w_2} \tag{6.17}$$

Here w is the weight fraction, and subscripts 1 and 2 designate the component in the blend. K_G is a constant, whose value is related to density and expansion coefficient and may be approximated using the Simha-Boyer rule [116]:

$$K_G = \frac{\rho_1 \Delta\alpha_2}{\rho_2 \Delta\alpha_1} \approx \frac{\rho_1 T_{g1}}{\rho_2 T_{g2}} \tag{6.18}$$

where $\Delta\alpha$ is the change in the expansion coefficient at T_g. When the difference in true density is ignored, i.e. $K_G = T_{g,1}/T_{g,2}$, the G-T equation reduces to the simpler Fox equation [117]:

$$\frac{1}{T_g} = \sum_i \frac{w_i}{T_{g,i}} \tag{6.19}$$

Hence glass transition of the ASD matrix depends on the T_g and the concentration of each component (drug, polymer, and surfactant), but will fall in middle of the ranged defined by the lowest and highest T_g of the individual components. The minor component with lower T_g is generally called a plasticizer, and that with higher T_g may be called an anti-plasticizer. In order to maintain a T_g of at least

Table 6.4 Polymers for amorphous solid dispersions

Polymer	MW (kD)	T_g (°C)	T_d (°C)	Notes
PVP K17	10–30	126–160	n.a.	K17, K25, K30 grades. Non-ideal thermoplasticity but may interact with drug strongly
Copovidone	45–70	106	230	Readily extrudable
HPMC	10–150	160–180	200	Non-thermoplastic but good crystallization inhibition
HPMC (extrusion grade)	80–550	115	250	AFFINISOL™ HPMC HME 15, 100, and 4000 cp grade
HPMCAS	55–90	120–135	<200	Readily extrudable wo plasticizer; Soluble at pH > 5.5 Good crystallization
HPMCP	37.9	143	n.a.	Crystallization inhibition; interaction with certain drug molecules
PEG/PEO	4–20	~45	n.a.	T_m ~ 60–63 °C; may crystallize during storage
Ammonio methacrylate copolymer	n.a.	55	n.a.	Eudragit® E; pH-dependent solubility
Polyacrylic acid	450	110	n.a.	Carbomer, carbopol 940
Soluplus®	90–140	~70	250	Good thermoplasticity. Designed for HME
Poloxamer 188	10–14	n.a.	180	T_m ~ 55 °C; may crystallize during storage
PVA, type LM 25	~49	~45	240	33–38% hydrolized
PVA, type LM 22	~48	~49	240	45–53% hydrolized
PVA, type 505	~126	~61	240	72–75% hydrolyzed, T_m ~ 154
PVA, type 4-88	~93	~67	240	88% hydrolized
HPC	80–1150	~105	n.a.	Non-thermoplastic, need plasticizer; Klucel EF to HF grade
Polymethacrylates	34	130	155	Not easily extrudable wo plasticizer

Data source: [61, 118–121]
n.a. not available

75–80 °C, attentions need to be paid to the choice of polymer, surfactant, and their concentrations. In addition, ASD formations can absorb a significant amount of moisture during manufacturing and storage due to the hydrophilic nature of many carrier polymers. Moisture is a great plasticizer ($T_g \sim -140$ °C) therefore can cause significant decrease of the T_g of an ASD product. The reduction in T_g by moisture sorption can lead to both physical and chemical stability issues for the drug product and need to be considered during formulation/process development. Packaging design may also offer an option in addressing the moisture sorption during product storage.

Table 6.4 lists some of the polymers and their relevant properties that are commonly used in manufacturing amorphous solid dispersions.

Drug solubility in the polymer is an important factor for carrier selection both for the HME and spray drying process. Crystalline solubility of the drug molecules in polymer represents a benchmark when considering the physical stability of an amorphous solid dispersion. When drug loading is below the crystalline solubility, the ASD is thermodynamic stable against crystallization, while the contrary is true when the solubility limit is exceeded. The extent of drug loading in excess of the solubility represents the degree of supersaturation of the drug molecule in the polymer, which adversely impact the crystallization tendency of the ASD. Typically, crystalline drugs have limited equilibrium solubility in common polymers used for ASD under the normal long-term storage conditions, unless strong interactions (e.g. H-bonding, ionic) exist between the drug molecule and the polymer. Solubility of the crystalline drug increases at high temperature, which is important for HME formulation and process development, where full melting/solubilizing of the crystalline drug by the polymer is required to ensure a robust HME process.

Drug solubility in polymer can be measured using the melting temperature depression method. In this method, physical mixtures of drug and polymer at different ratios are heated using a differential scanning calorimetry (DSC) until all the crystalline drug melts. The melting point represents the temperature at which the drug solubility equals to the drug concentration in the physical mixture, if equilibrium is achieved.

When crystalline drug is in equilibrium with a polymer melt, the chemical potential of the crystalline drug equals that of the dissolved drug in the drug-polymer system. The chemical potential of the drug in solution can be derived based on the free energy of mixing:

$$\Delta\mu_d = \mu_d - \mu_d^0 = \left(\frac{\partial \Delta G_{mix}}{\partial \phi_d}\right)_{\phi_p, T, P} = RT\left[\ln \phi_d + \phi_p\left(1 - \frac{1}{m}\right) + \chi_{dp}\phi_p^2\right] \tag{6.20}$$

Here μ_d^0 is chemical potential of the pure amorphous drug, which is the same as the molar free energy of the pure amorphous drug at the reduced melting temperature, T_m. The molar free energy of the amorphous form is related to that of the crystalline form as:

$$\begin{aligned}\mu_d^0(T_m) &= G^a(T_m)\\ &= G^x(T_m) + \Delta H_m\left(1 - \frac{T_m}{T_m^0}\right) + \int_{T_m^0}^{T_m} \Delta C_p dT - \int_{T_m^0}^{T_m} \frac{\Delta C_p}{T} dT\end{aligned} \tag{6.21}$$

Here T_m^0 is the melting temperature of the pure crystalline drug and $G^x(T_m)$ is the molar free energy of the crystalline drug at temperature T_m. At the observed melting temperature for a crystalline drug/polymer mixture, the chemical potential of the drug dissolved in the polymer equals to that of the pure crystalline drug, and we get:

$$-\frac{\Delta H_m}{R}\left(\frac{1}{T_m}-\frac{1}{T_m^0}\right)$$
$$-\frac{1}{RT_m}\int_{T_m^0}^{T_m}\Delta C_p dT+\frac{1}{R}\int_{T_m^0}^{T_m}\frac{\Delta C_p}{T}dT=\ln\phi_d+\phi_p\left(1-\frac{1}{m}\right)+\chi_{dp}\phi_p^2 \quad (6.22)$$

The drug-polymer interaction parameter χ_{dp} can then be estimated based on the melting-point depression data from the above equation. Once the χ_{dp} becomes known, one can estimate the solubility at other lower temperatures of interest (usually around the storage temperatures), assuming similar solubility in the in the rubbery state.

A simplified equation has been used in the literature [122–124] by ignoring the contribution from excess heat capacity:

$$-\frac{\Delta H_m}{R}\left(\frac{1}{T_m}-\frac{1}{T_m^0}\right)=\ln\phi_d+\phi_p\left(1-\frac{1}{m}\right)+\chi_{dp}\phi_p^2 \quad (6.23)$$

Experimental improvement was first made by Tao et al. [125], where cryo-milling of drug/polymer mixtures was applied to overcome the sluggish melting/dissolution kinetics under highly viscous environments. In addition, instead of using the onset melting temperature, the offset melting temperature was employed and extrapolated to zero heating rate to better mimic the equilibrium melting. Sun et al. [126] developed an annealing method that can attain the equilibrium solubility even better thus further improved the measurement accuracy. The drawback of the last method is that the experiments are tediousness and time-consuming. In the early stage screening, a more rapid method can be used, at the expense of scarifying some accuracy. It is justifiable because the information is primarily used for comparison purpose and for polymer selection during the ASD formulation development.

The miscibility of the ASD system is a key factor that may impact both drug product properties and product performance. As eluded above, drug loadings frequently exceed the crystalline solubility limit in many ASDs. A homogeneous glassy solution cannot be obtained if the drug loading further exceeds the miscibility limit. Above the miscibility limit, phase separation may occur spontaneously and may compromise product performance.

Just like liquid mixtures, the phase behavior of amorphous drug/amorphous polymer system ranges from miscible to partially miscible, depending on the composition and the intermolecular interactions. The Flory-Higgins lattice theory is applied to obtain the free energy of mixing of small molecule in polymer:

$$\frac{\Delta G_{mix}}{nRT}=\phi_s\,\ln\phi_d+\frac{\phi_p}{m}\ln\phi_p+\chi_{dp}\phi_d\phi_p \quad (6.24)$$

Here the symbols follow the same conventions as before (Eq. 6.14). Flory-Huggins parameter χ_{dp} represents the interaction between drug and polymer and $\chi_{dp}kT$ represents the enthalpy change of introducing one drug molecule to the

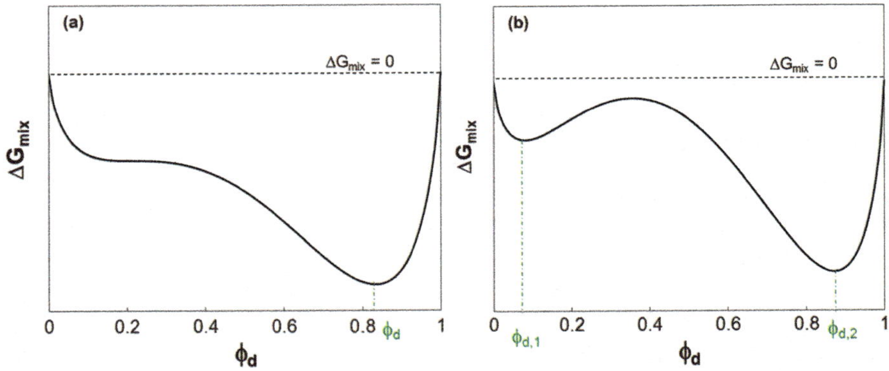

Fig. 6.20 Free energy schematics of drug-polymer mixing (left: completely miscible; right: partially miscible)

polymer. Figure 6.20 shows two scenarios: full miscibility and partial miscibility. Depending on the nature of polymer-drug interaction, the enthalpy of mixing can be positive (for example, in case of only van der Waals interactions), or negative (for example in the case of hydrogen bonding). Obviously, positive χ_{dp} causes an increase in enthalpy and free energy, which disfavors mixing. In addition, $\Delta G_{\text{mix}} \leq 0$ alone does not necessarily guarantee a homogeneous single phase. Similarly to the mixing of small molecules, miscibility gap usually exists, leading to the spinodal decomposition in certain range of the compositions, as shown in Fig. 6.20b. A daunting task during ASD formulation development is to pinpoint where the miscibility gap may lie at for the proposed ASD system.

6.5.2 *Optimize Oral Bioavailability of ASD*

The behavior of ASD during dissolution and the implications to oral bioavailability enhancement has been discussed in the first section. We get on board the ASD train with a goal to reap its potentials for improving oral absorption and we aim to make the best out of it.

Here are the three key points: (1) in the absence of bulk or solution crystallization, and bulk amorphous drug phase separation in the ASD when contacting with the medium (i.e. hydrophobic capture), dissolution of ASDs often results in liquid-liquid phase separation; (2) amorphous solubility represents the practical limit of solubility advantage of amorphous solid dispersions; and (3) the particle morphology and size distribution impact the dissolution rate from the amorphous drug precipitates upon LLPS and therefore the oral bioavailability. Hence, the ASD formulation needs to be able to achieve: (1) inhibition of drug crystallization during dissolution from solution or from the bulk solids, and prevention of amorphous drug phase separation from the bulk ASD during dissolution; (2) generation of LLPS;

and (3) optimization of the amorphous drug nanoparticles and maintaining them over the timeframe relevant to oral absorption.

6.5.2.1 Inhibition of Crystallization During Dissolution

While amorphous form is thermodynamically unstable in the solid state, the kinetics can be slow enough to afford viable long term product stability even without special measures. However, in solution state, the crystallization kinetics is generally much faster due to that facts the molecular mobility is much higher than in solid state. Crystallization tendency of a molecule in the solution state generally reflects the intrinsic properties of the molecule since mobility is no longer a limiting factor. From this perspective, the configurational entropy provides a good insight on the crystallization tendency of the drug molecule in supersaturated solutions.

Crystallization inhibition have been observed and studied for quite some time and in particular, various polymers have been found to effectively slow down the solution crystallization of various organic molecules. While the mechanism is still far from a complete capture, the hydrophobic interaction between the polymer and various crystal surfaces has been suggested a possible pathway. There is no universal rule on whether and why effective crystallization inhibition occurs or not for a specific drug/polymer system. Hence, the crystallization of a supersaturated drug solution and the inhibiting polymers need to be evaluated experimentally.

No special attention may be needed if the drug molecule can't crystallize easily in solution. Many complex drug molecules have hard time to crystallize, which work favorably for ASD development. However, if crystallization is possible over the relevant time scale, the formulator must consider this in the product design and incorporate appropriate agent for hindering the crystallization kinetics. Extensive inhibitor screening may be performed in vitro and the effective crystallization inhibitor is then selected, while considering other aspects of the formulation design and the processing development.

It would be the ideal situation if the polymer selected for the ASD carrier (HME or spray-drying) is also effective in crystallization inhibition. Given that the carrier polymer dominates in the ASD, the crystallization inhibition becomes more powerful. A different polymer can still be incorporated in the formulation. However, doing so may decrease the drug loading, increase the size of dosage form, and potentially may increase the pill burden. In addition, the effectiveness of crystallization inhibition needs to be high considering its limited concentration in the formulation and the constant refreshing fluids during in vivo dissolution in the GI environment.

Paclitaxel is widely used in the treatment of advanced breast and ovarian cancer. It is a large small molecule (MW = 853 Da), has low solubility (<1 μg/mL) and no ionizable functional groups. The drug is originally marketed as an intravenous formulation (Taxol®, Bristol–Meyers Squibb, BMS) containing 6 mg/mL of paclitaxel, 527 mg/mL of Cremophor EL (polyoxyethylenated castor oil), and 49.7% (v/v) of dehydrated ethanol. During the development of a SEDDS formulation,

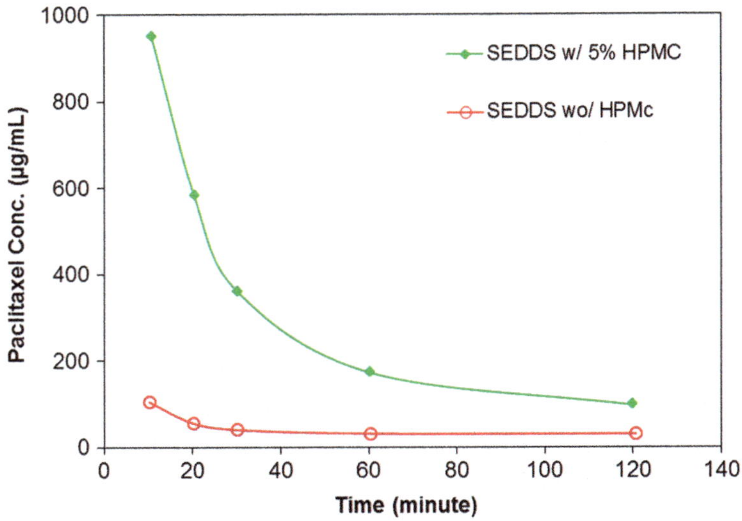

Fig. 6.21 Apparent in vitro paclitaxel concentration–time profiles from the SEDDS formulations with (S-SEDDS) and without HPMC. Data source from [127]

it was found (Fig. 6.21) that incorporating 5% w/w HPMC E5 in the SEDDS formulation significantly prolonged the supersaturation during an in vitro dissolution test of the formulation in 50 mL simulated gastric fluid [127]. The in vivo PK results [127] in rats confirmed that HPMC remained effect in delaying the crystallization during in vivo dissolution, resulting in higher oral bioavailability (Fig. 6.22).

HPMC offers an advantage that may not be reproduced by other polymer, if it is an effective crystallization inhibitor and a capsule dosage form is acceptable. That is, HPMC capsule may be used for encapsulating the product while HPMC itself is not incorporated into the bulk formulation. Figure 6.23 shows that HPMC capsules improved the exposure of drug X in human by fourfolds comparing the same formulation filled in gelatin capsules.

These above examples themselves are not ASD formulations, but the same principles apply. An effective crystallization inhibitor should be incorporated into the formulation to prolong supersaturation during release and improve oral absorption.

To screen polymer for effective crystallization inhibition and sustaining the supersaturation, a UV/Vis fiber-optic probe may be used to continuously monitor the apparent drug concentration–time profile from a supersaturated drug solution, like the setups in references [19, 20, 39]. Small apparatus, such as µ-Diss Profiler (pION, Billerica, MA), may be handy for small volume powder dissolution and crystallization inhibition study. It is worth to note that the drug nanoparticles may contribute to the absorption in solution therefore the drug-concentration profile is “apparent” in nature. The supersaturated drug solution may be generated by introducing a concentrated stock solution of the drug molecule in methanol, ethanol, DMSO or other appropriate hydrophilic solvents. The extent of supersaturation

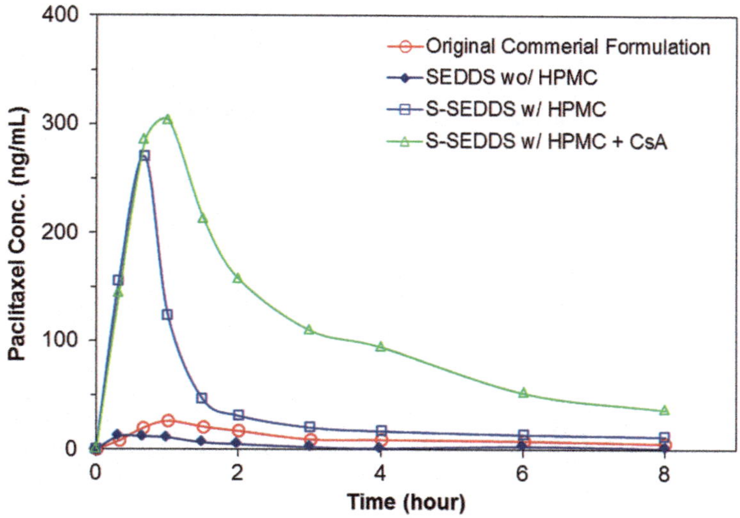

Fig. 6.22 Mean plasma concentration–time profiles of paclitaxel in rats after oral administration of 10 mg/kg paclitaxel using the four formulations indicated. Data source from [127]

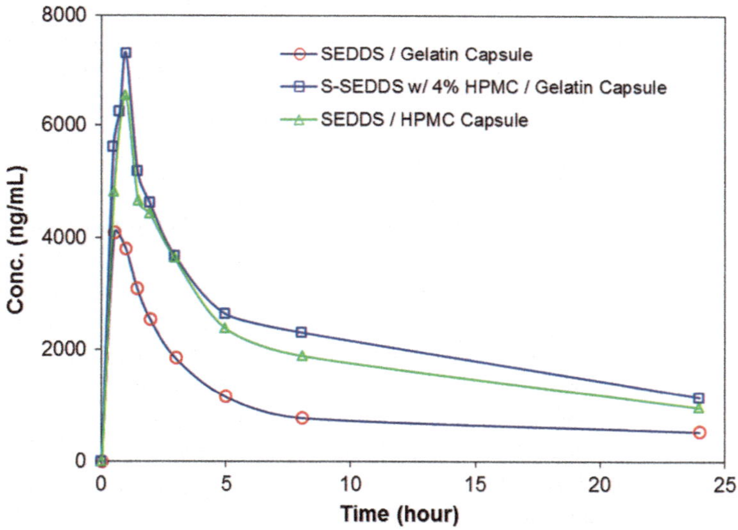

Fig. 6.23 Human bioavailability study results with three formulations of Drug X: formulated Drug X powder/hard gelatin capsule, an aqueous suspension, and a SEDDS formulation/softgel. Data source from [128]

should also be controlled to be relevant, i.e. in the vicinity of expected amorphous solubility. Figure 6.24 shows an example of this type screening [121]. The compound here is a weak acid, with p*K*a of 8.5. It has high melting point of ~270 °C and solubility below detection limit at pH < 8. The molecule has high tendency to

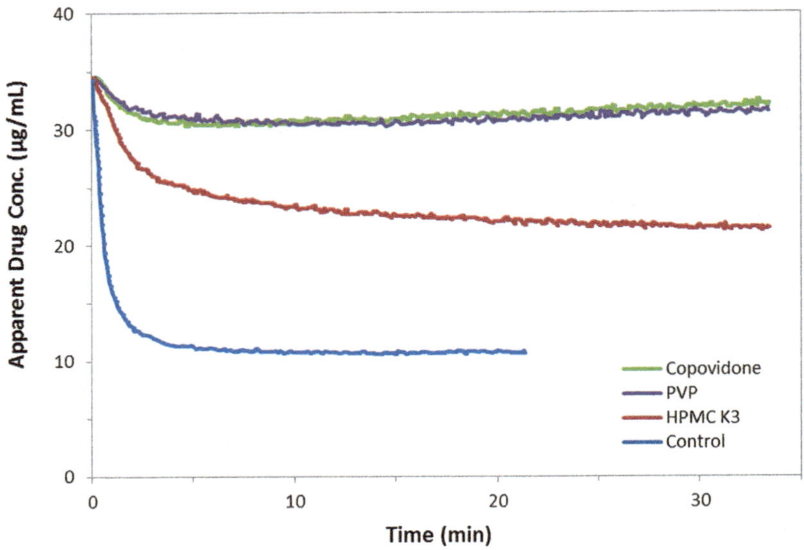

Fig. 6.24 Apparent drug concentration-time profiles during polymer screening from supersaturated solution of a model compound in pH 6.8 buffer. Data source from [121]

crystallize in solution. However, the crystallization can be significantly reduced or inhibited in the presence of 10 ppm of HPMC, PVP and copovidone. PVP and copovidone are most effective and are virtually similar in this case.

6.5.2.2 Optimize the Formation of Amorphous Drug Nanoparticles

Dissolution of properly design ASDs should facilitate the liquid-liquid phase separation both in vitro and in vivo and generate amorphous drug nanoparticles to enhance oral absorption. However, not all ASDs are capable of generating LLPS and enabling the formation of drug nanoparticles.

Water molecules prefer to interact with the carrier polymer over the drug molecules when ASD comes to contact with water. In a more quantitative treatment, the interaction parameters are very asymmetric, $\chi_{sd} \ll \chi_{sp}$, which promote phase separation. Depending on the competitive strength of the paired interaction among the drug, polymer, surfactant, and water, a drug rich phase may be formed facilitated by the unfavorable interaction with the water molecules and the hydrophobic drug-drug interactions. This bulk amorphous drug phase separation can occur quickly upon the ASD is in contact with the water, and is consistent with the theory of "hydrophobic capture". Once this bulk amorphous drug phase separation occurs, no LLPS would occur and no amorphous drug nanoparticles might be generated.

The phenomenon of hydrophobic capture is impacted by the properties of the drug molecule, polymer, surfactant, and their strengths of interaction with water, as well as the drug loading. While there are considerations during formulation design to mitigate this risk, it cannot be avoided a priori. Therefore, experimental screening provides a useful tool to eliminate formulations that are prone to the hydrophobic capture during dissolution.

The good news is that the bulk amorphous drug phase separation, the LLPS, and the formation of drug nanoparticles can be examined in the same set of screening experiments. Powder dissolution study can be performed on the ASD formulations. The apparent drug concentration can be conveniently monitored using the UV/Vis fiber-optics probe as a function of time; the particle size of the API precipitates may be examined using DLS; and the characteristics of the ASD powder during and after the dissolution may also be monitored by suitable techniques, such as Raman spectroscopy. Other experimental techniques, such as asymmetrical flow field-flow fractionation (AsFlFFF), optical microscopy, SEM/TEM, AFM, florescence probing, may also be invoked to further investigate and understand the dissolution and phase separation behavior of the ASD formulations, if necessary. Even visual observations can be a useful tool in differentiating the performance of different ASDs. The colloidal appearance during this type of dissolution testing often indicates a good sign.

The apparent drug concentration–time profile during powder dissolution can provide more quantitative insights on the quality of ASD. Ideally, ASD should generate apparent drug concentration exceeding the amorphous solubility limit quickly. The higher the apparent drug concentration, the longer the concentration is maintained, and the better is its performance. We understand that the solubilized drug concentration does not exceed the amorphous solubility limit. Therefore, any concentration above the amorphous solubility limit reflects the contribution from the amorphous drug nanoparticles. More drug nanoparticles are correlated with smaller drug nanoparticles and therefore increased contribution to the UV/Vis absorbance picked up by the probe. Hence, the apparent drug concentration–time profile not only indicate if LLPS occurs during dissolution, but also reveals the quality of the amorphous drug nanoparticles. Therefore, frequently the apparent drug concentration–time profile allows for formulation comparison and formulation decision directly.

Figure 6.25 provides an example of formulation screening with regard to polymer selection [129]. HPMCAS showed better capability in elevating and sustaining the apparent drug concentration while PVAP (polyvinyl acetate phthalate) could not sustain the supersaturation for longer than 60 min. Figure 6.26 shows an example of drug loading screening when HPMCAS was selected as the carrier polymer [129]. Increasing drug loading from 10% to 33% caused a decrease in the sustainable supersaturation. Similar screening can also be performed for surfactant optimization.

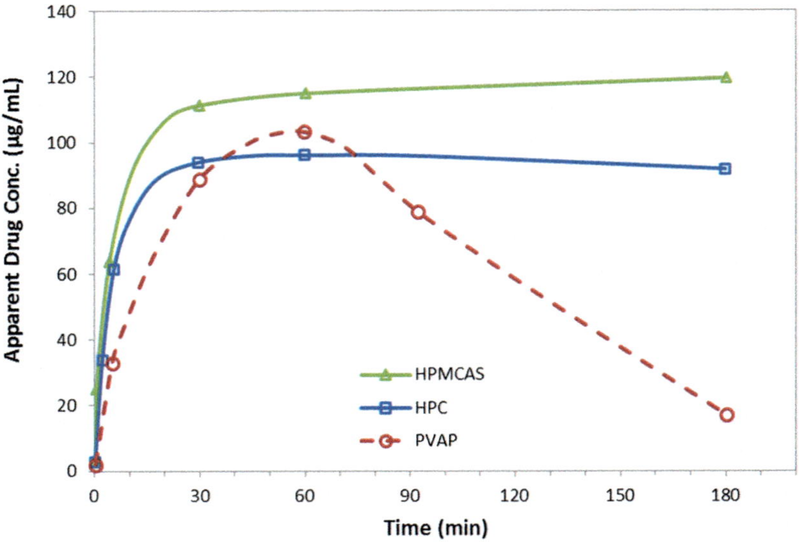

Fig. 6.25 Apparent drug concentration-time profiles obtained from dissolution of ASDs of a drug with various polymers at 50% drug loading. Data source from [129]

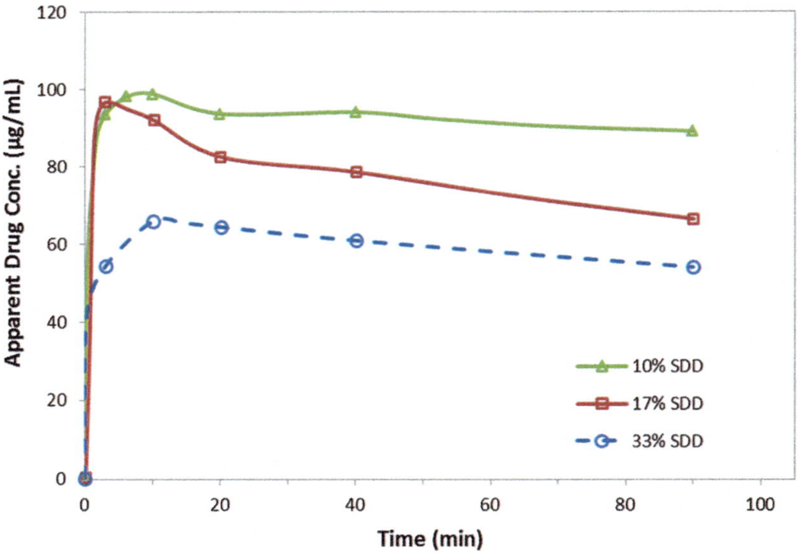

Fig. 6.26 Apparent drug concentration-time profiles obtained from dissolution of ASDs of a drug with HPMCAS at different drug loadings. Data source from [129]

6.5.3 *In Vitro and In Vivo Characterization*

In vitro and in vivo characterizations refer to characterizing the physicochemical, stability, dissolution, biopharmaceutical and pharmacokinetic properties of the drug molecule, the formulation, the dosage form, and the final product. The generated data are useful for formulation and process design, selection, interrogation, and optimization. It is not a single time point task. Instead, characterization intertwines with the entire drug product development at various stages and provide feedback to product and process development.

Basic physicochemical properties are characterized before onboarding the train of product development. These properties may include pK_a, log$P_{o/w}$, solubility in water and in pH buffers, solubility in common organic solvents, solubilization in the presence of surfactants and in lipid vehicles, pH-stability, and stability in the solid state. Solid-state properties including various solid-state forms (polymorphs, hydrate, solvate, amorphous), their crystallinity, crystal morphology, powder X-ray diffraction, moisture sorption, melting temperature, melting enthalpy, and melt stability.

Biopharmaceutical properties, such as permeability, substrate for various uptake and efflux transporters, first-pass metabolism, etc. may be directly measured (e.g. Caco-2 permeability and various biopharmaceutical screening using cell lines) or inferred from early animal PK studies and in-vitro drug metabolism pharmacokinetic (DMPK) screening. These data, together with the physicochemical properties of the drug molecules, form the initial basis of drug product development.

The early animal PK study dosed with neat drug substance, cosolvent solution, or simple suspension (such as neat drug substance suspended in 2% HPMC), together with the anticipated effective exposure levels, in consideration with the physicochemical properties, can provide a baseline assessment on the toughness of formulation development. Sometimes with certain manipulations of the solid state forms (e.g. salt, co-crystals) and proper product/process design (e.g. API particle size, surfactant, solubilization, complexation, granulation), a conventional dosage form may be adequate for certain BCS class II drugs. However, if significant efforts are required to achieve adequate absorption in animal studies, an enabling drug delivery technology is more likely necessary.

The properties of the amorphous form are needed to support the ASD development. Relevant information includes the glass transition temperature, the molecular mobility, configurational entropy, and the physical stability evaluation of the amorphous form. The easiness of crystallization may be assessed quickly from thermal analysis such as DSC based on molecular mobility and configurational entropy evaluation. In particular, if recrystallization event is observed during a DSC scan, physical stability can be a concern and need to be evaluated and mitigated during formulation design and process development. Fortunately, many drug candidates from today's combinatorial screening are poor crystallizers, may even have difficulty to crystallize from solution, and as a result, have less risk on physical stability.

Aqueous solubility of the amorphous form is a benchmark for oral bioavailability improvement. This number may be estimated from thermodynamics measurement of the crystalline and amorphous material (melting temperature, enthalpy, entropy) in conjunction with the moisture sorption isotherm [130]. It may also be determined experimentally by measuring dissolution of the amorphous form as in references [19], or by measuring the diffusion flux by varying drug concentration in the donor compartment similarly as in references [39, 41]. Similar experimental setups have been used to screen polymers for crystallization inhibition.

Carrier polymer screening is the next important step in developing ASD formulations. Many of the aspects are described in early sections. Drug loading capacity in the carrier may be an important factor in consideration of pill burden and market competitiveness, which can be started as a solubility screening. Various solubility screening method may be developed and each organization may have its own preferences. One method involves measuring solubility of crystalline drug in monomer or mixture of monomers that constitutes the repeating units of the polymer, such as *N*-vinylpyrrolidone for PVP, mixture of vinylpyrrolidone/vinyl acetate for copovidone, or low molecular weight oligomers of the carrier polymer, such as vinylpyrrolidone dimer. It should be noted that the solubility of the drug in polymer will be less than the value measured in the monomers or oligomers due to the entropic effect of the polymer (refer to Flory-Huggins lattice theory). However it provides at least a starting place. It should be further noted that no appropriate monomers/oligomers are readily available for some of the polymers such as HPMC, HPMCAS and other cellulose derivatives.

Solubility of the crystalline drug in polymer at high temperature is very relevant for developing hot melt extrusion process; because it determines the minimum processing temperature the crystalline/polymer mixture can be rendered as melt. As discussed before, solubilization temperature of a specific drug/polymer blend can be determined using the melting temperature depression method via a DSC. Solubilization temperature of a crystalline drug in a polymer/surfactant matrix can be measured similarly. This method provide a quick way to determine the minimum processing temperature needed for a certain drug load in a polymeric matrix that may include surfactants and other components.

Crystallization inhibition screening can be conducted using the method described earlier. The solution media is pre-dissolved with various polymers. Neat amorphous drug powders may be introduced, or even simpler, a concentrated drug solution in organic solvent such as methanol, ethanol, or DMSO, can be introduced to reach a desired supersaturation. The ability of the polymer to reach and maintain the amorphous solubility, or to delay the crystallization, can be evaluated from the apparent drug concentration–time profiles.

Similar experimental settings can be used to screen polymer, surfactant, and drug loading in order to optimize the LLPS and the formation of amorphous drug nanoparticles. ASD formulations can be prepared by solvent evaporation or melt extrusion method in laboratory settings. Dissolution of the ASD powders is then followed using a UV/Vis dip probe. The apparent drug concentration–time profiles are used to determine if LLPS occurs, to compare different formulations, and even to evaluate the size of the drug nanoparticles upon LLPS. The size of API nanoparticles may be visualized using an optical microscope or dynamic light scattering, or a particle size analyzer suitable for nanoparticles and colloids (such as Marven's Zetasizer). The information can also be used to further investigate whether bulk amorphous drug phase separation occurs during dissolution.

By this time a few candidate formulations should be available in terms of acceptable drug loading, glass transition temperature (>75 °C), and good in vitro dissolution behavior (both LLPS and size of API nanoparticle formation). These candidates may have different carrier polymer, different surfactant, different drug loading, or a combination thereof. If the physical stability is a significant risk, preliminary stability data should also be generated, initially under accelerated conditions. Formulations with significant signs of physical instability should be red-flagged.

The next step is to confirm the formulation performance in animal studies. Dog is often selected due to widespread availability but there may be reasons to go to other species. Oral bioavailability in dog or other animals is just used as preliminary verification of the formulation design and performance. Good bioavailability in animals may not necessarily translate into human bioavailability. However, if a formulation reflects poorly in animal studies, generally there is no good reason to move it forward to human study in the presence of other "better" candidates.

While the animal studies are being performed, longer term experimental stability data may be generated under typical long term storage conditions (25 °C/60% RH) or accelerated conditions (40 °C/75% RH). Meanwhile, analytical scientists may start the preliminary method development. The formulation scientists now can take a look at the manufacturing and develop a lab scale process for these candidates.

Once the performances of the formulations are confirmed in animals, the selected candidates can now progress to clinical development phase. Small scale GMP manufacturing processes are then developed to manufacture the clinical supplies for human PK study. Analytical methods are also developed to support the manufacturing, testing, and releasing of these supplies to ensure potency, purity, stability of the drug products.

Based on the acceptable human bioavailability, the properties of the drug product, and the nature of the manufacturing process, a formulation is then selected for full clinical and CMC development. However, if none of the candidates shows adequate absorption in human, one will have to go back to the drawing board, identify the potential causes for the poor in vivo performance, and start a new formulation cycle.

With the formulation finalized, the targeted effective plasma drug concentration, the overall dose strength can then be estimated. Pill size and pill burden may be adjusted to support the clinical program and final registration. Early phase clinical trials (Phase I, Phase IIa) usually prefer more flexible dose strengths in order to satisfy dosing flexibility. The focus of drug product development now moves to drug product understanding, manufacturing processes development, scale up, and optimization, in order to delivery adequate clinical supplies and to support the final market authorization application. Formulators, engineers, and analytical scientists work collaboratively to characterize the drug product/manufacturing process, and to address and mitigate any relevant issues in a satisfactory way for its approval and commercialization.

6.5.4 Process Development

Spray-drying and hot melt extrusion are the two major ASD manufacturing technologies and each has its own pros and cons. However, the selection of one manufacturing technology over the other is not a purely scientific decision. The knowledge and experience of the scientists, the engineers, and the organization as a whole, the existing equipment/facility capability, the experience and familiarity of the manufacturing organization, and the financial situation of the company, all can play a (sometimes determining) role on the final decision. It is not practical to work in a vacuum just to develop an ideal drug product and manufacturing process. However, it is the responsibility of the pharmaceutical scientists to consider the physicochemical, biopharmaceutical, PK, and PD properties of the drug molecules, the APIs, the formulation, to understand how factor may impact development success in a positive or negative way, and to work with relevant parties within the company and to propose a rational approach to drug product/manufacturing process development within the constraints he or she may have to live with. It is also worth to note that with the rise of many contracting manufacturing organizations, pharmaceutical scientists have more flexibility in choosing the right manufacturing technology to ensure a better chance of success.

Process development often starts in a laboratory environment on small scales because of the convenience of non-GMP nature. The goal of this initial process development is to provide a feasible process for manufacturing the product at a small scale to support the product characterization and to prepare for the expected clinical supply manufacturing. It may also provide a baseline understanding on what key process parameters/equipment settings may impact the manufacturing process and the properties of the manufactured product. These development activities are useful in that they provide an initial read on the potential critical equipment settings and critical process parameters. However, it is not advised to establish a full design space on the lab equipment, due to the differences in equipment vendor, equipment design, scale, geometry, heat transfer, mass transfer, etc., between the laboratory equipment and the commercial equipment, unless a scale-up/scale-down model is fully established.

Process scale up refers to the migration of a manufacturing process from the lab-scale to the pilot plant-scale and to commercial scale and usually involves equipment of increasing capacity and/or even different design. Sometimes it may not follow the exact same order, for example, a process may directly be scaled up to the commercial equipment, depending on the nature of the product, manufacturing process, volume projection, and facility/equipment availability. The pilot scale equipment usually resembles more of the commercial ones. Process studies on the pilot scale is generally more applicable to commercial equipment than on the lab scale, and is also preferred from the perspectives of material, resources, cost, and facility utilization. It would be very expensive to develop manufacturing process and establish an operating space directly on the commercial scale. Nevertheless, the operating space developed on a pilot scale should be rationally verified on the commercial scale prior to process validation and potential discrepancies (if any) should be addressed.

Regardless of the scale and equipment differences, it is mandatory that drug products manufactured on the commercial scale be equivalent to those used in the pivotal clinical trials to ensure the potency, purity, safety, and efficacy of the product, the latter of which are often manufactured at pilot scales. To be more specifically, critical quality attributes (CQAs) of drug products defined during the clinical and CMC development need to be maintained when the manufacturing process is transferred to the commercial scales.

As a personal note, a manufacturing process can be understood to consist of many individual events, in which a discrete element of the input material is exposed in certain timed sequence. The collective of those events in their timed sequence constitutes the history of this particular element. An entire batch material may be divided into many such small elements, each having the same history. This element can be called independent material domain. The independent material domain can be a single molecule or a single particle or a single tablet, if the process is imposing the same events to each molecule, particle, or tablet, etc. However, in most cases, the independent material domain is made up of an ensemble of molecules, particles, or tablets, and within which a distribution of history may exist. The independent material domain is defined as the minimum collection of the input materials such that all these domains have the same experiences during a process. As a somewhat simple example, in the process where tablets are coated by passing array of individual tablet through a spray zone one-by-one in the same fashion, the individual tablet is the independent material domain. In this example, every tablet has identical history.

The independent material domain may be difficult to define exactly for most processes. However, the concept is very useful for process transfer and scale up. If one can reproduce the trajectory of the independent material domain, all of its properties will be reproduced faithfully, and all product quality attributes will be maintained. This principle may be called "identical environment, identical exposure" and is a brutal force for process transfer and process scale up.

Of course, it is well known that not all history matter (or mater to the same extent). It is those critical events that will have meaningful impacts on the relevant properties of the independent material domain. The concept is very similar to the critical processes, and we can call them critical events/experiences. We move from the principle of "identical environment, identical exposure" to principle of "similar environment, similar exposure" to ensure the same critical histories are replicated during process transfer and process scale up.

6.5.4.1 Hot Melt Extrusion Process

The aim of a HME process is to produce extrudates that have no residual crystallinity, good homogeneity (glassy solution), and acceptable level of drug and polymer degradation during the manufacturing process and during the product shelf-life. These are also the major CQAs that may be impacted by the HME process.

HME process relies on both mechanical shear and thermal heat to achieve the above goal. Mechanical shear is impacted by the screw configurations and the process parameters (feed rate, screw speed, melt viscosity) and heat input/output is impacted by the barrel temperature, barrel cooling design, and the barrel heat transfer surface area. The mechanical shear and the heat have to work cooperatively, not too much (otherwise unacceptable degradation), not too little (otherwise remaining residual crystal or amorphous phase separations), and the Goldilocks principle applies. Therefore, an operating range exists for the HME process to obtain acceptable extrudates. The width of this range depends on the properties of the drug, polymer, the formulation, and the equipment. Sometime the formulation has to be tweaked to accommodate the process and sometimes the equipment setups and the process need to be modified to accommodate the formulation.

A progressive temperature profile is often applied for HME. The powder feeding zone is frequently set around the ambient temperature, but it gradually increases in the solid conveying area, and then elevates to the desired temperature at the melting zone. The temperature at the melting zone should be higher than the minimum thermodynamic melting/solubilizing temperature required for the particular drug loading (see Eq. 6.22 or 6.23), but is generally lower than the melting temperature of the pure crystalline form. The temperatures of the melt convey and discharge zone should be set to ensure stable flow into the die. Stability of the melt flow in the die is impacted by the pressure and the melt viscosity. While the pressure depends on the screw configuration in the conveying and discharging zone, the melt viscosity depends strongly on the temperature.

The screw configurations and different zone designs have been discussed previously. The screw configurations, together with the feed rate, screw speed, and melt viscosity (barrel temperature), determines the overall shear rate and shear energy. Therefore some of the parameters are confounded. Modification of the screw configurations may be necessary during HME process development as well as during scale up under situations where a change in the formulation is not allowed while process parameter alone is insufficient to achieve a robust process.

To ensure similar environment during HME process scale up, ideally the equipment similarity should be maintained. The design of the extruder should be based on similar principles in order to facilitate scale up and scale down among different equipment. This may be approached using the so-called geometric similarity that has been widely applied to a variety of other processes. The geometric similarity dictates: similar screw length/diameter (L/D) ratio, similar screw configurations, and similar zone length/screw diameter ratio. If the same L/D is maintained, the exact screw configurations at different scale may be reproduced with the same number of zones, the same zone length-to-screw diameter ratio, and the same screw length/diameter ratio for each element. Geometric similarity also requires the same screw outer diameter-to-inner diameter (D_o/D_i) ratio, and the same clearance between screw flight and the barrel, in order to maintain similar free volume, shear, and mixing capability.

There are a few classical scale-up strategies for HME process: degree of fill, heat transfer, and specific mechanical energy, depending on the critical nature of a particular HME process. The degree of fill rule is also called the volumetric scale up rule. It maintains the same free volume and with geometrically similar extruders, it also maintains the same degree of fill and shear rate. This rule is applicable when the feeder zone is full and the extruder is operated at its volumetric limit. However, as indicated before, twin screw extruder has good conveying capability and the screw speed is typically independent of material feeding (starve fed). Therefore, this rule is not typically invoked. When screw diameter ratio is the same, the rule can be written as:

$$\frac{Q_2}{Q_1} = \left(\frac{D_2}{D_1}\right)^3 \cdot \left(\frac{N_2}{N_1}\right) \cdot \left(\frac{\varepsilon_2}{\varepsilon_1}\right) \tag{6.25}$$

where Q is the feed rate, D is screw diameter, and N is the screw rotation speed, ε is the degree of fill.

The rule of heat transfer scale up concerns the extrusion process limited by the thermal heat transfer, which depends on the degree of fill, the barrel surface area, the temperature gradient between product and barrel, and the residence time. When the degree of fill and screw configurations are the same, heat transfer rate dictate the feed rate be proportional to the surface area of heat transfer, i.e.

$$\frac{Q_2}{Q_1} = \left(\frac{D_2}{D_1}\right)^2 \cdot \left(\frac{N_2}{N_1}\right) \tag{6.26}$$

Extrusion of amorphous solid dispersions generally depends on the shear to generate energy for the melting and mixing processes. Therefore, scale up based specific mechanical energy often result in more equivalent product quality attributes.

$$\frac{Q_2}{Q_1} = \left(\frac{D_2}{D_1}\right)^2 \cdot \left(\frac{\tau_2}{\tau_1}\right) \cdot \left(\frac{N_2}{N_1}\right) \cdot \left(\frac{R_2}{R_1}\right) \tag{6.27}$$

where τ is the torque (%), R is gearbox rating (kw/rpm).

To ensure "similar environment, similar exposure", another important parameter, the residence time and distribution (RTD), should be controlled to be equivalent between different scales. Ideally, each molecule should experience the same residence time under each different section (environment) during the extrusion. However, when geometric similarity is maintained, the overall residence time and distribution provides a good representation of the timing component of the history. RTD can be measured by introducing a tracer (e.g. a dye) at the feeding section and monitoring its concentration at the exit either online or offline.

6.5.4.2 Spray Drying Process

The spray drying process starts with atomization of the spraying solution into fine droplets. Droplet size of the spray defines the size of the dried particles. It also defines the length scales of the heat and mass transfer during drying and therefore the morphology and hence the final properties of the dried particles. Droplet size during atomization is impacted by formulation composition (drug, polymer, and other components), the solid concentration of the spray solution, spray solution viscosity and surface tension, nozzle design and geometry, spray rate, and other spray parameters (e.g. pressure, temperature).

Drying of the atomized droplet is impacted by the droplet size, the inlet and outlet temperatures, and the drying capacity of the inlet gas [64].

Spray drying is a very complex process and minor changes in spraying and drying conditions can result in different material attributes (e.g. density, morphology, flow, and compaction) of the spray-dried particles. It is nearly impossible at this stage to predict the droplet drying kinetics and the resulted particle attributes.

A principle for spray drying scale up is again to ensure similar history under similar environment. To be specific, the first step is to ensure similar droplet size and distribution upon spraying and atomization [131], and then subject these droplets under a similar drying environment to ensure similarity in the drying kinetics, so to ensure similarity in the final particle morphology and size distributions.

A concept called environmental equivalency (EE) has been used to ensure similar environment during aqueous coating [132, 133]:

$$EE = \frac{\omega_{wb} - \omega_{in}}{\omega_{out} - \omega_{in}} \tag{6.28}$$

Here ω is the absolute moisture ratio in kg of water per kg of dry air (kgw/kgda) and subscript wb designates the wet-bulb condition. The EE is a ratio between the overall drying capacity of the inlet air ($\omega_{wb} - \omega_{in}$) and the drying capacity actually consumed between the inlet and outlet. It is related to the relative rate of water evaporation and has been used as a relevant scale up factor for aqueous tablet coating. This environmental equivalency is similar to that defined earlier by Ebey [132] and the inverse of the EE is identical to the vaporization equivalency derived based on mass balance [134]. The key idea is that a similar spray droplet will dry similarly under a similar drying environment represented by the parameter environmental equivalency. This concept can be extended to the general air/solvent for spray drying and to ensure similar particle morphology upon spray-drying.

6.6 Concluding Remarks

Amorphous solid dispersion represents a state-of-the-art delivery technology for modern poorly water soluble drugs. Significant progresses have been made in the understanding of their physical, chemical properties, dissolution behavior and implications for oral drug absorption, and have led to the launch of many successful commercial products. This chapter provides a succinct review on the relevant dissolution behavior of amorphous solid dispersions that underlines their oral bioavailability improvements. Integrated with basic process understanding, together with considerations on the physical, chemical, biopharmaceutical, pharmacokinetic, and other relevant properties of the drug molecule and the formulation, a rational approach to amorphous drug product development can be realized to ensure maximized benefits to patients as well as good chance of business success. Nevertheless, developing amorphous drug formulation is a significant undertaking. It is not without challenges (e.g. bioavailability, physical stability, chemical stability, and manufacturing). However, with further advances in our scientific understanding and accumulation of institutional knowledge and experiences, the authors are optimistic in that amorphous solid dispersion drug product will become a mature or even a standard drug delivery technology.

References

1. Lipinski CA, Lombardo F, Dominy BW, Feeney PJ (1997) Experimental and computational approaches to estimate solubility and permeability in drug discovery and development settings. Adv Drug Deliv Rev 23:3–25
2. Loftsson T, Brewster ME (2010) Pharmaceutical applications of cyclodextrins: basic science and product development. J Pharm Pharmacol 62:1607–1621
3. Kola I, Landis J (2004) Can the pharmaceutical industry reduce attrition rates? Nat Rev Drug Discov 3:711–716
4. Pouton CW (1997) Formulation of self-emulsifying drug delivery systems. Adv Drug Deliv Rev 25:47–58
5. Neslihan Gursoy R, Benita S (2004) Self-emulsifying drug delivery systems (SEDDS) for improved oral delivery of lipophilic drugs. Biomed Pharmcother 58:173–182
6. Mistry RB, Sheth NS (2011) A review: self emulsifying drug delivery system. Int J Pharm Pharm Sci 3(Suppl 22):23–28
7. Gelperina S, Kisich K, Iseman MD, Heifets L (2005) The potential advantages of nanoparticle drug delivery systems in chemotherapy of tuberculosis. Am J Respir Crit Care Med 172: 1487–1490
8. De Jong WH, Borm PJ (2008) Drug delivery and nanoparticles: applications and hazards. Int J Nanomed 3:133–149
9. Kaialy W, Al Shafiee M (2016) Recent advances in the engineering of nanosized active pharmaceutical ingredients: promises and challenges. Adv Colloid Interface Sci 228:71–91
10. Hancock BC, Zografi G (1997) Characteristics and significance of the amorphous state in pharmaceutical systems. J Pharm Sci 86:1–12
11. He Y, Ho C (2015) Amorphous solid dispersions: utilization and challenges in drug discovery and development. J Pharm Sci 104:3237–3258

12. Vasconcelos T, Marques S, das Neves J, Sarmento B (2016) Amorphous solid dispersions: rational selection of a manufacturing process. Adv Drug Deliv Rev 100:85–101
13. Zhu T, Chiu Y, Doan T, Klein C, Chang M, Brun S, Hanna G, Awni W (2005) New tablet formulation of lopinavir/ritonavir is bioequivalent to the capsule at a dose of 800/200 mg. In: 45th Interscience Conference on Antimicrobial Agents and Chemotherapy (ICAAC), Washington, DC, Poster H-1894
14. Parks GS, Huffman HM, Cattor FR (1928) Studies on glass II: the transition between the glassy and liquid states in the case of glucose. J Phys Chem 32:1366–1379
15. Stagner WC, Guillory JK (1979) Physical characterization of solid iopanoic acid forms. J Pharm Sci 68:1005–1009
16. Corrigan OI, Holohan EM, Sabra K (1984) Amorphous forms of thiazide diuretics prepared by spray-drying. Int J Pharm 18:195–200
17. Elamin AA, Ahlneck C, Alderborn G, Nystrom C (1994) Increased metastable solubility of milled griseofulvin, depending on the formation of a disordered surface structure. Int J Pharm 111:159–170
18. Hancock BC, Parks M (2000) What is the true solubility advantage for amorphous pharmaceuticals? Pharm Res 17:397–403
19. Alonzo DE, Zhang GGZ, Zhou D, Gao Y, Taylor LS (2010) Understanding the behavior of amorphous pharmaceutical systems during dissolution. Pharm Res 27:608–618
20. Alonzo DE, Raina S, Zhou D, Gao Y, Zhang GGZ, Taylor LS (2012) Characterizing the impact of hydroxypropylmethyl cellulose on the growth and nucleation kinetics of felodipine from supersaturated solutions. Cryst Grow Des 12:1538–1547
21. Simonelli AP, Mehta SC, Higuchi WI (1970) Inhibition of sulfathiazole crystal growth by polyvinylpyrrolidone. J Pharm Sci 59:633–638
22. Holder GA, Thorne J (1979) Inhibition of crystallization by polymers. Polym Prepr (Am Chem Soc Div Polym Chem) 20:766–769
23. Oner M, Calvert P (1993) Influence of polymer architecture on crystal growth inhibition. Polym Mater Sci Eng 69:166–167
24. Agarwal P, Berglund KA (2004) Effect of polymeric additives on calcium carbonate crystallization as monitored by nephelometry. Cryst Growth Des 4:479–483
25. Inada T, Modak PR (2006) Growth control of ice crystals by poly(vinyl alcohol) and antifreeze protein in ice slurries. Chem Eng Sci 61:3149–3158
26. Vandecruys R, Peeters J, Verreck G, Brewster ME (2007) Use of a screening method to determine excipients which optimize the extent and stability of supersaturated drug solutions and application of this system to solid formulation design. Int J Pharm 342:168–175
27. Lindfors L, Forssen S, Westergren J, Olsson U (2008) Nucleation and crystal growth in supersaturated solutions of model drug. J Colloid Interface Sci 325:404–413
28. Raghavan SL, Trividic A, Davis AF, Hadgraft J (2001) Crystallization of hydrocortisone acetate: influence of polymers. Int J Pharm 212:213–221
29. Mehta SC (1969) Mechanistic studies of linear single crystal growth rates of sulfathiazole and their inhibition by polyvinyl pyrrolidone. Preparation and dissolution of high-energy sulfathiazole polyvinyl pyrrolidone coprecipitates, University of Michigan, Ann Arbor, MI, USA, pp 136
30. Sekikawa H, Nakano M, Arita T (1978) Inhibitory effect of poly(vinylpyrrolidone) on the crystallization of drugs. Chem Pharm Bull 26:118–126
31. Ziller KH, Rupprecht H (1988) Control of crystal growth in drug suspensions. 1. Design of a control unit and application to acetaminophen suspensions. Drug Dev Ind Pharm 14: 2341–2370
32. Ziller KH, Rupprecht H (1990) Control of crystal growth in drug suspensions. III. Isothermal crystallization in the presence of polymers. PZ Wiss 3:147–152
33. Ma X, Taw J, Chiang C-M (1996) Control of drug crystallization in transdermal matrix system. Int J Pharm 142:115–119

34. Wen H, Morris KR, Park K (2008) Synergic effects of polymeric additives on dissolution and crystallization of acetaminophen. Pharm Res 25:349–358
35. Konno H, Taylor LS (2006) Influence of different polymers on the crystallization tendency of molecularly dispersed amorphous felodipine. J Pharm Sci 95:2692–2705
36. Konno H, Taylor LS (2008) Ability of different polymers to inhibit the crystallization of amorphous felodipine in the presence of moisture. Pharm Res 25:969–978
37. Wattis JAD, Coveney PV (1997) General nucleation theory with inhibition for chemically reacting systems. J Chem Phys 106:9122–9140
38. Ilevbare GA, Liu H, Edgar KJ, Taylor LS (2013) Maintaining supersaturation in aqueous drug solutions: impact of different polymers on induction times. Cryst Growth Des 13:740–751
39. Alonzo DE, Gao Y, Zhou D, Mo H, Zhang GGZ, Taylor LS (2011) Dissolution and precipitation behavior of amorphous solid dispersions. J Pharm Sci 100:3316–3331
40. Van Eerdenbrugh B, Alonzo DE, Taylor LS (2011) Influence of particle size on the ultraviolet spectrum of particulate-containing solutions: implications for in-situ concentration monitoring using UV/Vis fiber-optic probes. Pharm Res 28:1643–1652
41. Raina SA, Zhang GGZ, Alonzo DE, Wu J, Zhu D, Catron ND, Gao Y, Taylor LS (2014) Enhancements and limits in drug membrane transport using supersaturated solutions of poorly water-soluble drugs. J Pharm Sci 103:2736–2748
42. Hsieh Y-L, Ilevbare GA, Van Eerdenbrugh B, Box KJ, Sanchez-Felix MV, Taylor LS (2012) pH-induced precipitation behavior of weakly basic compounds: determination of extent and duration of supersaturation using potentiometric titration and correlation to solid state properties. Pharm Res 29:2738–2753
43. Tachibana T, Nakamura A (1965) A methode for preparing an aqueous colloidal dispersion of organic materials by using water-soluble polymers: dispersion of β-carotene by polyvinylpyrrolidone. Colloid Polym Sci 203:130–133
44. Ilevbare GA, Taylor LS (2013) Liquid-liquid phase separation in highly supersaturated aqueous solutions of poorly water-soluble drugs: implications for solubility enhancing formulations. Cryst Growth Des 13:1497–1509
45. Bonnett PE, Carpenter KJ, Dawson S, Davey RJ (2003) Solution crystallisation via a submerged liquid-liquid phase boundary: oiling out. Chem Commun 698–699
46. Derdour L (2010) A method to crystallize substances that oil out. Chem Eng Res Des 88: 1174–1181
47. Veesler S, Lafferrère L, Garcia E, Hoff C (2003) Phase transitions in supersaturated drug solution. Org Proc Res Dev 7:983–989
48. Taylor LS, Zhang GGZ (2016) Physical chemistry of supersaturated solutions and implications for oral absorption. Adv Drug Deliv Rev 101:122–142
49. Almeida e Sousa L, Reutzel-Edens SM, Stephenson GA, Taylor LS (2016) Supersaturation potential of salt, co-crystal, and amorphous forms of a model weak base. Cryst Grow Des 16: 737–748
50. Tho I, Liepold B, Rosenberg J, Mägerlein M, Brandl M, Fricker G (2010) Formation of nano/micro-dispersions with improved dissolution properties upon dispersion of ritonavir melt extrudate in aqueous media. Eur J Pharm Sci 40:25–32
51. Kanzer J, Hupfeld S, Vasskog T, Tho I, Hölig P, Mägerlein M, Fricker G, Brandl M (2010) In situ formation of nanoparticles upon dispersion of melt extrudate formulations in aqueous medium assessed by asymmetrical flow field-flow fractionation. J Pharm Biomed Anal 53: 359–365
52. Frank KJ, Westedt U, Rosenblatt KM, Hölig P, Rosenberg J, Mägerlein M, Fricker G, Brandl M (2012) The amorphous solid dispersion of the poorly soluble ABT-102 forms nano/microparticulate structures in aqueous medium: impact on solubility. Int J Nanomed 7: 5757–5768
53. Harmon P, Galipeau K, Xu W, Brown C, Wuelfing WP (2016) Mechanism of dissolution-induced nanoparticle formation from a copovidone-based amorphous solid dispersion. Mol Pharm 13:1467–1481

54. Tadmor Z, Klein I (1970) Engineering principles of plasticating extrusion. Van Nostrand Reinhold, New York, pp 152–158
55. Breitenbach J (2002) Melt extrusion: from process to drug delivery technology. Eur J Pharm Biopharm 54:107–117
56. Chokshi R, Zia H (2004) Hot-melt extrusion technique: a review. Iran J Pharm Res 3:3–16
57. Breitenbach J (2006) Melt extrusion can bring new benefits to HIV therapy: the example of Kaletra® tablets. Am J Drug Deliv 4:61–64
58. Crowley MM, Zhang F, Repka MA, Thumma S, Upadhye SB, Kumar Battu S, McGinity JW, Martin C (2007) Pharmaceutical applications of hot-melt extrusion: Part I. Drug Dev Ind Pharm 33:909–926
59. Repka MA, Battu SK, Upadhye SB, Thumma S, Crowley MM, Zhang F, Martin C, McGinity JW (2007) Pharmaceutical applications of hot-melt extrusion: Part II. Drug Dev Ind Pharm 33:1043–1057
60. Douroumis D (2012) Hot-melt extrusion: pharmaceutical applications,1st edn. Wiley, Chichester, p 382
61. De Jaeghere W, De Beer T, Van Bocxlaer J, Remon JP, Vervaet C (2015) Hot-melt extrusion of polyvinyl alcohol for oral immediate release applications. Int J Pharm 492:1–9
62. Agrawal AM, Dudhedia MS, Zimny E (2016) Hot melt extrusion: development of an amorphous solid dispersion for an insoluble drug from mini-scale to clinical scale. AAPS PharmSciTech 17:133–147
63. Patil H, Tiwari RV, Repka MA (2016) Hot-melt extrusion: from theory to application in pharmaceutical formulation. AAPS PharmSciTech 17:20–42
64. Paudel A, Worku ZA, Meeus J, Guns S, Van den Mooter G (2013) Manufacturing of solid dispersions of poorly water soluble drugs by spray drying: formulation and process considerations. Int J Pharm 453:253–284
65. Sóti PL, Bocz K, Pataki H, Eke Z, Farkas A, Verreck G, Kiss É, Fekete P, Vigh T, Wagner I, Nagy ZK, Marosi G (2015) Comparison of spray drying, electroblowing and electrospinning for preparation of Eudragit E and itraconazole solid dispersions. Int J Pharm 494:23–30
66. Fong SYK, Ibisogly A, Bauer-Brandl A (2015) Solubility enhancement of BCS Class II drug by solid phospholipid dispersions: spray drying versus freeze-drying. Int J Pharm 496: 382–391
67. Singh A, Van den Mooter G (2016) Spray drying formulation of amorphous solid dispersions. Adv Drug Deliv Rev 100:27–50
68. Dobry DE, Settell DM, Baumann JM, Ray RJ, Graham LJ, Beyerinck RA (2009) A model-based methodology for spray-drying process development. J Pharm Innov 4:133–142
69. Paudel A, Loyson Y, Van den Mooter G (2013) An investigation into the effect of spray drying temperature and atomizing conditions on miscibility, physical stability, and performance of naproxen-pvp K 25 solid dispersions. J Pharm Sci 102:1249–1267
70. Vehring R, Foss WR, Lechuga-Ballesteros D (2007) Particle formation in spray drying. J Aerosol Sci 38:728–746
71. Çelik M, Wendell SC (2009) Chapter 5. Spray drying and pharmaceutical applications. In: Parikh DM (ed) Handbook of pharmaceutical granulation technology, 3rd edn. CRC Press, New York, pp 98–125
72. Mezhericher M, Levy A, Borde I (2010) Theoretical models of single droplet drying kinetics: a review. Dry Technol 28:278–293
73. Vicente J, Pinto J, Menezes J, Gaspar F (2013) Fundamental analysis of particle formation in spray drying. Powder Technol 247:1–7
74. Zhou D, Qiu Y (2009) Understanding biopharmaceutics properties for pharmaceutical product development and manufacturing I – oral absorption and the biopharmaceutics classification system. J Val Technol 15:62–72
75. Zhou D (2009) Understanding physicochemical properties for pharmaceutical product development and manufacturing – stability and excipient compatibility. J Val Technol 15:36–47

76. Zhou D, Qiu Y (2010) Understanding biopharmaceutics properties for pharmaceutical product development and manufacturing II – dissolution and in-vitro-in vivo correlation. J Val Technol 16:57–70
77. Benet LZ (2010) Predicting drug disposition via application of a biopharmaceutics drug disposition classification system. Basic Clin Pharmacol Toxicol 106:162–167
78. Wu C-Y, Benet LZ (2005) Predicting drug disposition via application of BCS: transport/absorption/elimination interplay and development of a biopharmaceutics drug disposition classification system. Pharm Res 22:11–23
79. Benet LZ (2013) The role of BCS (biopharmaceutics classification system) and BDDCS (biopharmaceutics drug disposition classification system) in drug development. J Pharm Sci 102:34–42
80. Amidon GL, Lennernäs H, Shah VP, Crison JR (1995) A theoretical basis for a biopharmaceutic drug classification: the correlation of in vitro drug product dissolution and in vivo bioavailability. Pharm Res 12:413–420
81. Zhou D (2003) Molecular mobility, physical stability, and transformation kinetics of amorphous and hydrated pharmaceutical solids. University of Minnesota, Minneapolis, MN, 270 pp
82. Zhou D, Zhang GGZ, Law D, Grant DJW, Schmitt EA (2008) Thermodynamics, molecular mobility and crystallization kinetics of amorphous griseofulvin. Mol Pharm 5:927–936
83. Zhou D, Grant David JW, Zhang Geoff GZ, Law D, Schmitt Eric A (2007) A calorimetric investigation of thermodynamic and molecular mobility contributions to the physical stability of two pharmaceutical glasses. J Pharm Sci 96:71–83
84. Zhou D, Zhang GGZ, Law D, Grant DJW, Schmitt EA (2002) Physical stability of amorphous pharmaceuticals: importance of configurational thermodynamic quantities and molecular mobility. J Pharm Sci 91:1863–1872
85. Knapik J, Wojnarowska Z, Grzybowska K, Tajber L, Mesallati H, Paluch KJ, Paluch M (2016) Molecular dynamics and physical stability of amorphous nimesulide drug and its binary drug-polymer systems. Mol Pharm 13:1937–1946
86. Szczurek J, Rams-Baron M, Knapik-Kowalczuk J, Antosik A, Szafraniec J, Jamóz W, Dulski M, Jachowicz R, Paluch M (2017) Molecular dynamics, recrystallization behavior, and water solubility of the amorphous anticancer agent bicalutamide and its polyvinylpyrrolidone mixtures. Mol Pharm 14:1071–1081
87. Kossena GA, Charman WN, Boyd BJ, Porter CJH (2005) Influence of the intermediate digestion phases of common formulation lipids on the absorption of a poorly water-soluble drug. J Pharm Sci 94:481–492
88. Custodio JM, Wu C-Y, Benet LZ (2008) Predicting drug disposition, absorption/elimination/transporter interplay and the role of food on drug absorption. Adv Drug Deliv Rev 60: 717–733
89. Shah N, Iyer RM, Mair H-J, Choi DS, Tian H, Diodone R, Fähnrich K, Pabst-Ravot A, Tang K, Scheubel E, Grippo JF, Moreira SA, Go Z, Mouskountakis J, Louie T, Ibrahim PN, Sandhu H, Rubia L, Chokshi H, Singhal D, Malick W (2013) Improved human bioavailability of vemurafenib, a practically insoluble drug, using an amorphous polymer-stabilized solid dispersion prepared by a solvent-controlled coprecipitation process. J Pharm Sci 102: 967–981
90. ICH Guideline (2016) Impurity: guideline for residual solvents, Q3C(R6), Step 4, 20 October
91. Scott RL (1949) The thermodynamics of high-polymer solutions. IV. Phase equilibria in the ternary system: polymer-liquid 1-liquid 2. J Chem Phys 17:268–279
92. Tompa H (1956) Polymer solutions. Academic Press, New York, p 325
93. Li J, Patel D, Wang G (2017) Use of spray-dried dispersions in early pharmaceutical development: theoretical and practical challenges. AAPS J 19:321–333
94. Davis MT, Egan DP, Kuhs M, Albadarin AB, Griffin CS, Collins JA, Walker GM (2016) Amorphous solid dispersions of BCS class II drugs: a rational approach to solvent and polymer selection. Chem Eng Res Des 110:192–199

95. Zhou D, Qiu Y (2010) Understanding material properties in pharmaceutical product development and manufacturing: powder flow and mechanical properties. J Val Technol 16:65–77
96. Kopp S, Beyer C, Graf E, Kubel F, Doelker E (1989) Methodology for a better evaluation of the relation between mechanical strength of solids and polymorphic form. J Pharm Pharmacol 41:79–82
97. Hancock BC, Carlson GT, Ladipo DD, Langdon BA, Mullarney MP (2002) Comparison of the mechanical properties of the crystalline and amorphous forms of a drug substance. Int J Pharm 241:73–85
98. Brungs MP (1995) Fracture and failure of glass. Mater Forum 19:227–239
99. Iyer R, Hegde S, Zhang Y-E, Dinunzio J, Singhal D, Malick A, Amidon G (2013) The impact of hot melt extrusion and spray drying on mechanical properties and tableting indices of materials used in pharmaceutical development. J Pharm Sci 102:3604–3613
100. Grymonpre W, De Jaeghere W, Peeters E, Adriaensens P, Remon JP, Vervaet C (2016) The impact of hot-melt extrusion on the tableting behaviour of polyvinyl alcohol. Int J Pharm 498:254–262
101. Surampalli G, Nanjwade BK, Patil PA, Chilla R (2016) Novel tablet formulation of amorphous candesartan cilexetil solid dispersions involving P-gp inhibition for optimal drug delivery: in vitro and in vivo evaluation. Drug Deliv 23:2124–2138
102. Joshi AB, Patel S, Kaushal AM, Bansal AK (2010) Compaction studies of alternate solid forms of celecoxib. Adv Powder Technol 21:452–460
103. Agrawal AM, Dudhedia MS, Patel AD, Raikes MS (2013) Characterization and performance assessment of solid dispersions prepared by hot melt extrusion and spray drying process. Int J Pharm 457:71–81
104. Grymonpré W, Verstraete G, Van Bockstal PJ, Van Renterghem J, Rombouts P, De Beer T, Remon JP, Vervaet C (2017) In-line monitoring of compaction properties on a rotary tablet press during tablet manufacturing of hot-melt extruded amorphous solid dispersions. Int J Pharm 517:348–358
105. Sarode AL, Sandhu H, Shah N, Malick W, Zia H (2013) Hot melt extrusion (HME) for amorphous solid dispersions: predictive tools for processing and impact of drug-polymer interactions on supersaturation. Eur J Pharm Sci 48:371–384
106. Lang B, McGinity JW, Williams RO (2014) Hot-melt extrusion – basic principles and pharmaceutical applications. Drug Dev Ind Pharm 40:1133–1155
107. Vogel H (1921) Das Temperaturabhängigkeitsgesetz der Viskosität von Flüssigkeiten. Phys Z 22:645–646
108. Fulcher GS (1925) Analysis of recent measurements of the viscosity of glasses. J Am Ceram Soc 8:339–355
109. Tammann VG, Hesse W (1926) Die abhängigkeit der viscosität von der temperatur bei unterkühlten flüssigkeiten. Z Anorg Allg Chem 156:245–257
110. Gibbs JH, DiMarzio EA (1958) Nature of the glass transition and the glassy state. J Chem Phys 28:373–383
111. Adam G, Gibbs JH (1965) On the temperature dependence of cooperative relaxation properties in glass-forming liquids. J Chem Phys 43:139–146
112. Hodge IM (1987) Effects of annealing and prior history on enthalpy relaxation in glassy polymers. 6. Adam-Gibbs formulation of nonlinearity. Macromolecules 20:2897–2908
113. Hodge IM (1994) Enthalpy relaxation and recovery in amorphous materials. J Non-Cryst Solids 169:211–266
114. Gordon M, Taylor JS (1952) Ideal copolymers and the second-order transitions of synthetic rubbers 1: non-crystalline copolymers. J Appl Chem 2:493–498
115. Couchman PR, Karasz FE (1978) A classical thermodynamic discussion of the effect of composition on glass-transition temperatures. Macromolecules 11:117–119
116. Simha R (1962) On a general relation involving glass temperature and coefficients of expansion of polymers. J Chem Phys 37:1003–1007

117. Fox T (1956) Influence of diluent and of copolymer composition on the glass temperature of a polymer system. Bull Am Phys Soc 1:123–132
118. Dow Pharma & Food Solutions. AFFINISOL™: HPMC HME for hot melt extrusion. 30 May 2017. http://msdssearch.dow.com/PublishedLiteratureDOWCOM/dh_0954/0901b803809543f4.pdf?
119. McPhillips H, Craig DQM, Royall PG, Hill VL (1999) Characterisation of the glass transition of HPMC using modulated temperature differential scanning calorimetry. Int J Pharm 180: 83–90
120. Baghel S, Cathcart H, O'Reilly NJ (2016) Polymeric amorphous solid dispersions: a review of amorphization, crystallization, stabilization, solid-state characterization, and aqueous solubilization of biopharmaceutical classification system class ii drugs. J Pharm Sci 105: 2527–2544
121. Gao Y, Gesenberg C, Zheng W (2017) Chapter 17. Oral formulations for preclinical studies: principle, design, and development considerations. In: Qiu Y, Chen Y, Zhang GGZ, Yu L, Mantri RV (eds) Developing solid oral dosage forms: pharmaceutical theory & practice, 2nd edn. Elsevier, New York, pp 455–495
122. Nishi T, Wang TT (1975) Melting-point depression and kinetic effects of cooling on crystallization in poly(vinylidene fluoride) poly(methyl methacrylate) mixtures. Macromolecules 8: 905–915
123. Hoei Y, Yamaura K, Matsuzawa S (1992) A lattice treatment of crystalline solvent-amorphous polymer mixtures on melting-point depression. J Phys Chem 96:10584–10586
124. Marsac PJ, Shamblin SL, Taylor LS (2006) Theoretical and practical approaches for prediction of drug-polymer miscibility and solubility. Pharm Res 23:2417–2426
125. Tao J, Sun Y, Zhang GGZ, Yu L (2009) Solubility of small-molecule crystals in polymers: D-mannitol in PVP, indomethacin in PVP/VA, and nifedipine in PVP/VA. Pharm Res 26: 855–864
126. Sun Y, Tao J, Zhang GGZ, Yu L (2010) Solubilities of crystalline drugs in polymers: an improved analytical method and comparison of solubilities of indomethacin and nifedipine in PVP, PVP/VA, and PVAc. J Pharm Sci 99:4023–4031
127. Gao P, Rush BD, Pfund WP, Huang T, Bauer JM, Morozowich W, Kuo M-S, Hageman MJ (2003) Development of a supersaturable SEDDS (S-SEDDS) formulation of paclitaxel with improved oral bioavailability. J Pharm Sci 92:2386–2398
128. Gao P, Morozowich W (2007) Chapter 13. Design and development of supersaturatable self-emulsifying drug delivery systems for enhancing the gastrointestinal absorption of poorly soluble drugs. In: Hauss DJ (ed) Oral lipid-based formulations: enhancing the bioavailability of poorly water-soluble drugs. CRC, New York, pp 303–327
129. Curatolo W, Nightingale JA, Herbig SM (2009) Utility of Hydroxypropylmethylcellulose Acetate Succinate (HPMCAS) for initiation and maintenance of drug supersaturation in the GI Milieu. Pharm Res 26:1419–1431
130. Murdande SB, Pikal MJ, Shanker RM, Bogner RH (2010) Solubility advantage of amorphous pharmaceuticals: I. A thermodynamic analysis. J Pharm Sci 99:1254–1264
131. Thybo P, Hovgaard L, Lindeløv JS, Brask A, Andersen SK (2008) Scaling up the spray drying process from pilot to production scale using an atomized droplet size criterion. Pharm Res 25:1610–1620
132. Ebey GC (1987) A thermodynamic model for aqueous film-coating. Pharm Tech 11:40, 42–43, 46, 48, 50
133. Strong John C (2009) Psychrometric analysis of the environmental equivalency factor for aqueous tablet coating. AAPS PharmSciTech 10:303–309
134. Reiland TL, Seitz JA, Yeager JL, Brusenback RA (1983) Aqueous film-coating vaporization efficiency. Drug Dev Ind Pharm 9:945–958

Index

M. Rams-Baron et al., *Amorphous Drugs*,
https://doi.org/10.1007/978-3-319-72002-9

F

G

H

I

J

K

L

Printed by Printforce, the Netherlands